COMMEMORATING WAR

Memory and Narrative Series
Mary Chamberlain and Selma Leydesdorff, series editors

Trauma: Life Stories of Survivors
Kim Lacy Rogers and Selma Leydesdorff, editors
(with Graham Dawson)

Commemorating War: The Politics of Memory
Timothy G. Ashplant, Graham Dawson, and Michael Roper, editors

Environmental Consciousness:
The Roots of a New Political Agenda
Stephen Hussey and Paul Thompson, editors

Memory and Memorials:
From the French Revolution to World War One
Mathew Campbell, Jacqueline M. Labbe,
and Sally Shuttleworth, editors

The Stasi Files Unveiled:
Guilt and Compliance in a Unified Germany
Barbara Miller

The Uses of Narrative:
Explorations in Sociology, Psychology, and Cultural Studies
Molly Andrews, Shelley Day Sclater, Corinne Squire, and Amal Treacher, editors

Narrative and Genre: Contexts and Types of Communication
Mary Chamberlain and Paul Thompson, editors

The Clash of Economic Cultures:
Japanese Bankers in the City of London
Junko Sakai

THE POLITICS OF MEMORY

COMMEMORATING WAR

TIMOTHY G. ASHPLANT
GRAHAM DAWSON
MICHAEL ROPER
EDITORS

Memory and Narrative Series

TRANSACTION PUBLISHERS
NEW BRUNSWICK (U.S.A.) AND LONDON (U.K.)

First paperback edition published in 2004 by Transaction Publishers, New Brunswick, New Jersey, by arrangement with Routledge.

This book is printed on acid-free paper that meets the American National Standard for Permanence of Paper for Printed Library Materials.

Library of Congress Catalog Number: 2004044942
ISBN: 0-7658-0815-3
Printed in the United States of America

Library of Congress Cataloging-in-Publication Data

Politics of war memory and commemoration.
Commemorating war : the politics of memory / Timothy G. Ashplant, Graham Dawson, and Michael Roper, editors.
p. cm. — (Memory & narrative series)
Originally published: The politics of war memory and commemoration. London ; New York : Routledge, 2000, in series: Routledge studies in memory and narrative.
Includes bibliographical references and index.
ISBN 0-7658-0815-3 (pbk. : alk. paper)
1. War and society. 2. Memory—Social aspects. I. Ashplant, T. G. II. Dawson, Graham, 1956- III. Roper, Michael, 1959- IV. Title. V. Memory and narrative.

HM554.P65 2004
303.6'6—dc22 2004044942

Contents

Contributors

T.G. Ashplant is a member of the Research Centre for Literature and Cultural History at Liverpool John Moores University. He has published on psychoanalysis and history, and the life-writings of working-class men and women in Britain; and is currently researching gender, class and politics in life narratives.

Ann Curthoys is Manning Clark Professor of History at the Australian National University, Canberra. Her recent publications explore radical political movements in Australia in the 1960s and 1970s, especially Women's Liberation, the movement for Aboriginal rights and the movement opposing Australian involvement in the war in Vietnam. She has published extensively on questions of nation and identity, and, jointly with John Docker, on the textual specificity of historical writing. With Mary Spongberg and Barbara Caine, she has recently begun the major task of compiling an international companion to women's history.

Graham Dawson was a member of the Popular Memory Group at the University of Birmingham and now teaches Cultural and Historical Studies at the University of Brighton. He is the author of *Soldier Heroes: British Adventure, Empire and the Imagining of Masculinities* (London and New York: Routledge, 1994); and a co-editor of, and contributor to, *Trauma and Life Stories* (London and New York: Routledge, 1999). He is currently writing a book on the politics of memory within the Irish Troubles and peace process, to be published by Manchester University Press.

Stephen Garton is Professor of Social and Cultural History at the University of Sydney. His most recent book was *The Cost of War: Australians Return* (Oxford University Press, 1996), and he is currently exploring the government of problem populations in Britain, Australia and America 1880–1930.

Elizabeth Jelin is a sociologist, and a Senior Researcher at CONICET (National Council of Scientific and Technical Research) and the University of Buenos Aires. She has done extensive research and writing on social movements, citizenship, human rights, gender and the family. She is a member of the World Commission on Culture and Development (United Nations and UNESCO).

Susana G. Kaufman is a clinical psychologist, and Professor in the Faculty of Psychology, University of Buenos Aires. Her main current research interests are in memory and trauma, and youth.

Yosefa Loshitzky is Senior Lecturer in the Department of Communication and Journalism at Hebrew University. She is the author of *The Radical Faces of Godard and Bertolucci* (1995) and *Conflicting Projections: Identity Politics on the Israeli Screen* (forthcoming) and the editor of *Spielberg's Holocaust: Critical Perspectives on Schindler's List* (1997). She is currently writing a book on the 'other' and the 'stranger' in European cinema. She has also written extensively on film, media and culture for a variety of journals and book anthologies and is on the editorial board of *Cinema Journal.*

Paulo de Medeiros has held the Chair of Portuguese Studies at the University of Utrecht since 1998. He has published articles on comparative literature, literary theory and modern Portuguese literature. He was guest-editor for a special issue of *Discursos* on 'Literature and National Identity' (1996), and was co-editor of *Portuguese Literary & Cultural Studies* (1998 and 1999), as well as of the first issue of *Ellipsis*, the journal of the American Portuguese Studies Association (1999) on the topic of 'Engendering the Nation'.

Bill Nasson is a Professor of History at the University of Cape Town. He has researched on the South African War 1899–1902, and on South African experience of the First World War. His publications include *Abraham Esau's War* and *The South African War 1899–1902*. He is an editor of the *Journal of African History*.

Petri J. Raivo is a PhD and docent in cultural and landscape geography working as a Senior Assistant at the Department of Geography, Oulu University in Finland. His current research as a Senior Fellow of the Finnish Academy (Project 4872) is concerned with the nationalistic landscapes of war and conflicts.

Michael Roper works as a social and cultural historian in the Department of Sociology at the University of Essex. His previous books include *Manful Assertions: Masculinities in Britain since 1800* (Routledge, 1991; co-edited with John Tosh) and *Masculinity and the British Organization Man since 1945* (Oxford, 1994). His current research is on the biographical shaping of management theory in the twentieth century, within the context of an interest in subjectivity, masculinities and life-story methods.

Peter Sjølyst-Jackson is a part-time lecturer in film studies and cultural theory in the Department of Communication, Culture and Media, Coventry University. He is currently writing a DPhil at the University of Sussex on problems around democracy in the interpretation of literary modernism and fascism. His research interests include psychoanalysis, deconstruction, postcolonial theory and film theory.

Jo Stanley is a writer and prize-winning cultural historian, specializing in women's work, especially at sea. She lectures at Leeds, Bradford and Manchester universities. Her books include *Writing Out Your Life: A Guide to Writing Creative Autobiography* (1998) and *Bold in Her Breeches: Women Pirates Across the Ages* (1996).

Preface

In recent years the issue of war memory and commemoration has been given an increasingly high profile in debates both inside and outside the academy. This volume contributes to those debates by exploring the relations of power that structure the ways in which wars can be remembered, across forms that range from public commemoration orchestrated by nation-states through to the personal testimonies of war survivors; and from the cultural memories of war represented in films, plays and novels, through to juridical investigations of wartime atrocities in courts of human rights. By the *politics* of war memory and commemoration, we signal the contestation of meaning that occurs within and between these various forms and practices, and the (unequal) struggle to install particular memories at the centre of a cultural world, at the expense of others which are marginalized and forgotten.

The book is international in scope and presents a series of case studies which explore the politics of war memory and commemoration in a diverse range of national contexts: Argentina, Australia, Britain, Finland, Israel, Norway, Portugal and South Africa. While we make no claims for global 'coverage', we have assembled a collection of essays with a genuinely global reach, from contexts which have the advantage of not commonly being considered in relation to one another. Our purpose in gathering these together is threefold. First, this collection extends international discussion of war memory and commemoration beyond the relatively well-trodden terrains of the Vietnam War and the two World Wars in Europe, encompassing lesser-known national histories and a wider range of wars. Second, it offers insights into the general processes involved in war memory and commemoration, informed by the distinctive concerns, political debates and traditions of intellectual enquiry specific to these particular national cultures. Third, by juxtaposing work from different contexts, it encourages comparative reflection and debate with the potential to generate fresh insights and perspectives.

The book is organized as follows. Part I, 'Framing the issues', consists of an essay in four sections by the editors, T.G. Ashplant, Graham Dawson and Michael Roper, which makes a critical intervention in debates about how the issues involved in war memory and commemoration are to be understood, and what is involved in the 'politics' of these practices. The first section presents a critique of three important paradigms in the study of war memory and commemoration: the 'state-centred' approach exemplified by the work of Eric Hobsbawm and Benedict Anderson; the 'social-agency' approach typified by the work of Jay Winter and Emmanuel Sivan; and the 'popular-memory' approach developed in studies by the Popular Memory Group at the University of Birmingham, and by Alistair Thomson. It argues the need for a theory able to integrate insights from each of these approaches, so as to identify the transactions and negotiations that occur between the various agencies involved in producing war memories: those of the state, civil society, 'private' social groups and individuals. In the subsequent three sections, the implications of this critique are developed in more detailed arguments, each of which addresses a particular dimension of the politics of war memory and commemoration. Thus, section 2 explores the processes whereby individuals and social groups articulate their war memories into narratives, the arenas within which they seek recognition for those memories and the agencies through which they act. Section 3 analyses what is involved in the politics of subjectivity and the complex mediations between representations of war memory and individual psychic responses. Section 4 examines how the national narrative of the nation-state is itself now increasingly subject to hegemonic framing at a global level. It focuses on the shaping of war memories through the political and human-rights narratives articulated through transnational agencies such as multi-state alliances and the institutions of international law; and concludes with a consideration of the transnational dynamics of change which are currently reshaping the articulation of war memories. The concerns of Part I of the volume are complemented by Part III, 'Debates and reviews', in which T.G. Ashplant presents readings of recent work on the politics of war memory and commemoration in Britain, France and Italy.

Part II consists of nine case studies. Each of these addresses issues raised in Part I in a variety of ways and from a range of different angles, refracted through the specific factors at work in its particular national or regional context. They are of interest both for the differences between them that emerge from these contexts, and for the common elements and issues that they share. In Part I we suggest ways of reading each of the chapters in relation to the concerns we explore there; indeed, our own thinking about these issues has developed very much in editorial dialogue with the contributors. The order of the chapters is loosely based on the analysis of the agencies and arenas of articulation, ranging from the most private to the most public, developed in section 2 of Part I. Thus, the essays are arranged in a sequence that begins with those that are more centrally concerned with the memory-politics of war as practised by the state and the political organizations that contest its power; moves on to those which explore the representation of war memories within a national culture and in particular cultural forms; and ends with those centred on the personal memories of individual survivors of war. However, the richness of all these studies is such that there are numerous other ways of reading them, and connections between them, which readers will no doubt discover for themselves.

Elizabeth Jelin and Susana G. Kaufman's study (in Chapter 2) of memories of the military repression in Argentina is placed first in the sequence because it is the most explicit exploration of the complex processes involved both in the construction of a public memory of conflict, and in personal remembrance among individuals. They analyse the struggle by the victims of state terrorism to secure public commemoration of the traumatic events under the military dictatorship; and explore the complex inter-relationships between private memories hitherto repressed, those made public within the legal processes which have been conducted, and the imperative to limit judicial and memorial processes in the interests of political stability, which has encouraged state-sponsored oblivion.

Bill Nasson (in Chapter 3) traces the decades-long process by which participant and civilian memories of the Anglo-Boer War were mobilized in support of an effort by Afrikaners unreconciled to defeat, to win state power within the new Union of South Africa, a political project which was crowned by the National Party electoral victory in 1948. Nasson further examines the diminishing political leverage of its narrative of courage in adversity in the struggle by extreme right-wing Afrikaners to prevent majority rule in the 1990s; and explores its potential future transformation into an element within a unified national narrative presenting all South Africans as having in their own ways fought and suffered.

Ann Curthoys (in Chapter 4) analyses the ways in which the commemoration of war in the Australian context is implicated in racial exclusion. She argues that, in this settler society, the 'forgetting' of colonial conquest was repeated in the exclusion from citizenship of Aboriginal ex-servicemen from both the First and Second World Wars. Her chapter focuses on the ways in which the established national narratives of war became a central target of civil rights campaigns from the 1960s as they sought to challenge this exclusion.

Petri Raivo (in Chapter 5) examines the centrality of the Second World War within the national memory of Finland, analysing its major themes in relation both to the forging of internal unity and post-war consensus, and to Finland's external relations with the USSR. Focusing particularly upon the inscription of memory within heritage landscapes such as cemeteries and battlefield sites, he traces a shift in the forms and function of war commemoration, from mourning and memorializing the dead in the immediate post-war decades, to the transmission of memory to the successor generations through heritage tourism since the 1980s.

Peter Sjølyst-Jackson (in Chapter 6) offers readings of the films, *The Cruel Sea* and *Ni Liv*, to compare the representation of war memories within the British and Norwegian post-war national cultures. He shows how each film attempts, through its portrayal of courageous masculinity, to negotiate the internal conflicts of its respective society during the Second World War; but argues that, both through the thematics of plot and through the repetition inherent in the very medium of film itself, wounds civic and psychic are shown to persist, despite narrative efforts at closure.

Yosefa Loshitzky (in Chapter 7) considers the significant role of films made by second-generation film-makers since the 1980s in questioning the dominant Zionist Shoah-memory in Israel, and in opening up new perspectives on the legacy and continuing effects of the Shoah. In a subtle and detailed analysis of one particular film, *Don't Touch My Holocaust*, and the play, *Arbeit Macht Frei vom Toitland Europe*, whose production it partly documents, she teases out the subversive potential of art which invites identification across divisions – between survivors and successors, *Ashkenazi* and *Mizrahi*, and Israelis and Palestinians.

Paulo de Medeiros (in Chapter 8) frames his essay within the context of the twenty-fifth anniversary of the 1974 revolution in Portugal, which ended half a century of dictatorial rule, and resulted in the ending of the colonial wars in Africa. He shows how the official remembrance of the revolution largely 'forgets' the colonial wars, and, through a detailed account of key novels of the wars, shows the role of literature in negotiating and confronting this official silence.

Focusing on a single autobiography, that of an Australian veteran of the First World War, Stephen Garton's essay (Chapter 9) also explores the role of literature in establishing 'counter-narratives' to official memory. He argues that the nature of the war experience, and its oppositions between home and front, civilian and soldier, laid the basis for the construction of a sectional memory amongst ex-servicemen in the inter-war period, which was simultaneously disenchanted and nostalgic.

Jo Stanley's essay (Chapter 10) – based on research at a treatment centre for British ex-servicemen who suffer from post-traumatic stress disorder (PTSD) – looks at PTSD as an example of 'involuntary remembering'. She explores the relationships of sufferers and their carers to the public memory of war, popular and official, and their participation – often, as part of the attempt to make memory biddable – in establishing alternative forms of public remembrance.

Finally, the editors would like to thank Dr Jan-Werner Müller and Dr Diana Wong for permission to cite their forthcoming publications; the Institute of Historical Research, University of London, for the use of meeting rooms; and Alison McDowall for help in preparing the index. T.G. Ashplant wishes to thank the School of Media, Critical and Creative Arts, and the Literature and Cultural History Research Committee, Liverpool John Moores University, for the award of study leave during the preparation of this volume; Sheena Streather and the staffs of the Aldham Robarts Learning Resource Centre and the Inter-Library Loans Unit, LJMU, and of the Upper Reserve and Upper Camera, Bodleian Library, University of Oxford, for facilitating access to a wide range of texts; and Dr Elspeth Graham for continuing support, advice and encouragement.

Part I

Framing the issues

1 The politics of war memory and commemoration

Contexts, structures and dynamics

T.G. Ashplant, Graham Dawson and Michael Roper

1 The study of war memory and commemoration

1.1 The re-kindling of interest in war memory and commemoration since the 1980s

During the last two decades there has been a proliferation of public interest and concern throughout the world in the various cultural and political dimensions and phenomena of war memory, and in the forms and practices of war commemoration. While this complex general development has yet to be adequately explained, a number of its key features can be clearly discerned. First, its most powerful, transnational manifestation has been the emergence into public visibility of the Shoah, through a variety of projects ranging from the establishment of new museums and the production of documentary and fictional films, to the campaigns to trace and bring to justice Nazi war criminals, and to restore the so-called 'Nazi gold' and other stolen property to the Jewish victims and their families. The debates surrounding the necessary remembrance, commemoration and reparation of Nazi genocide have had the most intense impact in the USA, Israel and Germany, but their resonance has been felt to varying degrees in all those parts of Europe where Jews were exterminated, and in countries throughout the world in which Jewish refugees made their homes, or to which their persecutors sought to flee.[1]

Second, social groups suffering injustice, injury or trauma that originates in war have become increasingly prepared to demand public recognition of their experience, testimony and current status as 'victims' or 'survivors'. Shoah survivors are prominent amongst such groups, which also include organizations representing veterans of combat from the two World Wars and other conflicts, as well as civilians whose lives have been damaged by war. Frequently, the demand for recognition also involves claims for material compensation and reparation requiring some form of official investigation in the political or legal spheres, thus acting as a further stimulant to public debate. As public recognition of the traumatic experiences undergone by survivors of war has increased, so the ageing of those who lived through the wars of the early and mid-twentieth century has added an urgency and poignancy to the endeavour of collecting their testimony and reflecting on its significance. For example, a sense of the pressing need to record the memories of survivors now nearing the end of their lives has reinforced the continuing debate over how if at all the Shoah may be represented.[2] Similarly, as the First World War generation dies out, at the very end of the span of living memory (in Britain today, it is believed there are still some 4,000 surviving servicemen now aged about 100),[3] there has been in recent years a resurgence of interest in that war and its meanings, and fresh imaginative responses from the grandchild generation, often explicitly addressing the question of cultural memory and the complexity of relations between past and present.[4]

A third factor is the increasing number and enhanced profile of anniversary commemorations to mark the beginning and ending of wars, and their key episodes. This is one component of a wider anniversary boom, fuelled and amplified by the public communications media, which seize upon forthcoming commemorative dates to stimulate cultural production of all kinds. Not only are commemorative ceremonies and other events reported in – and increasingly, staged for – the news media, but their contemporary meaning is scrutinized as well as celebrated in special publications, investigative reports and documentary features in which broader cultural and political significances are given a 'human face' by survivor testimony. In this way, war commemoration is transformed into a media event. This process is occurring world-wide, but the case of Britain, where the editors of this volume all live and work, can stand as an example. Here, a string of war anniversaries were propelled into high public visibility as media events during the 1980s and 1990s. A bonanza of fortieth- and fiftieth-anniversary dates commemorated the progress of the Second World War, from Dunkirk, the Battle of Britain and the Blitz in 1980 and 1990 through to the official state ceremonies marking D-Day in 1984/1994 and VE- and VJ-Days in 1985 and 1995. Besides affording occasions for retelling the mythic national narrative of 'our finest hour' to those already familiar with it, and for extending its public reception to a younger generation, these anniversaries also focused a national debate about the trajectory of British history since 1945, including the relationship of the UK to its 'European partners'.[5] In 1993 and 1998, the seventy-fifth and eightieth anniversaries of the Armistice ending the First World War saw a resurgence of public interest evidenced in an increase in poppy-wearing among the young, a popular, media-orchestrated campaign to restore the long-discontinued minute's silence on Armistice Day, and vast audiences both for a plethora of television documentaries, and for Sebastian Faulks' and Pat Barker's bestselling novels about the war and its memory.[6] In 1999, the ending of the twentieth century provided the occasion for media remembrance of the hundredth anniversary of the Anglo-Boer War, while major television documentary series – *War of the Century* (recounting the Nazi–Soviet confrontation of 1941–45) and *The History of the Twentieth Century* – reaffirmed the central importance given to the two World Wars in the way 'the story of the century' can be told from a British perspective.[7] The enhanced public visibility of these anniversary occasions has created opportunities for contesting as well as celebrating received memories. In 1995, for example, the national focus on VE-

Day was challenged by British survivors of the Japanese labour camps for prisoners of war in the Far East, who were still seeking governmental support for the campaign to secure reparation for their ill-treatment from the Japanese government.[8]

Finally, public concern with the memory of war has been stimulated by the collapse of the Soviet Union and the so-called 'Eastern bloc' in Europe (1989–91), the consequent ending of the Cold War and the realignment of power relations in what were once polarized spheres of influence throughout the world.[9] A new pattern of warfare has emerged in the 'post-communist' world of the 1990s, linked to what Jeanne Vickers has termed 'the explosion of ethnic strife', fought between peoples in the name of linguistic, religious or cultural affiliations and differences.[10] The disintegration of the Soviet Union itself generated an estimated 125 ethnic or minority disputes, and war broke out in Georgia, Moldova, Ossetia, Nagorno-Karabakh and, later, Chechnya (in 1994–96 and again in 1999–2000).[11] In Central Africa in 1994, the Rwandan genocide perpetrated by Hutu militias against the minority Tutsi tribe claimed an estimated 800,000 lives in 100 days, including a three-day massacre of 50,000 people in Murambi.[12] In the West, it has been the brutal ethnic conflicts in the former Yugoslavia which have made the greatest public impact and brought home the character of the new warfare. In Bosnia and Herzegovina, as many as 279,000 people are thought to have been killed between 1992 and 1995, and 2.7 million people (the great majority being Muslims) driven from their homes as refugees.[13] As the term 'ethnic cleansing' has entered popular discourse, so too has the possibility of recognizing the extent to which present-day enmities and antagonisms may feed off memories of past wars, and the cultural constructions of otherness encoded therein. In the case of the Serbo-Croatian War of 1991, the evident connections between current conflict and the unresolved legacy of the Nazi wars of national expansion and ethnic 'purification' have fed into renewed debate about the role of war commemoration and indeed war itself in the reproduction of national and ethnic identities.[14]

If these broad cultural and political developments provide a set of necessary contexts for a book on *The Politics of War Memory and Commemoration*, a further context is the proliferation of academic research and critical enquiry – particularly by historians – which has accompanied these developments, reflected upon them and itself contributed to the widening public interest in the phenomena of war memory and commemoration. Reviewing a clutch of recent studies of the war memorial, for example, Catherine Moriarty suggests that their 'simultaneous publication … is evidence of a surge of activity in the last twenty years which replaces the previous absence of scholarly literature'.[15] British art historians began serious investigation of war memorials 'as socially as well as sculpturally significant' during the 1980s; influenced on the one hand by Paul Fussell's earlier work on war experience and the 'literary means by which it has been remembered', and on the other by French studies of the relation between iconography, collective identity and '*les lieux de mémoir*' (sites of memory).[16] This new kind of cultural history challenged conventional British military, political and economic history-writing on the First World War through its emphasis on memory and meaning. At the same time, it helped to open up wider access to 'the space of remembrance', which hitherto had been 'the preserve of a few distinct groups' – traditional historians, students of war literature, war veterans, and the collectors of memorabilia.[17]

The process described by Moriarty is not restricted to the First World War or to the war memorial as a form. As 'memory studies' has emerged across a wide disciplinary spectrum, and developed into a field of interdisciplinary investigation with an international range during the 1980s and 1990s, war has proved to be one of its most productive and compelling subjects. Vast scholarly literatures produced mainly in the USA now exist on, for example, remembrance of the Shoah and of the American war in Vietnam; and the study of war memory is becoming an established element in national historiographies including those of France, Italy, Germany, Britain and Australia.[18] Increasingly, this work is beginning to push towards transnational, comparative studies, exemplified by Jay Winter's *Sites of Memory, Sites of Mourning* (1995); in Britain, the mid-to-late 1990s has seen a proliferation of academic conferences and the first books on this theme.[19]

The present volume is offered as a contribution to these developing debates, both theoretical and historical, about the significance of war memory and commemoration at the start of the twenty-first century. In referring in our title to the 'politics' of war memory and commemoration, we mean to signal the importance of retaining a live connection between the intellectual debates developing within the academic world and the wider contexts and constituencies where these issues are salient. A distinguishing feature of the most compelling 'academic' research on the cultures of war memory has always been its willingness to step outside the academy so as to engage critical theory and analysis in a more self-conscious dialogue with 'living memory'; that is, with cultural producers, political and civil rights activists, and those who have perpetrated military violence and been affected by it; for all of whom the remembering and forgetting of war is not an object of disinterested enquiry but a burning issue at the very core of present-day conflicts over forms of the state, social relations and subjectivity. All of the essays collected here, in different ways, are responses to vital concerns, and are informed by – and enter into dialogue with – specific debates and practices in the cultural politics of their own location. Each makes a contribution, not only to cross-cultural intellectual debate about the general issues involved, but also to the ways in which the remembering and commemoration of war is being understood, organized and contested within its specific context.

There is a danger in the push for internationalization at the level of theory and method – linked as it is to the growing interest in war and memory within the global academy – that the local or national specificity of the kinds of work represented in this volume may be abstracted, decontextualized and robbed of its potent connection to the cultural and political conflicts which have underpinned and inspired it. In introducing the studies which follow, then, we have wanted to insist on the importance of recognizing the particular political contexts and local 'applications' of their analyses, whilst also reading these essays for their contributions to the more general and theoretical debates about the processes and agencies that shape the common phenomena of war memory and commemoration. In this endeavour, our conception of the politics of memory has furnished a framework in which their various and varied, as well as shared, concerns can be inter-related. However, this has entailed developing an expanded notion of what the 'politics of memory' might involve, of the different agencies in play and of the various arenas in which they operate, if the full complexity of these inter-relations is to be grasped. The task of constructing a theoretical framework of this kind has required a critical engagement with existing work in the field, taking issue both with those perspectives which eschew *any* conception of a 'politics' of memory, and with those which construe it too narrowly or in an insufficiently mediated fashion. We turn next, then, to critical analysis of existing approaches.

1.2 *Politics or mourning? Current approaches to the study of war memory and commemoration*

War memory and commemoration have tended to be studied within one of two main paradigms. On the one hand, there can be found arguments which construe its significance as fundamentally political; that is, as a practice bound up with rituals of national identification, and a key element in the symbolic repertoire available to the nation-state for binding its citizens into a collective national identity. On the other, war memory and commemoration is held to be significant primarily for psychological reasons, as an expression of mourning, being a human response to the death and suffering that war engenders on a vast scale. Across the range of work that currently exists on this theme, there is a tendency to construe these two paradigms as if they were unrelated alternatives, and to focus analysis in terms of one or the other. In part these differences of emphasis are a product of disciplinary divisions within the emerging field of memory studies; between, for example, those historians influenced chiefly by political science, international relations or sociology; and those influenced by anthropology, cultural criticism or psychoanalysis. The differences can be exemplified by comparing the 'political' emphasis found in the widely influential work of Eric Hobsbawm and Benedict Anderson, with the 'psychological' emphasis in the work of Jay Winter (whose marrying of art and social history has emerged from that current of work on war memorials sketched by Moriarty), and in Winter's collaboration with Emmanuel Sivan.[20]

Hobsbawm and Terence Ranger's *The Invention of Tradition* (published in 1983) is concerned with the role played in modern societies by constructed versions of 'the past' – and of the continuity between past and present – in establishing social cohesion, legitimizing authority and socializing populations into a common culture.[21] Hobsbawm argued that these 'invented traditions' were 'highly relevant' to analysis of 'the "nation" with its associated phenomena:

nationalism, the nation-state, national symbols, histories and the rest'.[22] This argument meshed with Anderson's *Imagined Communities* (1983), in which 'the nation' is seen as a collectivity defined by its own ways of imagining itself and the identity of those who belong to it. Anderson locates at the centre of this process those 'ghostly imaginings' whereby the living generations feel their connection with the dead who belong to the 'same' national community, thus securing the nation's imagined continuity and transcendence of time.[23] For Anderson, tombs of Unknown Soldiers are the quintessential figures of such imaginings: evoking both the sacrifice that may be required from the citizen as the cost of belonging, and the means by which the nation-state persuades its citizens to die for it.[24] Besides opening up new ways of thinking about nationalism, these two books helped bring into focus the relations between the nation-state, the 'invented traditions' and 'imaginings' that give shape to national identity, and the forms and rituals of war commemoration, thereby generating a set of questions about the power of such commemoration to draw upon the sacrifice and loss occasioned by war as a means of preserving and reinforcing (or, it may be, challenging) dominant elites and ideologies within existing states.[25]

Jay Winter has explicitly criticized work that, in his view, over-emphasizes the political meanings of war remembrance, including those of nationalism: foremost among 'the distorting effects of a narrowly national approach', he argues, is its failure to acknowledge or address adequately the 'existential' function of mourning in commemoration.[26] In opposition to such work, Winter has attempted, as both author and editor, to position his own concerns as central to the way these issues are explored and understood. In *Sites of Memory, Sites of Mourning* (1995), his primary interest is in the role of First World War commemoration in translating individual grief into public mourning for the dead, stimulated by a universal human desire for psychological reparation of loss, in response to the traumatic impact of death in war. It is significant that his analysis is a comparative one that spans both sides of the conflict in Europe: his intention is to show that the work of mourning is a shared human impulse and knows no national boundaries. Winter's history reconstructs the variety of ways in which the bereaved come together to create forms and rituals to express and assuage the personal, psychological loss which they share on such a vast, collective scale. In *War and Remembrance in the Twentieth Century* (1999), he and Sivan explore these issues in a wider international context.

If the work of Hobsbawm and Anderson, on one hand, and Winter and Sivan on the other, is indicative of a tendency to separate the role of war commemoration within the political project of the nation-state, from its role in the work of collective mourning, these different emphases also lead to distinctive ways of theorizing the processes involved in remembrance and commemoration. Again (and despite protestations to the contrary), these tend to be mutually exclusive, with the former approach prioritizing the ways in which war memory is shaped centrally by the state, and the latter the work of remembrance performed by the agencies of civil society; as if the processes operating in each of the two domains were not inter-related and indeed constitutive of each other.

Thus, Hobsbawm, writing of Europe in the period 1870–1914, is careful to distinguish what he terms the 'official' and 'political' invention of tradition as practised 'by states or organized social and political movements', from its 'unofficial' and 'social' practice 'by social groups ... whose objects were not specifically or consciously political, such as clubs and fraternities'.[27] Yet in his analysis, the primary, shaping role of the state during this period is left in no doubt: '[T]he state linked both formal and informal, official and unofficial, political and social inventions of tradition ... [and, s]een from below, increasingly defined the largest stage on which the crucial activities determining human lives as subjects and citizens were played out.'[28]

In Winter and Sivan's work, the endeavour to refocus attention on hitherto neglected dimensions of remembrance does not deny the importance of the state but downplays it; indeed, in practice, it pushes it out of the frame of consideration. This can be seen in the theoretical introduction to *War and Remembrance in the Twentieth Century*, where they 'depart from those who define [collective memory] as the property of dominant forces in the state', signalling instead an intention 'to examine collective remembrance ... as the product of individuals and social groups who come together, not at the behest of the state or any of its subsidiary organizations, but because they have to speak out'.[29] Later, announcing that 'among these agents [of remembrance], we have chosen to concentrate on those coming from civil society because state agency and manipulation have been sufficiently well documented', they concede a role to the state whilst effectively dismissing its relevance to their analysis (and construing its effects, over-narrowly, as 'manipulation').[30] Winter's own contribution to the volume has 'agents of remembrance' operating, rather imprecisely, 'in the borderlands linking families, civil society, and the state'.[31] However, the essay's interest is exclusively in 'ordinary people' taking 'small-scale, locally rooted social action' within networks of family, local community, and 'fictive kinship'. [32] Although they have to 'deal with the state time and again', the sole substantive point made about this relationship is a reference to the struggle on behalf of the disabled 'to wrest from the state ... decent pensions and assistance in rebuilding their lives'.[33]

These dichotomies and polarizations – between an emphasis either on politics or on death and suffering, and between theoretical approaches that privilege either the state or civil society – have had a deleterious effect on the study of war remembrance. In reality, these various aspects of the phenomenon cannot be separated out quite so tidily. The politics of war memory and commemoration *always* has to engage with mourning and with attempts to make good the psychological and physical damage of war; and wherever people undertake the tasks of mourning and reparation, a politics is *always* at work. As Anderson goes some way towards recognizing, the commemorative rituals and patriotic rhetoric of the nation-state (or the nationalist movement) are involved precisely in making particular meanings about death in war: the 'noble sacrifice' of 'dying for your country'. Mourning activity in civil society takes place in the context of such 'official' meanings and understandings, which influence who can be commemorated and in what terms.

Indeed, in modern times, when the state usually assumes responsibility for regulating the treatment and disposal of the corpse, it also regulates what are acceptable practices of grieving and mourning. For example, Catherine Merridale has shown how, in war-torn Soviet Russia during the 1920s, the Bolsheviks launched an assault on the customs, rituals and supposed superstitions of traditional Russian funerary culture.[34] This led rapidly to their 'enforced abandonment' and replacement with officially sanctioned socialist alternatives, notably 'the practice of secular, scientific cremation'.[35] 'Soviet power', argues Merridale 'brought changes, many of which made mourning more difficult for those who survived'; indeed, for many people, 'the processes of loss and mourning were disrupted beyond recognition'.[36]

Unhelpful dichotomies obscure the full complexity and significance of the commemorative processes being addressed. As Alex King remarks, the primacy that is often given to 'a single psychological process or political intention in the formation of commemorative symbols leads … to an incomplete account of both their origins and their social functions'.[37] Under-conceptualized in both the state-centred and the social-agency approaches is the complicated relation between the various different agencies and arenas that are in play: those of individual memory, remembrance in civil society and national commemorative practices organized by the state. Two important detrimental consequences flow from this.

First, the very notion of a 'politics of memory' is impoverished. The 'politics of memory', according to Hobsbawm's analysis in *The Invention of Tradition*, is an 'exercise … in social engineering' from above: 'the history which became part of the fund of knowledge or the ideology of the nation, state or movement is not what has actually been preserved in popular memory, but what has been selected, written, pictured, popularized and institutionalized by those whose function it is to do so'.[38] From this perspective, the study of war memory is concerned with its official orchestration and embodiment in ceremonial and physical reminders, and with the patterns of inclusion and exclusion that determine which aspects of collective and individual experience are admitted to public recall and commemoration. However, such an approach may over-play the unity of social elites and tends to take for granted their capacity to touch off popular identifications. The impossibility of simply reading off popular concerns and emotional investments from the memorials and rituals of official commemoration is shown by David Lloyd's account of the construction and inauguration in Britain of the Cenotaph and the Tomb of the Unknown Warrior in 1919–20.[39] What his narrative vividly reveals is the process whereby the intended emphases of the political and military elites on a celebration of victory in war were displaced by large-scale and sustained popular pilgrimages for mourning the dead and consoling the bereaved – a displacement which ultimately had to be registered within the official ceremonies themselves. Hobsbawm himself has acknowledged more recently that nationalism 'cannot be understood unless also analysed from below, that is, in terms of the assumptions, hopes, needs, longings and interests of ordinary people, which are not necessarily national and still less nationalistic'.[40] Jay Winter's recent insistence that 'forms of official commemoration rested on a rich undergrowth of unofficial activity'[41] is an argument virtually indistinguishable from Hobsbawm's formulation here. Nevertheless, in Winter's case, it still does not lead to a sufficient recognition of the political dimensions of that activity on the part of the civic groupings and so-called 'ordinary people' with whom he is concerned.[42]

Second, both kinds of analysis are very weak on the ground of individual subjectivity, under-conceptualizing both the richness and complexity of personal memory, and the extent to which it is constructed through cultural practices of representation operating at the levels of civil society and the state. Precisely because state-centred approaches take for granted that the nation-state and its political movements secure the required identification of 'the people' whom they claim to represent, they cannot illuminate the key problem of how public commemoration achieves its subjective hold. Hobsbawm recognizes that 'official ideologies of states and movements are not guides to what is in the minds of even the most loyal citizens or supporters'; nor can it be assumed that 'for most people national identification … excludes or is … superior to, the remainder of the set of identifications which constitute the social being'.[43] Yet his own analyses do not even begin to get to grips with the psychic processes that identification entails.

Humanistic perspectives like Winter's also run the risk of proposing a grossly attenuated model of subjectivity, if for different reasons. In this case, the danger lies in the essentialist assumption of a universal psyche governed by eternal, unchanging emotional needs and desires, held to provide the basis for a common response to bereavement and thus to link humanity together across the nations and the ages. For example, whilst analysing the work of Käthe Kollwitz in *Sites of Memory, Sites of Mourning*, Winter slides from a deeply moving discussion of how Kollwitz's personal distress at the death of her son, Peter, in 1914 found expression subsequently in her memorial, 'Die Eltern', to a claim about the significance of war memorials generally in mediating bereavement. 'Is it fanciful', he asks, 'to suggest that rituals at war memorials, and in particular the reading of the names of the fallen, and the touching of those statues or those names, were means of avoiding crushing melancholia, of passing through mourning, of separating from the dead and beginning to live again?'[44] Such a claim about the supposed function of war memorials in handling the passage from melancholia to mourning risks collapsing the social specificity of Kollwitz's response to grief into a psychic universalism. Furthermore, it assumes that commemorative forms such as 'Die Eltern' straightforwardly express this universal response, and evoke it for viewers of the memorial, past and present, who empathize with Kollwitz's distress. There is no exploration, here, of the articulation of meanings within the process of production itself; nor of the various kinds of identification with such forms, and interpretations of their significance, which viewers actually make and have made. In effect, this argument takes the history out of commemoration.[45]

To conclude this critique, it follows that, in order to escape the dichotomies and polarizations analysed here, it is necessary to theorize the inter-relations between the elements which have been separated out in these competing models; thereby generating a more complex, integrated account of the interacting processes that link the individual, civil society and the state. This requires a redefinition of what constitutes the 'politics' of war memory and commemoration; a redefinition that is capable of embracing the operations of power in civil society as well as in the state; of recognizing the existence of a cultural politics surrounding representation and meaning-making; and of tracing the effects of these processes and conflicts from the social domain into that of the psyche, where they constitute a politics of subjectivity.

While this enterprise can benefit from the concerns and insights of both approaches critiqued above, further rich resources can be found in a third body of work on war memory that uses oral-history and life-story methods, and in the wider debates that surround this work.[46] The diverse traditions that constitute this field – which have been formative for the 'Memory and Narrative' series itself – are not always familiar to, or acknowledged by, practitioners working within the state-centred and social-agency paradigms.[47] However, the life-story paradigm often engages with their primary concerns albeit from a different perspective and starting-point: analysis is more likely to begin from personal memories as recounted in oral-history interviews or other forms of life story, than it is from 'official' commemorative occasions, or national narratives of war, or the war memories generated by artistic and other modes of cultural production. When confronted with the problem of getting to grips with the substance and formal mechanisms of personal war memories, researchers in other traditions of thought have tended to assume that these personal meanings can be read off from the artefacts or practices of public commemoration, or that the intentions expressed by their producers have been successfully realized in their reception, or that the range of possible responses is adequately represented by general public debate in the media.[48] By contrast, the starting-point for researchers working within the life-story paradigm is usually the meanings about war and its remembrance that people make for themselves and express in their own words and stories.

However, the debates concerned with critical interpretation of such accounts have long recognized the necessity of exploring their relation to wider cultural forms and contexts of remembrance. This has prompted theoretical consideration of the interaction between different agencies of remembering and forgetting, and the various arenas in which these practices occur. It is this interactive process – rather than individual remembering *per se* – that is most likely to be taken as the object of study. Unsurprisingly, then, much life-story research on war memories has been concerned with the relationship of individual subjects to national memory in the particular state to which they 'belong'. In this sense, such work has become a primary locus for investigating the 'view from below', identified by Hobsbawm as 'the area of national studies in which thinking and research is most urgently needed today'.[49] At the same time, since oral and life historians are in a position to explore in detail the operation of personal remembering and forgetting, they have also been able to make distinctive contributions to research on mourning, the effects of bereavement and those disturbances of memory associated with trauma. In those studies combining both concerns, we can find explicit attempts to move beyond the 'politics or mourning' dichotomy.[50]

To exemplify this third paradigm, we can point to the work of the Popular Memory Group and its subsequent development by the oral historian, Alistair Thomson.[51] Central to this work is the distinction between 'public representations' and 'private memory'. The former term refers to those representations of the past that 'achieve centrality' within the public domain, where their institutional propagation by the national and local state, the culture industries or the public media ensure their scope to make public meanings for vast audiences.[52] Private memory, by contrast, refers to 'the more privatized sense of the past' which circulates among particular social groups 'in the course of everyday life'.[53] The interaction between these aspects of memory is understood in Gramscian terms, as a hegemonic process of ideological domination and resistance. 'Private memories', the Group argues, 'cannot ... be readily unscrambled from the effects of dominant historical discourses. It is often these that supply the very terms by which a private history is thought through.'[54] By the same token, the power of dominant memories depends not simply on their public visibility, but also on their capacity to connect with and articulate particular popular conceptions, whilst actively silencing or marginalizing others.

The popular-memory approach has informed analyses of the processes of memory as these operate in a number of specific national contexts.[55] In Britain, work by Dawson and West has shown how the dominant memory of the Second World War, representing Britain as an 'island race' united under a 'totemic leader', Sir Winston Churchill, was re-evoked in the rhetoric of the Conservative Party in the early 1980s.[56] In particular, the then Prime Minister, Margaret Thatcher, 'reworked the meanings' of 1940 into a moment when the essential qualities of the British people were expressed, as in the great days of Empire once celebrated by popular imperialism. The reassertion of this imperial British identity found its most strident expression in the prosecution of the Falklands/Malvinas War of 1982, which was scripted in Churchillian terms as a defence of 'our British way of life' against the fascist threat of invasion, and thereby contributed to the marginalization of alternative social-democratic and radical leftist memories of the Second World War. Dawson and West argue that a necessary condition of this mobilization of Conservative popular memory was the 'ubiquitous presence' of public representations of the war from which it could draw: television and film productions, museum and historical tourist sites, military airshows, toy soldiers, advertisements, the vast output of the publishing industry, and the monuments and ceremonies of official war remembrance. A further condition was the war's significance in the personal memories of those who lived through it, and their narration of that experience in the private contexts of the ex-servicemen's reunion or the family gathering.[57] Their analysis points to the 'deep popular purchase' of dominant memory, despite the fact that private memories are always 'potentially ... disconfirming of the dominant public forms'.[58]

In Australia, Thomson's *Anzac Memories* (1994) drew on and developed the popular-memory approach in a study of the founding myth of white Australian nationhood which was forged through the 'baptism of fire' at Anzac Cove and Pozières during the First World War.[59] Based on oral-history interviews with Great War veterans, it sets out to explain the hegemonic power of the Anzac legend to make national meanings about the wartime experience of Australian soldiers (as celebrated in Australian commemorative culture and in popular representations such as the film, *Gallipoli* [1981]); and 'how and why it worked, or sometimes didn't work, for veterans'.[60] The complex entanglement of public and private memory is key to Thomson's approach. The public legend itself is analysed in the process of its historical formation and contestation. Thomson identifies a variety of agents involved in this process, from official war historians and returned-servicemen's groups of left- and right-wing persuasion which literally fought in the streets in 1919, through to the film-makers, anti-Anzac Day feminist protesters, oral historians and elderly diggers of the 1980s. The private memories of the veterans are similarly understood as in process, shifting in relation both to the changing forms and meanings of the public legend, and to the identities of the men themselves at different stages of their life course. Thomson's point in analysing these relations is to show how individual subjectivity is negotiated in relation to the Anzac legend. While this has helped many veterans 'to compose memories of their war ... that they can live with', those aspects of the diggers' experience that are 'not recognised and that cannot be articulated through the public narratives are displaced or marginalised within individual memory', causing alienation, silence and 'internalised trauma'.[61] Furthermore, Thomson shows how the subjective purchase of these narratives, as widely recognized and validated constructions of Australian identity, extended not only to the veterans themselves, but also to those who welcomed their return, mourned their loss and commemorated their contribution to the story of Australia.

Thomson's analysis utilizes an expanded conception of the politics of memory, to demonstrate how subjectivity articulates with the contestation of memory involving agents that operate both in civil society and in the state. This argument, and other work that draws on the popular-memory approach, addresses fundamental problems with the state-centred and social-agency approaches.[62] First, the concepts of public and private memory provide a means of showing the ways in which individual experience is always structured and understood through cultural narratives, including those of the nation-state. In this way, it has moved beyond the limitations of state-centred analyses which are unable to explain how official commemoration and memory achieves its subjective hold. Second, the popular-memory approach – contrary to social-agency arguments – demonstrates the ways in which expressions of grief, far from being 'outside' politics, are always framed within dominant narrative forms and genres, and may be mobilized on behalf of conflicting political positions.

All three approaches discussed in this section have been influential in developing our own understanding of what is entailed in the politics of memory. However, it must also be recognized that their analyses of war memory and commemoration, like those of the contributors to this volume, are products of a particular historical context, and responses to its characteristic political and cultural concerns. Hence the predominant focus on the national context – a mark of both the state-centred and popular-memory approaches – is an aspect of that broader interest in the exercise of power within the liberal-democratic states of the West, which was a preoccupation of Marxist thinking for twenty years after 1968. In focusing on the 'internal' relations of power, much of this work neglected the 'external' pressures of transnational power relations, brought to bear on the nation-state as it were 'from above', as it organizes its relations with other nation-states, and with a range of international institutions from NATO and the Warsaw Pact to the United Nations. Even for so committed an internationalist as Hobsbawm, analysis of the 'invention of tradition' is conducted entirely without reference to those inter-state transactions which, in other kinds of analysis, he would consider to be constitutive factors. In Thomson's work, the fact that Australian national identity was founded upon an assertion of independence from the 'mother country' ensures that British–Australian relations are placed at the heart of his analysis of the Anzac legend; yet only cursory consideration is given to the Australian state's support for American nuclear policy and to its involvement in the Vietnam War, both of which were central to 'internal' conflicts over the memory of Anzac during the 1980s.[63] By contrast, Winter's particular kind of transnational universalism can be contextualized with reference to the resurgence of pan-European liberalism stimulated by the dismantling of the Berlin Wall in 1989; and to the erosion of 'national' autonomy in the name of greater 'European' political unity consequent on the Treaty of Maastricht (1991). If, as we believe, the nation-state remains central to the politics of war memory in the early twenty-first century, it must be recognized that its role is currently undergoing radical restructuring by the processes of globalization. It follows that, to be useful, new theoretical frameworks for making sense of war memory and commemoration must also be sensitive to the dynamics of change that are operating transnationally, and the ways in which these impact within national contexts.

* * *

In the rest of Part I, we present our own analysis of what is involved in studying the politics of war memory and commemoration. The argument is threefold. Section 2 draws on elements from each of the approaches critiqued above, to produce a more nuanced and mediated model of how the politics of war memory and commemoration works. We explore the processes whereby individuals and social groups articulate their war memories into narratives, the arenas within which they seek recognition for those memories and the agencies through which they act. In section 3 we analyse what is involved in the politics of subjectivity, and show the complex mediations between representations of war memory (including forms of commemoration and mourning) and individual psychic responses. The question of survivor testimony is considered in its relationship both to the pre-existing cultural forms (or 'templates') which structure memory, and to the memory-work of successor generations. Section 4 examines how the national narrative of the nation-state is subject to hegemonic framing by the political and human rights narratives articulated through transnational agencies (such as multi-state alliances and the institutions of international law). Analysing the framing of memory by hegemonic master-narratives of the two World Wars and the Cold War, we conclude with a consideration of the transnational dynamics of change which are currently reshaping the articulation of war memories.

2 The struggle to articulate war memories

The nation-state is, we are suggesting, central to the politics of war commemoration.[64] This emphasis on the state need not mean, however, an inevitably state-centred model. A theory of hegemony allows examination of the complex interactions and political/cultural relations that constitute public memory as embodied both in sites and rituals of commemoration, and within national 'mythologies' or public 'legends'. State formations, political parties or movements, and other social agents are all involved in constructing versions of the national past and national identity, selecting from or reworking the repertoire of national stories and symbols to fashion effectively useable public memories for their particular ends and purposes. The dominant national narrative is always contested, the focus for internal conflicts. Even those contributions to this volume which broadly adopt the state-centred paradigm, focusing on the construction of official memory rather than its reception, are alert to the ways in which it may be internally riven and incapable of achieving the political ambitions of its architects. The politics of war memory and commemoration is precisely the struggle of different groups to give public articulation to, and hence gain recognition for, certain memories and the narratives within which they are structured. The history of war memory and commemoration involves tracing the outcomes of particular struggles, as represented both by those memories which are publicly articulated, and by those which have been privatized, fragmented or repressed.

Within this political/cultural process, it is helpful to distinguish several aspects of the struggle to articulate war memories: its narratives, its arenas and its agencies. Narratives of articulation refer to shared formulations within which social actors couch their memories. They extend, in social range, political power and symbolic potency, from hegemonic official narratives cementing international alliances or binding together a nation-state, through oppositional or other sectional accounts of subordinated experiences striving for public recognition, to locally common or shared, and even individual (sometimes fragmentary), memories. Such narratives draw on the language of wider discourses. Pre-eminent among these are the discourses of national identity. Other modes of discourse (for example religious, political and – increasingly – human rights) partially overlap with such national discourses, both contributing to them and offering resources for challenging them. Social groups endeavouring to extend or modify current regimes of war memory may engage with already existing and widely disseminated narratives, or be compelled to elaborate newly emergent ones.

Arenas of articulation refer to those socio-political spaces within which social actors advance claims for the recognition of their specific war memories, and for whatever other benefits they seek to derive from such recognition. They range, in social breadth and political importance, from the networks of families or kinship groups, through those of communities of geography or interest, to the public sphere of nation-states and transnational power blocs. The articulation of memory involves struggles to extend, or alternatively to limit, the arenas within which specific memories are able to circulate, and hence make claims for recognition.

Agencies of articulation refer to those institutions through which social actors seek to promote and secure recognition for their war memories. They encompass the official bodies of the nation-state, the organizations and movements of civil society (where such exists) and more localized face-to-face groupings. These aspects of the articulation of war memory are distinguishable only analytically; they describe what are, in any given case, different dimensions of a single social practice. Nevertheless, we would suggest that analysing specific instances of war memory and commemoration in these terms can help to clarify its politics, by specifying which social groups, via what agencies, are the promoters of a particular narrative addressed to which arenas.

Within a given society, a diverse range of social groups, as well as individuals, strive to articulate their memories of war. Some of these groups have a prior existence independent of war (national or ethnic minorities, political, civic or religious constituencies); others are brought into existence by war itself (veterans, the disabled, war widows, displaced or annexed populations). They vary in their access to political and cultural power, and hence in the resources they command to develop and broadcast their narratives. They may advance claims for the recognition of their memories in any or all of the social arenas. They may work through the agencies of the state, of civil society, or of more informal groupings. The following sections 2.1 and 2.2 examine, first, the narratives through which, and the arenas within which, memories are articulated and, second, the agencies which perform that task.

2.1 Arenas and narratives of articulation

This section will examine the process of articulation by tracking it through a range of arenas from the bottom up. Starting with the formation of individuals' war memories, it will trace the ways in which these can be transformed into the shared/common memories of a social group and then be projected into a public arena, whether promoting a new sectional or oppositional narrative, fitting within or modifying an existing dominant national narrative, or connecting with a transnational narrative. It will be seen that this process requires a constant engagement – whether negotiation or contestation – with the nation-state, its agencies and its narratives.

War memory, from this perspective, is first of all the possession of those individuals, military or civilian, who have experienced war. The crystallization of that experience into war memories is in part shaped through pre-existing cultural narratives, both those specific to that individual and their immediate social environment (such as family myths), and those which circulate through wider social discourses (such as national or generic – for example, warrior – myths).[65] But the memories thus laid down are also formed from the specific, immediate personal experiences themselves; these experiences are not simply the product of, nor totally absorbed into, extant or newly created social discourses, but exist always in (a potential) tension with them.[66] The movement from individual remembering to state commemoration is a transition from direct personal to cultural memory; but these modes of memory are related. Only if it continues to engage with individual memories does cultural memory retain its political effectiveness, while personal memories are in part shaped by pre-existing national and local cultural memories.

Such individual memories may never be articulated in any wider arena. Some may not require expression, being private but not suppressed. Others may be too painful to bear, even when they are apparently consonant with public narratives. As Jo Stanley's striking phrase (in Chapter 10) 'involuntary commemorations' indicates, war memories may escape the individual's control, so that family members and partners become enforced witnesses of, even sharers in, continued suffering. Her account of British war veterans suffering from PTSD, based on interviews with both veterans and their carers, shows how painful memories may continue to haunt those whose service, in a war widely accepted as necessary, is recognized. Medical support services seek to help such veterans find better mechanisms for coping with their memories, by providing a variety of therapeutic techniques and safe arenas for their expression, or by encouraging them to develop their own.

Many war memories, however, both seek and find expression. When individuals can express and compare their memories with the experience of contemporaries, can begin to formulate a shared language and identify common themes, then what may be termed 'shared' or 'common' memories emerge.[67] These are the memories, retained by larger or smaller groups, of the same (shared) or directly cognate (common) wartime events; examples would include the survivors of a single campaign or massacre, or the victims of a common occurrence such as aerial bombing or widespread rapes in newly conquered cities. They are most commonly formulated, preserved and transmitted through face-to-face groupings, ranging from family and kinship networks and gatherings of old comrades to local communities and interest groups. Already in this arena, the transformation of direct personal memory into cultural memory has begun. Such is the intensity of war experience, and of the memories in which it is retained, that it may be transmitted to a survivor's family (spouse, partner or children) so forcefully that they too, who did not live through the events, nevertheless become bearers of a common memory.[68]

The transformation of individual into shared/common memories can, however, be blocked for various reasons. Apart from the consequences of state repression, considered below, particular individual experiences may be (or seem) singular, and – especially if not consonant with official narratives – be withheld from articulation and become isolated. Thomson has shown this to be the case for those First World War Anzacs whose individual memories did not fit the official picture of the heroic Australian which was already being constructed during the war itself. Such individuals felt forced to remain silent, sometimes at considerable personal cost.[69]

Common memories may also be suppressed with the partial collusion of those most directly involved, both through a personal and community sense of shame, and through fear of the political repercussions of bringing them into public debate. A now widely discussed example is the extensive rapes by Soviet soldiers of women in Germany, Austria and Hungary in spring 1945. In the case of Germany, Attina Grossmann has traced the complex process of dissociation by which 'silence' was constructed. In the immediate aftermath of the fall of Berlin, the rapes were widely known. The legal and medical authorities responded by permitting abortions for women who became pregnant as a result. Grossmann points out that the women themselves also talked and wrote about their experiences at the time. Soon, however, since full expression within the political arena risked damaging political consequences, discussion was to be repressed, the events 'forgotten'. In the Western sector, such memories would have weakened the attempts to re-masculinize German society. In the emergent East Germany, the rapes proved a major obstacle to seeing the Russians as socialist models and mentors.[70] Norman Naimark analyses a brief official attempt by the German Communist Party in 1948–49 to discuss this still festering issue publicly under the rubric 'About "the Russians" and about Us'. His analysis of the verbatim transcript of a large and lengthy meeting shows how rape was alluded to – by speakers from both platform and floor – constantly but only obliquely, transposed usually into complaints about thefts of bicycles or watches. The political stakes were too high, however, and the discussion was rapidly closed off.[71] Thereafter, a set of memories which had briefly entered the political arena was returned to the shared/common or the personal. In Austria a similar sequence obtained; official recognition of the crisis resulted in toleration of abortion over the next few months, and then the events were 'forgotten'.[72] Thereafter, in all three countries, though the rapes attained no explicit public articulation, 'silent knowledge' of them was pervasive. The women affected 'locked away' or repressed memories of the fear and trauma they had suffered.[73] It may take a later change of political circumstances such as the collapse of communism, and/or a new development within civil society such as the emergence of the feminist movement, to create a social and narrative space within which shared memories can be articulated by political or cultural organizations into sectional memories claiming recognition. For the women who had been raped in 1945, for example, this happened after the disintegration of the Soviet bloc in 1989.[74]

A particular dimension of war experience, though not suppressed or repressed, may not easily find expression within wider narratives. Where there appears to be such a complete dissociation between official and common memory, the latter may be preserved in the face of public neglect by being acknowledged and shared in private, never acquiring public valency. Thus for many years British civilian internees in Singapore (mostly women) had no immediate means of publicly reckoning with their war experiences. National memory celebrated the defeat of fascism in Europe, yet they were part of a different theatre of war, and prisoners rather than active citizens for the duration of the conflict. Bernice Archer has noted how the keeping of private mementoes by ex-internees – in one case, a teaspoon used in the internment camp – might act as a focus of memory. Here the spaces of private and public commemoration remain separated. The absence of public forms to which survivors could relate their experiences, gives the objects associated with 'private memory' a particular poignancy. Domestic artefacts used in the wartime camps became, within the setting of their post-war families, commemorative objects embodying their courage and survival, around which they could narrate their experiences to their families.[75]

Shared or common memories enter the public arena when they are articulated in some cultural or artistic form, or into a political narrative, whether sectional or official/dominant. The term 'sectional' is used here to refer to those memories which, though they have achieved the level of open public articulation, have not yet secured recognition within the existing framework of official memory. Sectional narratives, to achieve political purchase, have to be articulated in the double sense of 'linked together' and 'expressed'. Even at its early stages, a sectional narrative shapes the individual and common/shared memories from which it is composed, selecting some and excluding others, highlighting key themes and framing them within its preferred narrative tropes. Only when memories have been woven together into a narrative which is both widely held and publicly expressed do they have the potential to secure political effects. Such publicly articulated sectional memory may be subordinate, if accorded only limited or partial recognition; marginalized, if simply neglected or not deemed worthy of recognition; or suppressed, if treated by the nation-state as incompatible with the parameters of the dominant narrative. On the other hand, it may become oppositional, depending on the extent to which its demands for recognition can be accommodated within the hegemonic frame, and the degree of socio-political mobilization it can achieve.

Consequently, the relationships between sectional narratives of war, and hegemonic official narratives, are very varied. If the experiences represented within a sectional narrative are felt to be distinct from, but not necessarily inimical to, those expressed in official memory (for example those of a national or ethnic minority within the state), then that sectional narrative (or elements of it) may be relatively easily incorporated into the official narrative, its authors, too, accorded recognition and esteem. If, however, those experiences have hitherto been actively marginalized or excluded, the social actors mobilizing around such a sectional narrative may be compelled to challenge or undermine the official narrative by claiming a (more prominent) place within it. Such demands for inclusion may expose a profound contradiction within or between dominant narratives, where failure to recognize the involvement of marginalized groups in war draws on a longer history of silence or omission. Thus, in the case of Aboriginal ex-servicemen discussed by Ann Curthoys (in Chapter 4), local exclusion from Anzac Day parades and from veterans' organizations reiterated the white settler myth in which Aborigines were thought to have simply 'disappeared'. The notion of white Australian victimhood, implied in the Anzac legend of the 'birth of a nation', could only be sustained by the continued symbolic and material exclusion from public memory of the war service of those who had earlier been the victims of colonization. Here we see how closely the politics of war commemoration – who counts as an ex-serviceman and who does not – are linked to wider issues of social exclusion and citizenship.

In all societies, different social groups have a differential power to make their meanings and memories central and defining. The weaker and more marginalized have less access to the agencies of either state or civil society, and less capacity to influence prevailing narratives or project their own narratives into wider arenas. This is apparent in relation to both race (the struggle for recognition of Aboriginal ex-servicemen just cited), and class. Thomson's memory-biography of Fred Farrall, a working-class digger, reveals that only when he joined the Communist Party in the 1930s did he acquire access to a language separate from the dominant legend of Anzac in which to express his revulsion from the war, and thereby free himself from some of its continuing effects on his life.[76] Such differential access to power is also very marked as regards gender. In the context of national imaginaries, dominant memory is often centred around the idealized figure of the masculine soldier.[77] This may have the effect that the meanings and memories of others who have sacrificed and suffered – above all women – are relegated to the margins. Stephen Garton (in Chapter 9) shows this process in action in the structure of Raymond Lording's autobiography, as the heroism of the author and his fellow Anzacs, both in war and in the peace-time struggle for recovery and rehabilitation, is defined through a silencing of the (always potentially suspect) Others – women especially, but also non-combatants and the older generation – who together made up the home front.

An excluded group which feels completely alienated from official memory may, if strong and potentially cohesive enough, seek to mobilize its counter-memories into an oppositional narrative, mounting a full-scale challenge for political power. An example of this is unreconciled Afrikaner memory of the Anglo-Boer War (1899–1902), as Bill Nasson portrays it (in Chapter 3). Following their defeat, the Afrikaners took divergent political paths. While some allied themselves with the victorious British when the Afrikaner republics were absorbed into the newly formed Union of South Africa (1910), helping consolidate the new state, others, unwilling to accept the legitimacy of their defeat and its political consequences, set about building up a separate sub-national Afrikaner identity. As part of this process, they collected and publicized memories of the war, of courage and cunning in the field and stoic endurance in the farmsteads and concentration camps in the face of a brutal British enemy and traitors within the Afrikaner camp itself. In this way individual and shared/common memory was translated into public sectional memory. The resultant oppositional narrative helped sustain the growth of a political movement which, some forty-five years after the end of the war, would triumphantly take over state power with the (Afrikaner) Nationalist Party victory of 1948.

In the modern era, it has been the nation which has been the prime arena for the articulation of war memories and the mobilization of commemoration, since war has been central to its identity and symbolic continuity. In Benedict Anderson's influential formulation, the nation secures its symbolic continuity through time, and mobilizes the willingness of current generations to die in its defence, by interpellating them as members of an imagined community which transcends death. War commemoration is a vital moment in that process of interpellation.[78] Both fighting for and dying for one's country help to solidify national identity, one's own or that of one's family and immediate community.[79]

The term 'official memory' refers to those dominant or hegemonic narratives which underpin and help to organize the remembrance and commemoration of war at the level of the nation-state. After each new conflict, such official memory is re-articulated with the aid of pre-existing war narratives, which provide a national repertoire of useable images, plots and figures. Official memory is expressed most explicitly via permanent memorials, and through a calendar of ceremonies (annual and anniversary) which repeatedly recall key wartime events, and meditate on their meaning. But it also exists within what Graham Dawson has termed the national imaginary, a political unconscious filled with resonant imaginative figures and scenarios from the national past.[80] Such is the power of official memory to bestow recognition on those whose experiences it embraces, that where possible claims about the meaning of war, by oppositional voices seeking to advance their own interests, or by individuals asserting the dignity or defending the necessity of their own wartime actions, will be couched within that framework.

Existing elites strive to re-work dominant national narratives when necessary to accommodate memories of new wars, while managing and/or repressing sectional memories which are (or threaten to become) oppositional. We may now trace the articulation of memory in the opposite direction, from the top down, observing how the promoting of official narratives and the marginalizing or repressing of alternatives may have the effect of accommodating, constraining, reshaping or silencing sectional, shared/common and even individual memories. The extent to which existing elites within the nation-state achieve success in this struggle to maintain its hegemonic narratives depends both on the degree of pre-existing internal division and conflict, and on the nature of the war concerned. In a Britain which had come through the First World War victorious and largely united, the government and state agencies had considerable scope to respond to popular pressure in crystallizing its narrative of war remembrance. As David Lloyd shows, the original Cenotaph was intended as a temporary monument for the dead, and a saluting point, at the Peace Day victory parade of July 1919. However, popular reaction, expressed and perhaps amplified by the press, marked it out as the key site of the day, converting it into a sacred space hallowed by reverence and intense emotion. Consequently, the government rapidly accepted the demands which followed for it to be rebuilt as a permanent memorial.[81] Despite this, initial plans for its unveiling in November 1920, in conjunction with the entombing of the Unknown Warrior in Westminster Abbey, reverted to the theme of celebrating victory and the armed forces. Places at both ceremonies were reserved primarily for current servicemen. The outcry which followed, from both press and ex-servicemen's organisations, forced the government into a complete reversal. While ex-servicemen were now included in the procession following the coffin, the central place in the Abbey was given to the bereaved. This proved to be a key element in establishing an official war narrative centred more on the duty, loss and grief of ordinary soldiers and citizens (women as well as men) than on victory and triumph.[82]

It is civil wars (or the social or ideological conflicts unleashed during wars against other nations), and metropolitan struggles against liberation movements, which have proved especially difficult for nation-states to commemorate in ways which do not require the suppression of sectional memories. National narratives, even in these circumstances, have great potential assimilative power, as is shown in Petri Raivo's account (in Chapter 5) of Finnish commemoration of the Second World War. He notes that the custom of returning the bodies of dead soldiers for burial in their own locality, which had originated during the Civil War of 1918, was now adapted to convey a double message. Externally, it emphasized Finland as a Western Christian nation, which did not leave its dead in the hands of the Soviet atheist invader. Internally, building on but intensifying a process of reconciliation begun in the late 1930s, it stressed the unity of all Finns. This emergent discourse of unity began the process of retrospectively absorbing into national memory the defeated Reds of the Civil War; their dead, too, now gained memorials, commemorated as having fallen for their convictions. Even this very powerful national narrative, however, cannot contain all citizens. Those on the left who opposed the national war effort (and its consequent alliance with Germany during 1941–44) later erected their own memorials to those executed for their political activities; but these have been omitted from popular guidebooks and histories. In this case, the character of the war (invasion by a powerful neighbour, from which Finland had only recently secured its independence, and which was itself the heartland of communism) was sufficient to make possible a wide measure of national unity. Other nations invaded and occupied during the Second World War were left with more complex internal divisions, only partially sealed over in wartime and post-war narratives.[83] Still more fractured was the situation in Spain at the end of the Civil War. So determined was Franco's victorious Nationalist regime to efface any oppositional memory of what had been at stake in the war, that they treated half the nation virtually as a foreign conquest, drawing on the pre-existing discourse and iconography of the late medieval *Reconquista* of Moorish Spain by the Catholic monarchs to construct a memorial narrative of triumph.[84]

The complexities of liberation struggles, especially where the colonized territory has been legally incorporated within the metropolis, are illustrated by the Algerian War of 1954–62. This brought down the French Fourth Republic, came near to detonating civil war and ended in defeat, leaving an embittered minority of former settlers unreconciled to their loss. In France, no agreed national narrative of the conflict emerged. In the aftermath of war De Gaulle sought to refocus national energies towards modernization. He relied for legitimation on memories of the Resistance; whereas to remember the divisive Algerian War might recall the more recent, politically ambiguous, circumstances of his return to power.[85] Consequently, some 3 million French conscripts involved in the war faced marginalization. Martin Evans has examined how the main veterans' organization, the FNACA, set about generating a public counter-narrative of their experiences in the face of virtual silence about the conflict. At first, this effort was couched within the existing French regime of official war memory and commemoration. Neither site nor date was readily available for commemoration. Since it was essentially a guerrilla war, fought on what had become foreign soil, there were no obvious places of pilgrimage. Lacking an agreed day of remembrance, the FNACA chose the day of the Algerian ceasefire in March 1962; a politically resonant choice, since the French government had never officially recognized the conflict as a 'war', and thus its conscripts were not initially granted the status of '*anciens combattants*', with the benefits this brought. When their demands for official commemoration on this date were blocked, they inaugurated a strategy of commemoration from below, by local authorities. This too, however, met with only very limited success, since official silence had the support of ex-settler and far-right hostility to any commemoration of what they viewed as a betrayal. Denied significant recognition even after the socialists came to power in 1981, thwarted in their efforts to gain recognition within existing national narratives of war, they turned instead to the emergent transnational discourse of trauma and human rights. They directed their efforts towards educational and reconciliation work, building a sense of self-worth by bringing what they regarded as the lessons of their experience – the need to avoid manipulation by politicians, and to promote racial tolerance and the avoidance of armed conflict – to later generations. They were able to draw on both the work and the paradigms of American Vietnam veterans' organizations, to develop a narrative of French conscripts as victims, both of the pointless war itself and of post-war neglect.[86]

The struggle to maintain hegemony occurs also within alliances of nation-states. As has been noted above, the dominant nation or coalition can ensure that its definitions of war, and its patterns of commemoration, are imposed upon allies and enemies alike. So, during the Cold War, America and Russia alike developed narratives linking their victories of 1945 with their current political goals, enforcing these on their respective alliances.[87] Since the mid-1970s, and more rapidly since 1989, there has emerged a transnational discourse of trauma, victimhood and human rights. Social groups which have been denied recognition for their claims within the arena of their nation-state (such as the French conscript veterans) have begun to construct narratives drawing on this discourse. At present, formal international arenas to which such narratives may be directed are very unevenly developed. But such is the political strength of some of the groups now mobilized around such narratives, for example Jewish and other forced labourers under the Nazis, that they have begun to achieve some significant measure of redress denied them when addressing a single nation-state.

2.2 Agencies of articulation

This section will examine the process of articulation through a range of agencies from the top down. It will start with the efforts of the nation-state, both liberal-democratic and authoritarian, to maintain dominant national regimes of war memory and commemoration, looking especially at the difficulties posed by particular types of civil and liberation wars. It will then trace the roles of organizations in civil society in both cooperating with and challenging the state. Finally, it will examine more informal agencies, and family and kinship networks, which bring together individuals and mediate between them and more formal institutions of state and civil society. The activities of the nation-state impact on all other agencies, reaching down to the very process of memory-formation by individuals and within families; but even an authoritarian state cannot exercise complete control over war memories, and more typical is a process of contestation within which various agencies seek recognition for the experiences of those social groups acting through them.

Different kinds of state power can be examined in terms of the regimes of official memory and forgetting which they institute, and their relative importance in relation to other civil or private agencies. In striving to preserve and adapt existing national narratives around war memory and commemoration, the nation-state plays a leading role. It puts in place a wide range of public institutions (war graves commissions, museums, libraries, archives, official histories) to maintain war memorials and transmit the narratives surrounding them. It may also assume responsibility for the welfare of the returned servicemen who fought in its name, and of the bereaved relatives of those who died. The armed forces themselves have a particular need for war commemoration to consolidate the institutional identity both of the services as a whole, and of component fighting units (such as regiments). Diana Wong points out that it was the Singapore armed forces which led the restoration of official war memory there, because of the need for military history to help build *esprit de corps*.[88]

At the same time, the liberal-democratic nation-state with its political movements also relies heavily upon, and often responds to, initiatives originating from various components of civil society. As Adrian Gregory has shown, crucial developments leading to the evolution of the central national sites and rituals of British First World War commemoration often emanated from individual or semi-official sources, to be adopted and sometimes reshaped through a process of negotiation between interested parties (including the armed forces, veterans' organizations and the Churches) brokered by the government. In the consolidation of the annual cycle of ceremonies over the next decade, and their modification to accommodate changing attitudes to the war, key roles were played by both the state broadcasting service and the (fiercely competitive) commercial daily press.[89] Alex King's complementary study of British local war commemoration reveals the close interplay between the local state, the Churches (both established and nonconformist), agencies within civil society such as artists' professional organizations and civic pressure groups, the local press and community leaders in determining the form taken by a given town's or village's memorial. King emphasizes the commitment of memorial committees, modelling themselves on the practices of municipal government, to inclusiveness; this meant that the practical purpose and symbolic meanings of proposed memorials were subject to extensive debate before they were adopted.[90]

Official war commemoration comes under pressure when significant sections of society challenge, or feel excluded from, dominant narratives. Much recent work has focused on the ways in which the experiences of subordinated groups may take form as sectional memory through the construction of 'counter-monuments', in a process of political negotiation of the meanings expressed in public commemoration. To put it in Pierre Nora's terms, such work explores the dialogue between sites of 'dominant' and 'dominated' memory, and the political conditions which make such dialogue possible.[91] Two variables in the setting of such dialogue seem crucial.

The first is the nature of the state. To what extent does it permit pluralistic debate, to what extent does it impose a singular reading of the national past? How flexible are its strategies for maintaining hegemony, and how far can they permit modification of dominant national memory? In an openly divided liberal democracy, such as America during the fierce political debate over the value and meaning of the Vietnam War, there may be a visible public dialogue between dominant and dominated memory. Kristin Ann Hass has suggested that, at the end of the twentieth century, it has become increasingly difficult in mass democratic societies to construct a unified public memory of war. Thus America's role in the Vietnam War is commemorated by not one but three memorials on Capitol Hill. The main one, the Veterans Memorial, was not instigated by the state, but sponsored by a veterans' association set up specifically for the purpose of breaking the silence in public memory over Vietnam. It simply records the names of the dead by the dates of their death. Such was the furore it aroused when plans were first released that a more traditional, heroic memorial was also built with the support of the political right. Later a Vietnam

Women's Memorial was added.[92] Here, the commemorative site itself, through its diverse memorials, encodes the political conflicts over the meaning of the war.

At the other extreme, authoritarian regimes may seek to impose rigidly a single meaning on war commemoration, excluding any alternative narratives. The Soviet Union enforced a monolithic (martial, male, Russian) mode of commemoration of the Great Patriotic War of 1941–45, which, argues Catherine Merridale, left no space for other perspectives on the war (such as those of women, Jews or other nationalities). This intense celebration of victory through endurance also masked a profound silence about the other, hugely costly, Russian wars of this century, from the First World War through the Civil War to the Afghanistan War of the 1980s.[93] Following the conquest of Singapore in 1941, the Japanese occupation regime engaged in what Wong suggests was a notable level of symbolic violence. Their politics of memory embodied both an obliteration of the colonial past (including place names and colonial memorials), and a refocusing on the Japanese present (including the adoption of Tokyo time and the Japanese calendar) and on the new Singapore (with days of commemoration to mark the Japanese declaration of war against the Allies, and the British surrender).[94]

The second crucial variable, which may intensify especially an authoritarian state's attempts to control memory, is when war has taken the form of an internal conflict. This may give rise both to extreme efforts by the state to impose silence in the face of continuing national division, and later to the most bitter political struggles as social groups mobilize against that silencing. Richards' study of Spain in the decade after the Civil War traces the maintenance of a state of 'cold' civil war. The immediate post-war years were marked by continuing physical repression and near-starvation of the defeated Republican population. The Nationalist regime owed its claim to legitimacy substantially to the victory of the military rising; so all alternative perspectives (symbolized by references to 'the time of the Reds') had to be systematically de-legitimized, their public representation repressed, their memory obliterated. Only the dead victors were to be commemorated; while a new spatial and temporal organization was forced upon the population.[95] In the starkest example in this volume, Elizabeth Jelin and Susana Kaufman (in Chapter 2) chart the contradictions of a later stage in this process, when an authoritarian regime which had triumphed in civil conflict has been replaced by a still not fully secure democracy. They examine efforts, following the fall of the military regime in Argentina, to secure commemoration of those who were the victims of state terrorism. The holding of public judicial hearings, and their subsequent publication, initiated a space for deaths and disappearances to be acknowledged and recorded. Given the still minimal public commemoration of the victims, however, struggles continue over whether and how the sites of repression should be publicly marked. Consequently, despite the political changes, the traumatic effects of the regime show in the continued separation of official from oppositional memory, and in its failure to engage with significant elements of common and individual memory. The continuing refusal of official commemoration may serve, they suggest, not to disperse but to intensify efforts to remember and record, undertaken now by agencies in civil society representing victims or their families.

Below the level of the nation-state and its political movements are various components of civil society, including veterans' and victims' organizations, and cultural, community and artistic movements, which in recent years have been the subject of much scholarly attention.[96] Some of these necessarily work in close cooperation with institutions of the state, while others seek to mobilize support to modify or overthrow dominant narratives.[97] Close and unforced cooperation seems to prevail in Finland, where Raivo's account (in Chapter 5) shows local communities and veterans' organizations, in a memorialization of the Second World War which commenced during the war itself, working consistently within an established framework of what has been termed 'the great national narrative of the Finns'. Their efforts, together with the informal activities of families, and the formal endeavours of state structures such as national – and later reserve – military service, function together in what Raivo terms a 'socialization process for the nation'. Moreover, as the generation which fought the war approaches the end of its life, and their grandchildren enter adulthood, agencies of civil society together with the state have developed new initiatives, such as the construction of replica landscapes and heritage centres; these continue to transmit memory of the war, and the national narrative it underpins, but through the rather different channels of education and entertainment.

In contrast, examples of oppositional mobilization include: the Australian Aboriginal activists of the 1960s, Freedom Riders inspired by the American Civil Rights movement (discussed by Curthoys in Chapter 4); the American Vietnam veterans' organizations in the 1970s; and the organization of French conscripts involved in the Algerian War, FNACA. This organization (drawing in part on the Vietnam veterans' experience) engaged with the state to demand both practical rights (lobbying for proper recognition of the traumas they had suffered, and adequate medical and financial compensation), and official commemoration of their forgotten war.[98]

The different ways in which histories of war are written and disseminated can serve to illustrate the range of relationships between the state and the organizations of civil society. Many states sponsor the writing and publication of official war histories. These may seek to include a wide range of shared/common memories within their narrative. As Thomson has shown in his study of the Australian official history of the First World War, its author C.E.W. Bean strove to produce an account which would remain true to the experience of the troops, while helping to consolidate the emerging narrative of Australian nationhood in which its contribution to that war was central.[99] No such narrative, however, can be all-embracing; and alternative memories may be mobilized into sectional or oppositional narratives. As Garton shows (in Chapter 9), the main Australian veterans' organization, campaigning in the 1930s for a new type of pension for long-term unemployed veterans, chose to publish Lording's autobiography, since it expressed what he (and they) felt Bean's narrative failed to represent adequately – the more negative emotions left by the war. Nasson (in Chapter 3) describes how, in the aftermath of the Anglo-Boer War, academic historians, working in parallel with popular journalists, helped to gather individual and common war memories, and weave them together into a oppositional counter-narrative of heroism in defeat supporting the political fight-back of unreconciled Afrikaners.

In a less institutional way, literary or artistic productions, even single works, may operate to draw together shared or common memories into a wider narrative within civil society. In the early post-junta years in Argentina, when the insecure democratic government sought to bring an end to legal measures against the military regime, the publication of the testimonies to and findings of its own investigative commission, under the title *Nunca Más* (Never Again), served to link up, by stating publicly, hitherto fractured and divided memories. Jelin and Kaufman's interviewees (in Chapter 2) testify to the power of this book in crystallizing their own memories and hence their understanding of what they had lived through during the years of repression.[100]

In a transitional space between the formal organizations of civil society and the informal networks of family and kin, there exist what Jay Winter has termed 'fictive kinships', referring to particular groups of survivors, small-scale agents who form what he calls 'families of remembrance'. He cites as an example the French Union of Disfigured Men after 1918, who came together in a quasi-adoptive self-help relationship to supplement the inadequacies of state support.[101] Such veterans' organizations, bringing together those whose identical or cognate experiences can be articulated in shared/common memories, have always been among the major agencies of unofficial war commemoration. Stanley (in Chapter 10) notes that British veterans suffering from PTSD are often most comfortable 'taking refuge' in a services environment, where memories may be shared with fellow veterans and trained staff who are familiar with the realities of war. Commemorative narratives may also be rehearsed through existing groupings with a pre-war framework of shared values, such as local face-to-face communities, whose networks may be at once public and familial. Barry Doyle has explored the ways in which the Free Church community in Norwich sought to give meaning to the death in battle in 1917 of one of their business, religious and political activists, Harry Jewson, a man seen as the leader of the next generation. As was typical of the provincial nonconformist bourgeoisie, various members of his large and prominent local family were also key activists in the city's religious and political life; and their felt sense of familial identity extended beyond kin into the wider community of chapel-goers. Doyle's account suggests that in this context personal mourning alone was insufficient, and only by retelling the story of his life and death in terms of their common religious and political ideals could they attempt to make sense of his death.[102]

Finally, of course, families and kinship networks play a crucial role in connecting and then transmitting the overlapping memories of individuals. Winter has recently examined war remembrance as an event in family history, through examples of stories and books produced in such familial settings.[103] Again, though, the nature both of the state and of the war being remembered may have a crucial impact in shaping even these most basic agencies for the preservation and circulation of memories. When an authoritarian state seeks to enforce a dominant narrative of a civil conflict, the intensity of repression may reach down to disrupt even the habitual mechanisms for transmission of memory between generations within the family. In post-Civil War Spain, as depicted by Michael Richards, so intense was the effort to promote the values of the victorious Nationalists, and de-legitimize those of the Republicans, that families were frightened to voice their memories openly even to their own children. The 1940s and early 1950s became known as the years of silence.[104] Jelin and Kaufman's investigation (in Chapter 2) of attempts to retrieve and make public memories of the period of state terrorism in Argentina from 1976 to 1983 show in detail the impact of extreme repression. Here, too, family transmission of memories became disrupted through fear and the necessities of survival. The parents of one of their interviewees, Julia, could shield her as a child from the sounds and sight of killing, but not tell her from what she was being protected. Only since the restoration of democracy can the generations begin to discuss the nature of the recent past they lived through, and make sense of memories which were fragmented or sealed away. Such studies testify to the repression or fragmentation of individual memory through explicitly political activities involving the disruption of existing mechanisms for the formation of shared memories.[105]

Nevertheless, even where there is direct and pervasive control of war commemoration from above in authoritarian societies, this may serve to preserve private memory as a counter-force, rather than successfully harnessing it for national political purposes. The emphasis of official Russian commemoration of the 'Great Patriotic Dead' of 1941–45, Catherine Merridale has argued, aimed to censor consciousness of the 10 million or so earlier deaths caused under Stalin's Revolution by collectivization, famine and the purges. Yet 'private' responses to such commemoration could not be controlled. Merridale suggests, on the basis of her own and others' interviews, that some of those who lived through these traumas may have responded to triumphalist celebration of the war dead by remembering losses due to other conflicts: the commemorations may have re-awakened consciousness of events expressly not recorded in public memory.[106] Even in a totalitarian regime, the private work of mourning may in fact revive the very sectional memories whose destruction is sought by official commemoration.[107] Jelin and Kaufman's account (in Chapter 2) of the proliferating struggle by groups in Argentinian civil society to commemorate the sites of repression, against the governmental and military wish to draw a line under the past, raises the question whether, in an emergent democracy, attempts to silence oppositional memories may have the paradoxical effect of stimulating them. Emergent memory becomes hydra-headed.

Even where it is a case of suppression supported by unofficial violence, rather than by state repression, the countervailing weight of official memory, if intense and long-lasting enough, may be such that it threatens to repress, isolate or fragment – at worst to efface – individual private memories. Such was the case in the Irish Free State with the families of Catholic returned servicemen from the First World War. Jane Leonard has noted how, in the context of the resurgence of the Troubles from the 1960s, her interviewees began to doubt whether their memories of commemorative events of the inter-war years – such as the British Legion poppy appeals held in churches – had actually taken place. So successful had been the combination of state-sponsored amnesia and direct action by the IRA and other nationalist organizations in obliterating the British symbols of commemoration, that an earlier, more ameliorative stance towards commemorating participation in the war became unthinkable.[108]

In a less authoritarian context, where public remembrance of war has been absent or discouraged, but no attempt made to enforce an official narrative, family, kinship and similar networks may help to preserve memories until changing political circumstance allow them to re-emerge. In Singapore, Wong has shown how, both under the restored British colonial regime after 1945, and during the early years of independence, no public acknowledgement of the Japanese conquest of 1942 was made. Yet despite this official downplaying or suppression of public memory of the war, in popular memory it remained a profound watershed, especially through locally based and family remembrance. Indian families preserved memories of taking part in the struggle for Indian independence, while the Chinese population kept alive the memory of their war dead through their annual clan association rituals. The latter remembered especially Japanese reprisals against the Chinese population in 1942 and later. During the 1960s, the Chinese Chamber of Commerce established a war memorial, where it held commemorative rites every year. Only when Singapore had become more confident of its own security as an independent city-state did the government move to record this suppressed aspect of its past, building in part on these practices from kinship networks and agencies of civil society.[109]

The complex interrelationships between different agencies, and their ramifications through different dimensions of the politics of war memory, can be illustrated from Stanley's study (in Chapter 10). One of her interviewees, Frank Capacity, in his efforts to come to terms with his own painful memories of being a Far East prisoner of war in the Second World War, has helped form an organization working for commemoration, the Kwai Railway Memorial 'Three Pagodas Group' (an agency of civil society, working within a narrative aiming to prevent future wars through education). This has involved unsuccessful contacts with the agencies of the nation-state, when he unavailingly pressed the British Prime Minister to help finance a memorial. Simultaneously, he has addressed himself to kinship and fictive kinship (veteran) networks, photographing the graves of dead comrades which he saw on his own journey of remembrance and presenting them to surviving relatives.

Finally, though for analytical purposes it is valuable to distinguish between narratives, arenas and agencies for the articulation of war memories, consideration of a specific case – Yosefa Loshitzky's account (in Chapter 7) of the work of second-generation playwrights and film-makers in Israel – will serve as a reminder that these are all aspects of the same social process. In the newly independent Israeli state, memories and remembrance of the Shoah – at first marginalized because of their associations with a powerlessness and victimhood against which the new state defined itself – later became a central symbolic articulation of the need for such a state and its continuing security. The plays and films which Loshitzky explores work aesthetically within the arena of civil society to challenge current hegemonic Israeli narrative of the Shoah, in several ways. They counter monolithic representation by asserting that the Shoah is not simply a catastrophic event in the history of the Jewish people as a whole, but in the lives of a multitude of specific families and communities with their own histories to be recorded. This draws attention to the many different social and ethnic groups within contemporary Israel, with diverse relationships to the Shoah. Those involved in these productions have uncovered and explored histories previously preserved within family and kinship groups, including their own personal stories. Representation of such private and individual histories in public artistic media, working against the appropriation of the Shoah into a national political narrative, helps transform shared or common into sectional memories. These artists have also engaged with the metaphorical relationships between the Shoah, the position in Israeli society of the *Mizrahim* (Oriental Jews, who did not directly experience it) and of Israel's Palestinian citizens, and the current Israeli– Palestinian conflict, questioning the monopolization of the Shoah within a single narrative, and foregrounding its complex impact on different groups within Israel.[110] Hence, artistic groups in civil society, drawing on individual and shared memories which they help connect up into sectional memories, may help to generate a dissenting or oppositional narrative which can challenge dominant official memory.

3 Subjectivity and the cultural forms of war memory

In introducing the range of agencies involved in the articulation of war memory, we have stressed that the experiences and memories of individuals – both survivors and their successors – need to be understood as analytically distinct from the narratives which such agencies endorse. Yet at the same time, of course, the relationship of individuals to these public forms of war remembrance is intimate and complex. Public narratives may bring deeply felt psychic needs into play, because of the ways in which they resonate with personal experience. Alternatively, individual psychic reactions to war and its aftermath may provide a powerful stimulus to the creation of new forms of remembrance. Indeed, the very process of symbolizing loss in a public form may, as the social-agency approach recognizes, be a means of psychically negotiating the personal aftermath of war. This section considers such interactions from the perspective of individual subjectivity. It focuses on the ways in which subjective responses to war are shaped by existing cultural forms of expression, and what is involved subjectively in the struggle to establish new forms and meanings. Taking the view 'from below', it illuminates the nature and range of individual psychic investments in the production of, and participation in, war commemoration.

A problem with both the state-centred and social-agency approaches is that they do not consider closely enough the ways in which subjectivity is constituted through the narratives and practices of war remembrance. State-centred approaches are more interested in the operations of the state agencies themselves, than in the ritual or physical forms upon which such agencies draw in creating new commemorations. The efficacy of pre-existing forms – the resonance of images of the soldier-hero often used in official monuments to war, to take one example – is under-analysed. If, on the one hand, state-centred approaches ignore the narrative forms which make up the stuff of 'invented traditions', on the other hand, their emphasis on the state means that little is said about the subjective needs and desires which are mobilized by these official activities. Eric Hobsbawm, in acknowledging the limitations of a state-centred focus on the mechanisms of state ritual, pointed out the importance of these kinds of identifications. Invented traditions, he argued, are only effective because they 'meet a felt – not necessarily a clearly understood – need among particular bodies of people'.[111] Although Hobsbawm asks the historian to be aware of gaps between 'official' ideologies and 'popular' consciousness, his own sights remain firmly locked onto the former, as we saw in section 1.

The social-agency approach might initially appear to be strong in relation to precisely these kinds of issues. It certainly takes as central, questions of both form and subjectivity. After all, the point of Winter's *Sites of Memory, Sites of Mourning* was to establish the robust nature of traditional commemorative languages after the First World War, precisely because these were the forms with which individuals identified most easily in coming to terms with personal loss. Christian imagery of sacrifice and redemption, and spiritualist ideas of the return of the dead, proved more able than modernist forms to express grief. As Winter puts it, 'the strength of what may be termed "traditional" forms in social and cultural life, in art, poetry, and ritual, lay in their power to mediate bereavement'.[112] However, this analysis assumes a universal human response to grief. It rests on a view of survivors as psychically undifferentiated and, as a result, it generalizes rather too freely about the ways in which individual subjectivity is implicated in particular public narratives of remembrance. And whilst Winter certainly demonstrates that traditional forms remained ubiquitous well into the twentieth century, he does not tell us much about how subjective understandings and responses to the First World War were themselves constituted through these forms. He assumes that individuals drew upon and adapted traditional languages in order to express themselves, but that those languages had little efficacy of their own in conditioning personal responses to war. In the social-agency approach, the symbols of remembrance are seen as reflecting rather than constructing subjective meaning. As a consequence, it neglects the fact that individual subjects come to identify their experiences through the pre-existing narratives fashioned by the agencies of the nation-state and civil society. Finally, the social-agency approach lacks an account of the ways in which the mourning of survivors may complicate politics. The psychic and emotional charge attached to direct memories of war gives those who are designated as survivors particular authority in the process of articulating public memory. As we shall see below, attention to subjectivity alerts us to a further important dimension in the politics of war memory: that is, the psychically charged nature of the struggle between survivors and their successors, as the direct memory of war passes into cultural memory.

3.1 The templates of war remembrance

As section 2 has suggested, responses to war are formed in relation both to personal experience and to pre-existing narratives. The latter may relate to personal history – as a family story about a father's or grandfather's war, for example, which influences the attitude of the children to later conflicts. Or they may circulate within the wider arena of the nation. In either case, part of their subjective hold may lie in the way in which, in the remembrance of one war, they draw upon key 'pre-memories' or 'templates'. These templates, consisting of cultural narratives, myths and tropes, are the frames through which later conflicts are understood. They are akin to what Peter Burke refers to as the 'schema' of memory: they are 'associated with the tendency to represent (or indeed to remember) one event or person in terms of another'.[113] Thus individuals expressing their direct personal memories of war may invoke elements drawn from experiences and representations which originate before the lifetime of current living populations. Even the 'eyewitness' memory of war, then, is constructed both from personal experience and in relation to pre-existing cultural templates. These templates provide horizons of representation through which later conflicts are understood. They may be more generic, as in the Western film, which Alessandro Portelli tells us provided a narrative resource through which Italian partisans in the Second World War remembered key battles. Just as in a Western, many described confrontations with the fascists in terms of a 'primal scene' in which the outlaw reaches for his gun first, but is beaten to the draw by the hero.[114] The power of such templates to shape personal memory often derives from the psychic charge embedded in their meanings. Dawson, writing about stories of British imperial soldier heroes – from the novels of Sir Walter Scott to biographies of Sir Henry Havelock or T.E. Lawrence – argues that these provide masculine 'figures for identification' that may become internalized within the psyche as the basis for the composure of personal identities.[115] In this way, 'the national past lives on ... in the psychic lives of succeeding generations'.[116]

The memory of heroic victory or suffering endured in a previous war may act as the template through which later conflicts are understood. Such templates tend to function at both a psychic and political level to intensify the response to later conflict. As such they provide the latent sources which later popular mobilizations (such as that in support of the Falklands/Malvinas War in Britain) have drawn upon. The right-wing, nationalist-militarist interpretation of

Germany's defeat in 1918 as a 'stab in the back', which fuelled the Nazi aggression of the 1930s and 1940s, tapped into powerful and widespread personal memories of courage in the trenches or endurance on the home front. Such templates may of course invoke a much longer time span than living memory. Deep-seated memories of colonial, ethnic, religious and national wars may intersect with later conflicts in powerful ways. The Serbian invocation of the battle of Kosovo in 1389 is a core memory of national identity transmitted purely through cultural memory. Or consider the Northern Ireland loyalists' invocation of the Siege of Derry and the Battle of the Boyne in the Williamite War of 1689–91. The cultural memory of this long-distant war has been transmitted to contemporary Unionists through the annual rituals of Orange commemoration; it is their participation in this commemoration, not the war itself, which constitutes their personal memories.[117] In regions of so-called 'ethnic' conflict such as these, the roots reach back deeply into the historic past. One template may be superimposed on another, itself a rewriting and re-remembering of some previous war. In former Yugoslavia, wartime events which are within the horizon of personal memory, including internecine struggles between and amongst Serbs and Croats, communists and fascists, are condensed with earlier memories of wars against the Ottomans and others (stretching from the early twentieth back to the fourteenth centuries). In Ireland, the loyalist cultural memory of resistance at the Siege of Derry encoded the importance of vigorous 'defence' to prevent a repetition of the massacres of Protestants in the 1641 Rising. These fears, as well as the heroic qualities of the defenders, structured a template which was invoked in subsequent wars over two centuries, from the United Irishmen's Rebellion of 1798, to the independence conflict of 1912–22, and on to the present 'Troubles'.[118] In the process, as Raymond Williams grasped in his concept of the 'selective tradition', the distinct meanings of earlier historic episodes will be re-worked and may be entirely transformed or erased by subsequent hegemonic pressures; but wherever their traces remain, these may also, in the right conditions, be recovered in the service of 'alternative or opposing practical continuities'.[119] Such traces hold out the possibility of redrawing the lines of connection between past and present, and rewriting the histories which have stitched a people, or a state, or a region, into its particular position within a dominant global order. Hence, questions of 'national memory' in the newly emergent states of the 1990s, or in states subordinated but peripheral to the Cold War polarization, may involve a return to wars prior to 1939–45: an archaeology of memory in the service of a refashioned national narrative. However, the political possibilities for this depend upon the traces having been preserved in cultural forms which have retained their psychic efficacy, and thus their potential to organize subjective identifications.[120]

The contributors to this volume attest to the pervasive legacy of such templates. Writing of the defeat of the Afrikaners in the Anglo-Boer War of 1899–1902, Bill Nasson, for example, traces (in Chapter 3) how the memories of that 'historical trauma' came to 'fertilize the future'; their power to 'kindle consciousness' of a distinct identity proving a 'defining legacy' of the Afrikaner people throughout the century. These 'embers of 1902' could still be 'stoked' by Afrikaner hardliners during the transition from apartheid in the 1990s, in ultimately futile efforts to mobilize against the transition to black majority rule. Now the template is being re-worked yet again, in a very different effort to forge a new national history. The imprisonment of blacks and Afrikaners in British concentration camps is being linked to the apartheid struggle itself in order to underscore the quest for freedom as the 'common historic destiny' of all South Africans. Such reinterpretations retain a political efficacy because they are able to galvanize subjective identifications across a wide constituency. Another example of the way in which such templates operate at a psychic level is given by Petri Raivo in his discussion of war memory in Finland. Raivo shows how the commemorative practice of burying the dead in local graveyards, instigated in the civil war of 1917, functioned in the 1939–45 war as a means of galvanising the troops. The return of the dead to Finnish soil was worked into a powerful catch-cry to 'Never abandon your companion in arms'. In the context of a largely agrarian, Lutheran population, the practice of local burial drew upon deeply embedded religious and rustic ideals. It was thus a template through which, in these later wars, soldiers affirmed their loyalties to each other and to the nation, and their enmity towards the atheist Bolshevik.

Templates of more remote wars, then, do not draw upon the direct memory of survivors. Yet these 'pre-memories' must be considered as no less powerful than survivor narratives in their capacity to galvanize public responses to conflict. The cultural templates associated with wars past provide the raw materials for psychic identifications through which subsequent conflicts are understood and support or opposition mobilized. Whilst these templates may be consciously manipulated by political elites, to some extent this misses the point. Their efficacy lies in the fact that they circulate in cultural spaces which antecede, and thus are part of the constitution of, personal memory. Where subjectivities have been shaped by such templates, people *feel* the significance of these past events to be deeply personal.

3.2 The micro-politics of aesthetic form

Responses to war thus always employ well-established narrative templates drawn from earlier conflicts and related images circulating within the culture. Yet this is not to say that these templates necessarily determine later subjective responses to war. What matters is which particular templates are adopted by which agencies, and for what ends. In addition, new representations of war may reject or provide a critical commentary on old templates, or recycle them in ways which transform their original meanings, thus invoking different kinds of subjective identifications. Templates do not always work in the service of a dominant national narrative, as we shall see below. The extent to which they are subject to contestation depends on the particular political context. Their analysis may require a more local approach by the historian than is supplied by the national canvas of the invented tradition. Such an approach might ask how particular designs and sites of commemoration are chosen and why; and how forms are contested both during their construction and afterwards as sites of memory. However, these issues need to be understood in terms of the range of identifications which are brought into play by new aesthetic forms, and the extent to which these provide a basis for a broader, more inclusive politics. Such questions constitute what we call the micro-politics of aesthetic form.

These aspects of war memory tend to have been under-emphasized in recent debate about forms. Rather, the emphasis has been on the extent to which war overwhelms or radically alters existing modes of representation. For example the place of the First World War in the chronology of modernity has been hotly debated. The argument of Paul Fussell's *The Great War and Modern Memory* was that the unprecedented scale of destruction in this war could not be represented within traditional languages of war.[121] Focusing largely on the published work of British war poets and memoirists such as Sassoon, Graves, Blunden, Owen and others, he argued that the experience of trench warfare had itself forced a shift in the modes of war representation, from epic to ironic, euphemistic to realist, heroic to disenchanted. The First World War was thus a cultural watershed, breaking with traditional literary modes and establishing the forms of 'modern memory'. Later writers such as Modris Eksteins largely supported this view of the war as a radical break.[122] An alternative position – represented by Winter – is that, far from the First World War making existing cultural forms bankrupt, it gave them new life.[123] Modernist detachment was not the only available form, he argues; existing genres continued to be enthusiastically deployed and adapted by survivors throughout the inter-war years. However, *Sites of Memory, Sites of Mourning* does not so much refute the central place of war in the trajectory of modernity, as simply relocate the watershed. In Winter's narrative it was the Second World War which 'put an end' to traditional languages of commemoration and mourning.[124]

The focus of such work on large-scale shifts in the narratives of war memory – its locking on to the metanarrative of the birth of the modern – has obscured more contextualized discussion of aesthetic forms. New aesthetic forms of commemoration can have the effect of bringing dominant narratives into question, introducing alternative understandings of war and its effects into the national arena. Even while artistic media may seem to be operating within the terms of conventional public fictions, they may still create spaces for the representation of otherwise hidden dimensions, the unforgivable and unforgettable or the unassimilated. Peter Sjølyst-Jackson shows this in his discussion of post-war mainstream film in Britain and Norway in Chapter 6. Such films did not work in a simple way to reinstate conventional narratives of masculinity and the nation after the traumas dealt by the war. His reading of *The Cruel Sea* suggests that the film explores the psychic costs of war, particularly the persisting damage it may do to those involved in fighting it; albeit with the consequence that the film has been placed outside the canon of British war cinema. The film's most powerful after-image is the screams of British sailors killed by their own side – screams which may be shut away but never completely silenced. Norway, like Britain, had a widely shared consensus about the meaning of the Second World War. But, because of the closeness of elements of Nazi Aryan ideology to the National-Romantic imagery central to Norwegian national identity, there was a powerful need to locate the source of darkness outside the Norwegian family and community. The plot of their most admired war film, *Ni Liv*, in Sjølyst-Jackson's reading, achieves this: all true Norwegians, the film implies, collaborate to help the resistance hero escape. However, even here disruptive elements remain. The film shows that the hero's flight has depended on assistance from the Sames, an ethnic group who were subjected to a concerted attempt at cultural assimilation in Norway during the 1950s when the film was made. In representing this group, *Ni Liv* brings its status as 'Other' and its marginalization to the consciousness of the audience.

Aesthetic forms which question the national narratives in this way may even act as a stimulus to new commemorative practices. Such is the case with the Vietnam Veterans Memorial at Capitol Hill. As Kristin Ann Hass shows, this veteran-sponsored memorial challenges existing ideas of nationhood and the place of war within it. The design, by the Chinese-American Maya Lin, eschewes national or militaristic symbolism. Etched into the hillside, it grows in height along its length, reflecting the mounting toll of casualties each month. This 'black gash of shame and sorrow' contrasts with the monument which faces it, sponsored by the political right. 'The Three Fighting Men' is a 'strong, highly masculinized and heroic' figurative image of American soldiers advancing in the jungle.[125] Hass's analysis suggests the range of emotional reactions which may be touched off by these particular public forms. Whereas the traditional imagery of the fighting man encourages a pose of respectful passivity among viewers, the open form of Maya Lin's wall encourages an active relationship with the dead; for example through the bringing to the wall of objects which symbolize personal loss. Popular practices such as these, Hass argues, draw on a distinct vernacular language, derived for example from Mexican and Italian Catholic funerary practices. Yet it is the design of Maya Lin's memorial which facilitates this: it literally constructs a space underneath the names of the dead where 'dominated' memories and representational forms are expressed.

Our contributors also suggest the contested nature of form, and the ways in which aesthetic representation may prepare the ground for political expression. Jelin and Kaufman show how, in response to official pressures in Argentina to quell the popular tide of remembrance surrounding state terrorism, the turning of sites of disappearance into sites of commemoration has become a focus of popular activism. Protest art – for example, the painting of murals on the site of the El Olimpio concentration camp on the eve of the twentieth anniversary of the dictatorship's accession to power – became a means of pressing for more permanent commemoration, the turning of the building into a museum of *Nunca Más* (Never Again). Our contributors also suggest that aesthetic expression may provide possibilities for understandings not yet within the political domain. De Medeiros shows how in the Portuguese context it is novels which, in the face of official silence surrounding the end of the colonial wars and dictatorship in 1974, have attempted to address collective amnesia. Similarly, Loshitzky remarks that, in Israel, 'art preceded purely political debate'. It was film and theatre which, in the 1980s, initiated critical assessment of the dominant Shoah memory. Loshitzky shows how the cultural productions of the second generation open up consideration of what that memory privileges, such as the experiences of survivors and their children, and what it marginalizes: the experiences of *Mizrahim* Jews, and particularly of Palestinians, whose suffering is hidden by the centrality of the Shoah in Zionist memory. Her analysis of the film, *Don't Touch My Holocaust*, and the play, *Arbeit Macht Frei vom Toitland Europa*, explores how they force the audience into the position of the Other in relation to dominant Israeli Shoah memory: as Nazi collaborator, second-generation survivor or Arab Israeli. The focus of film and play on the internal and individual meanings of the Shoah provide a means of disrupting institutionalized forms of commemoration. It counteracts the 'banalization' of the Shoah by forcing the audience to confront its own memories and repressions. Such representations – whether in film, literature or drama – interrogate dominant war narratives and, through this, open public spaces within which larger understandings may develop.

A comparison of the Israeli situation which Loshitzky comments upon, and that in Portugal, raises questions about the differential power of aesthetic forms to mobilize a wider audience. In part, perhaps, the capacity of cultural productions to touch off wider responses depends on the extent to which a range of alternative narratives is already in circulation. This may be why it is that in Portugal, where novelists work within a context of near silence, aesthetic movements are not able to reach back into the political domain, but operate within a separate arena which – almost uniquely – registers the psychic traumas of the colonial wars. By contrast, the cultural production of second-generation Israelis gains a wider resonance – in part, Loshitzky explains, through its critical commentary on earlier, 'eyewitness' Shoah narratives. The study of such representations, then, requires an understanding of both the templates through which they are composed, and the political contexts of their creation and circulation. It must also recognize their power to touch off unconscious identifications as part of an ongoing dialogue with other individual, sectional and national narratives of remembrance.

3.3 Witnessing: trauma, mourning and the psyche

If the templates of memory provide a means through which individual psychic responses to war are mediated, at the same time the very experience of war itself also stimulates lasting psychic reactions amongst those whom it touches. Yet whilst remembrance may be motivated by such reactions, or may stimulate them, these psychic responses are nevertheless far from unitary. This section focuses on the nature of survivor narratives, and the range of psychic processes associated with individual remembering and forgetting.

Recent literature on war and memory has brought to the fore the question of how far public commemoration can be seen as a means of addressing individual experiences of trauma or grief. Such questions tend to be ignored in work which operates within the top-down conception of war commemoration as invented tradition. Jay Winter in particular has warned that historians have focused too much on the 'exuberance of the state' in organizing commemoration, and not paid enough attention to the symbolic work of ordinary people in assuaging grief.[126] In *Sites of Memory, Sites of Mourning* and subsequent essays he explores the function of activities ranging from the telling of stories by family and fictive kin, to the construction of war memorials, as a means 'to convert trauma into misfortune'.[127] His approach to the history of war commemoration is characterized by a focus on human motivations of 'sadness and dignity' – what Winter and Sivan describe in *War and Remembrance* as the 'evanescent expressions of the force of civil society itself' – rather than on the coercive force of political authority.[128] We would largely accept the criticism of approaches which merge such expressions within a 'state-bounded space of hegemony and domination'.[129] Shifting the focus from the level of the state enables the social-agency approach to show the extraordinary range of activities through which kin and fictive kin, and local associations and networks, addressed the psychological aftermath of war. Public war commemoration clearly has a central relationship to the grief and mourning of survivors as one 'subjective need' which invented traditions may mobilize.

At the same time there are a number of problems with the way in which the social-agency approach treats mourning. First, it creates a false dichotomy between politics and mourning. As we have seen, politics informs both the very templates which are available for the expression of grief, and the arenas in which it operates. Second, the social-agency approach presumes a 'typical' process of grieving, in which the shared activities of remembrance assuage each individual's grief. At one level the formal framework of the social-agency approach acknowledges the limitations of such a model. Winter and Sivan point out that the work of mourning may never end and may not be recuperative.[130] Nevertheless, empirical work in this vein tends towards the assumption that shared symbolic forms mediate individual grief. Such approaches draw on the Freudian notion of mourning as a form of psychic 'working through', in which the monopolizing of energy by the pain of loss, and lack of interest in the outside world, is overcome by the gradual detachment of the ego from the lost object. It is distinguished from melancholia, a pathological form of mourning in which the identification with the lost object remains, and the loss is continually revisited. Commemoration, in the social-agency approach, is seen as a means of enabling and expressing the reality of loss which is associated with 'normal' mourning.[131]

The example of Käthe Kollwitz illustrates these limitations in the social-agency approach. In both *Sites of Memory, Sites of Mourning*, and *War and Remembrance*, Winter argues that Kollwitz's work can be understood in terms of the impulse to translate individual loss into shared forms. Yet as Regina Schulte has powerfully shown, the grief that Käthe Kollwitz felt at the death of her son Peter in October 1914 was not contained either by the range of her 'private' remembrance or by her war memorial at Roggevelde. Kollwitz wrote nothing in her diary for nearly two months after Peter's death, and in the first entry after resuming it, conceived the idea of a memorial to him. In her case public commemoration was bound up with private grief from the very start. In the following years she created 'cultic' spaces in his bedroom, with constantly replenished wreaths and flowers: her grieving did not pass neatly from the private to the public. Moreover, an important theme in her mourning was the failure of the patriotic mode, linked to the facing of her own complicity in Peter's death (in a reversal of gender stereotypes, she had endorsed his joining-up, against the wishes of her husband). Her memorials became a 'political gesture, ... a protest against the appropriation of mothers and sons, and also grandsons, to lend meaning to war'.[132] In Winter's account, however, Kollwitz's memorial expresses a trajectory from the private grief of the nuclear family, to the metaphoric embracing of *all* the nation's sons.[133] Such a stance might be considered problematic on many levels. As Schulte reminds us, it artificially separates the more political motivations of artistic activity from the psychic work of mourning. It is based on the assumption that the private memories of the bereaved become objectified through ritual activities commemorating the dead. It further assumes that the psychic states of the bereaved artist and the viewer are commensurate, and that both, through memory work in the public sphere, become reintegrated into society, thus avoiding the repetitive grieving associated with melancholia.

The social-agency approach thus tends to impose a model of 'normal grieving' which ignores the range of individual psychic responses to death. In some cases it may be psychically necessary not to let go of the dead. This was the position of Walter Benjamin, for example. As Martin Jay has shown, Benjamin, writing in 1932 at the high-point of Great War commemoration, argued that the blows of trauma could not and should not be cushioned. The use of redemptive symbolism to 'work through' risked premature forgetting. The 'violence of amnesia' could only be avoided by the ritualized repetition of melancholic states, through what Jay calls the 'trope of troubled burial'.[134] In Benjamin's words:

> He who seeks to approach his own buried past must conduct himself like a man digging. This confers the tone and bearing of genuine reminiscences. He must not be afraid to return again and again to the same matter; to scatter it as one scatters earth, to turn it over as one turns over soil.[135]

The contributors to this volume similarly indicate that there are situations in which there is a psychic imperative not to 'work through' from melancholia to mourning. This is perhaps most striking in Stanley's chapter on veterans suffering from post-traumatic stress disorder (PTSD). Sufferers are encouraged to work at creating their own private remembrance – for example, lighting candles for dead mates or using drama-therapy to enact their burial; constructing collages which represent visually their trauma and how they feel; even returning to the sites of their traumatic experience. These are the means of making memory biddable rather than involuntary. Yet many ex-servicemen are reluctant to 'let go' of their memories in this way. The very fact that this memory remains private and unassimilated is important to them: it shows that they are keeping faith with the dead.

The social-agency approach perhaps also overstates the capacity of commemoration to achieve working through. The examples in this volume emphasize that public remembrance of war may lead to psychic 'unbinding' as well as to mourning. Thus Stanley suggests that war films such as *Saving Private Ryan*, or remembrance day ceremonies, may trigger trauma over again. In his account of post-war representations of the Second World War in film, Sjølyst-Jackson focuses on its 'retroactive' capacities, that is, its ability to re-evoke the memory of war amongst audiences. As such it was an ambivalent medium. It could both offer the promise of release from memory and disrupt its working through. *The Cruel Sea*, for example, had a double effect. On the one hand it portrayed typically British and imperial characteristics of steady nerves under pressure and responsible judgement in difficult moral circumstances. On the other it confronted the viewer with unassimilated trauma. In showing the haunting of the living by the dead, it commented on the incapacity of public memory to 'put the lid' on private grief. From the perspective of individual memory Jelin and Kaufman make a similar point, insisting that, no matter how prolific or inclusive the public commemoration, it can never fully deal with individual memories of trauma, necessitating the ongoing performance of small acts of private commemoration.

If the use of commemoration to 'work through' private grief may sometimes be neither possible nor desirable, on occasion forgetting may be psychically necessary. Some of the ex-soldiers whom Stanley describes do not 'work through' traumatic memories, so much as set them aside. The drama-therapy offered to PTSD sufferers helps achieve this; for example in the act of psychically placing such memories in a box for safe-keeping. If not exactly 'forgotten', they are more contained than worked through. Alternatively, there may be strong internal impulses to simultaneously remember and forget.[136] The novelists discussed by de Medeiros, in exploring their characters' struggles with their memories of the Portuguese colonial wars, represent an internal split between the desire to bury the traumatic past on the one hand, and to connect private memory with historical memory on the other. The split is a 'necessary strategy' to ensure individual survival, the traumatic nature of the experience being incapable of containment through remembrance. Thus, whilst there might be important political and psychic motivations for the continual re-remembering of war, in the attempt to admit what de Medeiros calls its 'hauntings', equally, there are situations in which such hauntings are unbidden and uncontrollable, and can only be handled through forgetting or the displacement of memory into other cultural activity. Both processes may occur simultaneously. Neither response approximates the social-agency model of 'working through', however: sometimes forgetting from the very start may be psychically necessary, just as, in Benjamin's terms, to be melancholic may be the only way to keep faith with a past that must not be forgotten.

The relation between grief and commemoration assumed in the social-agency approach is one in which the private pain of past experience is alleviated through being symbolized in shared forms. This view of commemoration as a kind of therapeutic reflex presumes an organic relationship between the individual, the agencies of civil society and the nation-state. It takes the politics out of mourning. In so doing, it projects a contemporary emphasis on the recuperative function of narrative back on to both the commemorative activities and the psychic realities of people in the past.

3.4 Postmemory and generational transmission

The process of transmission of war memory from the direct experience of survivors to the 'cultural memory' of their successors, further demonstrates the complexity of subjective relations to war memory. We have seen that the individual experience of war is handled in psychically diverse ways by survivors. Equally, however, the memory of war shapes the subjectivities of survivors and their descendants in different ways. Moreover, the period of transition between the first generation and its successors may, as Winter and Sivan suggest, be 'a very difficult one', as survivors seek to ensure that 'their' version of war is not forgotten, whilst successors struggle with the conflict between acting as the 'trustees' of survivor memory, and reassessing this legacy and their own relation to it.[137] Broadening the focus from the agency of the nation-state to that of civil society allows recognition of the politics involved in this transmission. In particular, it highlights the struggle of successors – who are always at some level 'outside' the direct memory – to negotiate the authority of survivor narratives. A focus on generational transmission enables us to see how successors seek to secure recognition for their own shared memories within public arenas.[138]

The resurgence of interest over the past two decades in war and its meanings, we noted earlier, partly reflects this phenomenon, as remembrance of the two World Wars passes from direct memory to the second and third generation. This is both a private and public process, the children and grandchildren of survivors engaging in activities that translate individual memory and family stories into museum exhibits, film, history and fiction. As the span of living memory reaches its end, the eyewitness experience is often given renewed authority and public space. The British television series, *Veterans*, for example, timed to coincide with the eightieth anniversary of the end of the First World War, was based entirely on the testimonies of men aged 100 or over. Its significance was less as a history of the events of war – their testimonies were not especially unusual – than as a reminder that the First World War is only just within direct memory, and rapidly passing entirely into cultural memory. The commemoration of such passings is not uncommon. It stems from the recognition that, as Jelin and Kaufman explain here, the past is not automatically passed between generations, but must be actively transmitted so that later generations 'accept that past as meaningful'. The proliferation of Shoah memory since the 1970s has had much to do with the question of how its legacy can be defined, and how transmitted to succeeding generations as the eyewitnesses pass from the scene.[139] Indeed, as Loshitzky notes in Chapter 7, the very trajectory of the culture of Shoah remembrance may reflect this, from the 6 million commemorated as an amorphous mass in the post-war ceremonies of the Israeli state, towards a desire for the multiple nature of the experience to be registered.[140] The placing of the individual story at the centre was one means of maintaining, in Friedlander's words, 'the concreteness of death and despair' in the face of historical and institutionalized memory.[141] Annette Wieviorka makes a similar point about the changing function of personal testimony, from archival purposes of documenting the practices of genocide – making events known – to the present impetus to 'keep them before our eyes. Testimony is to be a means of transmission to future generations.'[142] Thus the increasing receptiveness to the survivor's voice is linked to the process of generational transmission.

This transmission, although its mode is primarily familial, crosses back and forth between public and private arenas. It can be a largely private matter; as for example in the action of Horacio, a survivor of Argentina's dictatorship recorded by Jelin and Kaufman, who feels obliged to remind his children every time they pass by the sites of detention that 'here's where they killed and made people disappear'. But it can also move from the familial to the public sphere, as the children of survivors seek to keep alive this memory. This is shown by Winter in his discussion of how kin and fictive kin often translate events in family history into public remembrance. This search for memory is also evoked in Sebastian Faulks' fictional *Birdsong*, whose central character, Elizabeth, is moved by the sixtieth anniversary of the Armistice to reconstruct, through reading letters and histories of the Great War and visiting battlefield sites, the experience of a grandfather she never knew.[143] Of course Faulks's narrative itself functions to fill the gaps and silences in both family and public memory: it graphically reconstructs the experience of trench warfare for generations who have lost their link with the survivors. In the plays and films of the second-generation in Israel, as Loshitzky notes here, the 'private' memories of directors and actors are made the material of public memory as part of a process in which the second generation reinterprets the Shoah. The 'Tower of faces', a gallery of pre-war family photographs taken in the Lithuanian town of Ejszyszki and now in the Washington Holocaust Museum, also testifies to the crossing back and forth of this generational transmission between the public and the private. The photographs, collected by the granddaughter of the town's Jewish photographers, depict the everyday activities of its Jewish population, many of whom were killed in a pogrom during the war. As Hirsch observes, the conventionality of the family photograph provides a powerful way for viewers to bridge the gap between themselves and the victims: 'We reanimate the pictures with our knowledge of daily life.'[144] Association with one's own family memory enables successor generations to identify with the victims, and heightens the emotional response to their violent deaths.

The expression of the survivor's voice has often found its most poignant expression among the second generation, as Loshitzky observes here. At the same time, the process of moulding public memory that occurs with this passing is often fraught with difficulty. Institutionalized forms such as the memorial or remembrance day might just as often work to divest as to remind subsequent generations of the obligation to remember. The memorialization of war in physical and visual forms may create the illusion that the past will not be forgotten, whilst also springing from 'an opposite and equal desire to forget'.[145] The transmission of memory through such forms of institutionalized remembering may be particularly important, however, in national contexts – such as Portugal or Argentina discussed here – where official policies have promoted forgetting. In Argentina for example, acknowledgement of the dictatorship's effects at the national level has been constrained by the political pressure to limit further indictments. Whilst this increases the need for the experience of violence and repression to be recorded at the level of civil society, at the same time Jelin and Kaufman argue that the process of transmission must allow for reinterpretation: 'There is no way to close off new readings of old stories.' Successor generations may wrestle with the way in which the eyewitness experience blocks out or marginalizes other ways of remembering. They may impose a critical perspective on the survivors even as they seek to prevent the acts of violence to which they were subjected from being forgotten. As Hartman has remarked, 'the "generation after" ... struggles against as well as for Holocaust remembrance'.[146] Whilst aspects of traumatic experience can be passed on (sometimes explicitly, more often perhaps in symbolically coded form) from the survivors to their immediate descendants, the authority given to the first-hand memory makes public recognition of this secondary trauma difficult. The presumed truth and legitimacy given to those who suffered personally creates a danger that new meanings of the past cannot easily gain recognition, limiting the enlargement of the 'us' of memory.

The tensions surrounding this succession may show themselves in struggles over who counts as a witness, or what counts as memory. In Portugal, as de Medeiros shows, there are ongoing debates over who has the authority to write about the colonial wars in Africa. Its ex-veteran novelists face a tension between a perception of the combatant as the only true survivor of war, and the only 'empirical' memory as that of combat; and the desire to widen the circle of memory. Some survivors eschew secondary understanding as in any way constituting 'memory'. For them it is important that the particular quality of personal memory is not confused with the 'vicarious' memory conveyed in cultural myths about war. For Samuel Hynes, a historian of the First World War and veteran of the Second, 'memory' always entails a mental process of '*now* reaching back to *then*'. By implication, the understanding of others' wars, such as that which Hynes brings to bear in his study of First World War myths in literature and art, *A War Imagined*, must never be confused with the individual psychological process of recalling and making sense of eyewitness experience. As he insists, 'I have a clear and quite detailed set of images of that war in my head ... but I don't *remember* it as I remember my own war.'[147]

Hynes's attempt to guarantee the sovereignty of memory for the eyewitness, distinguishing it from the representation of other people's wars, illustrates the difficulties posed by the process of generational transmission. His attitude, whilst couched in the terms of academic debate, is not uncommon amongst survivors. In Chapter 9, Garton reports that Australian veterans after the First World War cultivated what Fussell has called the '*versus* habit' in extreme ways.[148] They retained a deep-seated scepticism towards all those, particularly wives, lovers and politicians, who were not at the front. Even whilst the national myth of Anzac proclaimed them as founders of the nation in the inter-war years, they cultivated an exclusive memory based on the disavowal of home and nostalgia for the mateship of war. Similarly, the veterans suffering from PTSD that Stanley reports on here largely reject the memory-work done by secondary groups. Many feel alienated from homecoming ceremonies and from later public commemorations. This scepticism towards other people's representations of war is not restricted to veterans of battle itself, nor to the contestation of public memory between generations. It may also be a characteristic of family memories of exile and war. Miriam Glucksmann has recently written of visiting the apartment block in Berlin where her mother – a Jewish refugee who arrived in Britain in 1938 – lived prior to the war, and the excitement of being able to locate in bricks and mortar the childhood she'd heard so much about. On telephoning her mother with the news that she had just been in 'her' house, she was cross-questioned as to which particular flat she had been shown around. On hearing the answer, her mother replied 'But that was the wrong flat. I lived on the first floor.' Glucksmann remarks that this summed up 'My mother's and my relationship in a nutshell.'[149] The exchange testifies not only to the personal character of their relationship, but to the social character of relationships between witnesses and the second generation: metaphorically in the same realm of memory, yet never in an identical place; the efforts of the children to connect always confronting the fact of their absence from the event itself.

Survivors often seek to preserve the inviolable truth of witnessing from the 'almost memory' of others. Yet the question posed by the second generation is, can such a clear distinction be made between other people's wars and our versions of them? The depth of mourning they witness in parents and family friends itself constitutes a powerful memory.[150] It is this quality of absence and presence that characterizes what Marianne Hirsch has called 'postmemory'. Postmemory addresses the experience of being separated in time and space from the war being remembered, yet of living with the eyewitness memory. As Hirsch observes, the children of Shoah survivors 'grew up dominated by narratives that preceded their birth'. At the same time 'their own belated stories are evacuated by the stories of the previous generation, shaped by traumatic events that can be neither fully understood or re-created'.[151] Their mourning is always unfinished:

the act is a secondary one, the lost object is not their own. Saul Friedlander writes movingly of this dilemma. As he puts it:

> a distance – impassable, perhaps – separated me from those who had been directly caught up in the tide of events, and despite all my efforts, I remained, in my own eyes, not so much a victim as – a spectator. I was destined, therefore, to wander among several worlds, knowing them, understanding them ... but nonetheless incapable of feeling an identification without any reticence.[152]

This quality of being at one remove gives a particular impetus to the artistic discourses of postmemory. The second generation 'creates where it cannot recover'.[153] In the absence of direct memory, artistic forms supply the material of imagination. The fact that the past must be reconstructed from stories rather than from direct knowledge encourages artistic creation.[154] The qualities of postmemory which Hirsch writes about can be clearly seen in Loshitzky's contribution in Chapter 7 of this volume. The theatre and film practitioners whom she discusses have no direct experience of either the Shoah or the founding of the state of Israel, yet their memories testify to the legacy of these events, not least in the present-day moral ambiguities posed by the Israeli–Palestinian conflict. They use their own experiences as second-generation Shoah survivors, *Mizrahim* Jews and Palestinians to explore what Hirsch calls 'exile from the space of identity'.[155] Productions such as these view the eyewitness experience from the outside and, in so doing, construct more inclusive forms of Shoah remembrance. Postmemory, then, signals the shift from narrative based on direct memory to cultural productions which explore what it means to live under the shadow of past wars. It is constantly negotiating events and experiences which are outside personal experience, but which nevertheless shape the subjectivities of the 'outsiders' in profound ways.

3.5 Historians, witnesses and the politics of commemoration

One of the characteristic features of survivor testimony is that it offers itself as unmediated, 'raw' experience. Indeed, this is one of the features which gives it authority. At the same time, of course, such memories must first be composed into a narrative before they can be projected into a public arena. Furthermore, once composed, the circulation of survivor memories within public arenas involves a wide range of social agencies – from publishers to broadcasters. Often, if not typically, this process is undertaken by descendants on behalf of survivors, as part of the translation of direct experience into cultural memory. In this translation, the 'professionals' of war memory such as historians, archivists and museum workers play a key role. Of course, historians have long been instrumental in creating templates of war memory, fashioning the memory of past wars into national or sectional narratives. For example, Nasson shows here how the rewriting of the Anglo-Boer War in children's textbooks was central to the inter-war Afrikaner project of nation-building. Although historians might be said to be part of a 'second-order elite', they have long exerted a crucial influence in the fashioning of national imaginaries.

However, given their role in gathering, interpreting and giving an audience to survivor memory, such professionals could be said to have taken on added significance in the modern context of war commemoration. The recent rapid proliferation of academic literature on war and commemoration begs the question of what its relationship is to other, more 'popular' productions of war memory. Sometimes this relationship is direct, as with the use of historians as 'experts', consultants and anchors for media events commemorating anniversaries. The period surrounding the fiftieth anniversary of the end of the Second World War in Britain was marked by, for example, Ian Kershaw's history of the Nazis, whilst Paul Fussell, author of *The Great War and Modern Memory*, used his double role as veteran and expert on memory in narrating his series on the allied experience of combat in the Second World War. A crop of television documentaries appeared to coincide with the eightieth anniversary of the end of the First World War during 1998: Jay Winter's award-winning series *Total War*, funded by the Imperial War Museum, and Stephen Humphries's *Veterans: The Last Survivors of the Great War*, based on the oral testimonies of men in their late 90s and early 100s. Historical work has also figured centrally in recent literary representations of the Great War: Elaine Showalter's work on the history of shell shock, figured through the relationship between W.H.R. Rivers and Siegfried Sassoon, informs Pat Barker's trilogy and the film *Regeneration.*[156] As historians have made war commemoration in the past their object of study, so too, they have become important agents in the shaping of its future.

This relation is arguably of unprecedented importance because the specific *form* of memory-production that is favoured in modern war commemoration is the personal testimony. It constitutes the dominant genre of 'modern memory' of war. Its intensely individuated meanings differentiate it from the monument, remembrance day or other 'official' and collective markers of memory. Projects involving its collection and preservation have become ever more ambitious and technologically complex. The national war museums in Britain, Australia and other countries, for example, not only use clips of oral testimony and video interview in their exhibits, but have become primary sites for the generation and archiving of personal war memory from both world wars, including the home front as well as combat experience. This process of generating personal testimony has been most fully developed in the case of Shoah memory. The Fortunoff Video Archive for Holocaust Testimony, founded in 1981, had amassed a collection of 3,400 witness accounts by 1995.[157] Steven Spielberg's Visual History of the Shoah Foundation comprises the most ambitious attempt yet. In 1995 alone it collected 20,000 testimonies, with a projected total of 50,000 by the millennium.[158] *Shoah*, Lanzmann's ground-breaking film of 1985 not only featured first-hand testimony, but offered a critical commentary on the very notion of witnessing.[159] The personal testament has arguably become the end-of-century equivalent of the war memorial as a symbol of war. These mass collections of individual testimony have in themselves become, in Nora's terms, '*lieux de mémoire*' for the public representation of war.

This modern commemorative culture often draws on methods developed in the 'new social history' of the 1960s, especially oral history. But if the emphasis is placed on the experiences of ordinary people, it is the academic who most often acts, as Ken Plummer puts it, as the coaxer and shaper of war stories.[160] We elicit them, we edit them and we give them a public audience. Arguably, the historian has a place in the commemorative culture of the late twentieth century as privileged as that of the war poet or monument designer in mid-century. As such we must ask questions about the social uses to which such testimony is put. Into what kinds of commemorative service is the personal testimony elicited by the oral historian being pressed? What areas of historical understanding are privileged and marginalized in this process? How is the relation between public narratives of war and private experience reconfigured by such memorialization?

In Jelin and Kaufman's chapter for this volume on the politics of commemoration in Argentina, it is clear that the work of oral historians plays a crucial role in keeping memory alive given the pressure to close off public debate about the atrocities perpetrated by dictatorship. They argue that whilst the private memory of victims can never be fully assimilated within public commemorative forms, nevertheless oral history is an important means of keeping the dialogue open. A similar spirit pervades Alistair Thomson's *Anzac Memories*. Funded partially by the Australian War Memorial, Thomson's work is nevertheless aimed squarely at dominant public narratives of the Great War as the making of Australian nationhood. *Anzac Memories* reflects the emergence of 'counter-narratives' from the 1980s which questioned the celebratory association between war and nationhood. It simultaneously shows the power of national legends in shaping individual memory, and seeks, by highlighting those memories which have been marginalized or excluded, to articulate new public meanings of the event. Thomson works outwards from the role of the oral historian as advocate towards the fashioning of a new, more inclusive public history of the place of war and war memory in Australian society.

Such attempts to transform public commemoration through giving the voice of survivors a public audience can be problematic, however. Annette Wieviorka shows how the memory of the Shoah has shifted since the 1950s from that of 'survivor' to that of 'witness'. In the immediate post-war period, *Yizkor* books,[161] diaries and other recordings were motivated by the survivors' internal need to tell their story, and by the shared desire to transmit a collective culture of 'knowledge, gestures, habits, forms of organization' – the very frameworks of memory – which risked loss. Yet such memory did not have much impact on the wider society: it circulated largely within the separate ethnic groups of European Jewry. What replaced this from the 1970s was a 'social imperative', in which private experience has been introduced to an ever-widening public arena, encouraged by the use of oral history, video, television and other interview techniques. Paradoxically, the effect has been to individualize Shoah memory. It is the display of individual emotion which gives the testimony of the survivor authority – whether they are talking to the camera, to schoolchildren, or at the very site of their internment.[162] As Loshitzky remarks (in Chapter 7), the effect of this hegemonic narrative in Israeli society has been to elevate victimhood 'to the level of "civil religion"'.

The fixing of public commemoration on personal experience in general, and the focus of narration on victimhood in particular, is not restricted to the representation of the Shoah. Indeed, it has become a widely accepted means for survivors to seek recognition and pursue political claims. In his work on the French Algerian War, Martin Evans shows the pivotal role of documentary makers such as Bernard Tavernier in lifting the taboo on recognition of this 'undeclared war' in French national memory. Interestingly, Tavernier's film, released to coincide with the thirtieth anniversary of the war's end, seeks redress through its unrelieved focus on the hidden personal suffering of the conscripts: their physical disabilities and continued emotional trauma. Both the film and the activities of the main ex-servicemen's organization, the FNACA, sought, through commemoration focusing on individual trauma, to constitute 'a distinct minority group in special need of consideration'.[163]

Such attempts to create new commemorative spaces through the bringing of survivor narratives to a public audience raise many difficulties. It makes critical judgement difficult, and risks turning the historian into a mere conduit for the pursuit of claims based on victimhood.[164] Sensitivity to the grief expressed in personal testimony can threaten to overwhelm and obliterate understanding of the politics of war memory. This can be seen for example in the recent efforts by ex-Nazis who were interned by the Soviets in the Nazis' own concentration camps after the end of the war to bring attention to their plight. They draw on the tropes and imagery of Jewish concentration camp survivors to insist that *their* suffering should receive equal attention to that of the Jews. Oral historians working in such a situation may face a dilemma: in their attempts to document the 'hidden history' of Soviet prisoners in the camps, they may at the same time be confronted with the interviewees' desire to influence public opinion, by establishing their personal suffering as commensurate with those of Jewish survivors.[165] Alessandro Portelli notes rather similar tensions between the claims of survivors and the need for critical historical understanding in his powerful account of the changing cultural memory of the Nazi massacre at Civitella in Italy in 1944. Oral historians had been prominent in helping to organize a conference on European memory of Nazi crimes, partly as a means of giving restitution to Civitella's neglected memory of the massacre. However, in so doing they had helped to elevate a particular version of events – a sectional memory – into an 'official' local memory in which the massacre was seen to have been caused, not by the Nazis, but by the Resistance fighters who had supposedly provoked them. The ideological nature of this memory proved difficult to expose or challenge because to do so seemed deeply disrespectful to the mourning of the survivors. At the conference, Portelli observes, 'reverent contemplation – no doubt justified by the dramatic quality and emotion and pain of the witnesses –

prevailed over analysis and interpretation'.[166] The historians' personal experience of witnessing the grief of survivors – first as interviewers and then as co-organizers of the conference – limited their capacity to interrogate the accounts of survivors. In effect they had become unwitting accomplices in the resurgence of right-wing politics in the mid-1990s, and its questioning of the post-war assumption that Resistance fighters had 'won' the war for Italy.[167] Portelli's work illustrates the extent to which oral historians, in giving a voice to survivors, are inevitably immersed in the politics of war commemoration. Critical judgement is not easy to sustain in a public culture which has a heightened sensitivity towards the personal experience of trauma.

We also need to ask what other possible understandings of war are excluded by the focus on victimhood. Wieviorka argues, for example, that intimacy and the expression of personal emotion becomes the point of recording, displacing other forms of historical analysis. Martin Evans observes a similar problem with French documentary films of the Algerian War. Such films powerfully evoke the conscripts as the victims rather than the perpetrators of violence. At the same time they do not consider the effects of the war on Algerian society. The pain of French conscripts who had to witness torture keeps on the margin the pain of the Algerians who were tortured.[168] De Medeiros makes a similar point (in Chapter 8) about Portuguese veteran narratives of the colonial wars. In focusing on the personal psychic scars of their war, they effect a double process of blocking out of alternative understandings. The privileging of veteran memory defines the 'survivor' in a way which threatens to displace other kinds of war experience, such as those of partners and families, for example. In addition, the effect of the veteran narratives – sometimes despite the author's explicit intentions – is to leave the impact of violence on the colonized, not just during the 'wars', but across the centuries of Portugal's colonial rule, outside the frame of understanding. Powerful images of veterans re-living traumatic memory may promote the 'forgetting' of other victims, and threaten to supplant historical analysis. What Wieviorka terms the 'compassionate pact' between witness and hearer drives out understanding of the wider realities of social relations.[169] This difficulty is not restricted to novelists and film-makers: in Emmanuel Sivan's discussion of the commemoration of Jewish dead in the fifty years of Arab–Israeli conflict since 1948, an emphasis on the private and therapeutic functions of remembering leads to an almost total neglect of the arena of the nation-state. The humanist notion of grief as mediated through collective practices such as the creation of booklets of commemoration by family and kin, means that the relationship of such practices to official Zionist memory of the Arab–Israeli conflicts is not examined.[170] Yet this relationship surely demands attention since, as Yael Feldman has pointed out, the post-war generation of Jews 'grew up under the soothing images of heroic partisans', to the point where even the Shoah experience was marginalized.[171] Not only does Sivan contend that commemoration of the wars is overwhelmingly voluntary, 'spontaneous' and thus wholly autonomous from the agency of the nation-state, but also that its effect has been to 'put in relief the human price of war and thus mitigate the tendency towards its glorification'.[172] His emphasis on personal loss, and the marking of the familial–civic relation as the primary arena of analysis, creates an untenable separation between Zionist politics and the mourning of survivors.

Thus, in recent years, the survivor testimony has become a form of dominant memory in its own right. This has important implications for the way in which we consider the relation between the dominant public narratives of war memory which are produced through the agencies of the nation-state and discussed in section 2, and the politics of subjectivity explored here. Because of their emotional resonance, survivor testimonies are often mobilized in support of these dominant narratives, as we see, for example, in the way that personal testimony sustains the British image of the Second World War as the 'people's war', or how it is used to support Israeli narratives of Zionism. Yet, even when survivor testimonies articulate alternatives to the dominant national and transnational narratives, still their voices may be privileged in ways that obscure and marginalize the voices of others who, like the second generation, have lived with the shadows of war and whose identities have been shaped by their connections with the survivors. Whilst we have argued for closer attention to the ways in which these 'private' and 'public' politics articulate, the determinations that operate between them are complex, multi-faceted and often contradictory. As such, accounts which seek to analyse these politics must recognize their inter-relations without collapsing the two together. The contributions in this volume illustrate the difficulty of this task. Jelin and Kaufman (in Chapter 2) argue for holding open the disjuncture between 'public politics' and 'the personal' as valuable and necessary, as it is within the latter realm that the memory of loss is most keenly preserved, in the face of public imperatives to forget. De Medeiros (Chapter 8) is interesting for the way he holds the two aspects in creative tension, beginning with a discussion of silences about the colonial wars at the level of the national culture, but tracing the exploration of memory into the psyches of the novelists' characters, where complex and ambivalent responses to the past are operating. These discussions retain a recognition of the autonomy and yet the mutual construction of 'personal' and 'public' memory, and the politics of this relation. In the absence of such a recognition, we have argued here, the professional producers of memory and interpreters of the meanings of war, may on the one hand become unwitting agents in the privileging of survivor testimony, or, on the other, adopt a perspective which makes assumptions about, and thus overdetermines, the effects of public narratives on individual subjectivity.

4 Naming wars, framing memories

As we have argued in previous sections, the sovereign national state plays a key role in the politics of war memory and commemoration. The legacy of war – and the ongoing political vitality of war memory and commemoration – is intimately related to the very formation of many modern nation-states, and to the basis of enduring conflicts, identities and traditions of political life within others. In section 2, we identified the hegemonic process whereby the official national narrative promoted by state agencies operates so as to 'frame' war memories articulated from below, in forms which serve the interests of that nation-state. Such framing establishes horizons and structures which condition the meanings assigned to any particular war. It may be based upon cultural templates of remembrance derived from wars of the past (as discussed in section 3.1); but it is also responsive to changing political circumstances. As the state meets the challenge of internal conflict over war memory, or engages afresh in new hostilities, the significance of previous wars may also be reworked retrospectively, in accordance with the political requirements of the present. This hegemonic framing of memory is a selective process in which the nation-state exercises its power to recognize and incorporate within its national narrative only certain war memories, whilst others are officially marginalized or forgotten.[173] Indeed, the willingness or refusal of the state to recognize and name particular kinds of violence as 'a war' in the first place, together with its bestowal of names upon particular wars, are fundamental to the construction (and contestation) of the national narrative and official memory.

In section 4, we consider the politics of naming wars and framing memories. However, our concern here is with the hegemonic process as it operates, not just 'internally', within the relations between the nation-state and civil society, 'private' social agencies, and individuals, but also within transnational arenas. These are constituted by the relations between any one individual nation-state and others; between the nation-state and a range of transnational agencies (from multi-state alliances and the institutions of international law, through to non-governmental organizations [NGOs] and the globalized culture industries); and between the nation-state and supranational agencies, most notably the United Nations (UN) organization.

Section 4.1 explores how the naming of 'war' and of particular wars by the nation-state is both structured and circumscribed, on the one hand by the definitions and regulatory principles of international law concerning war and human rights, and on the other by the hegemonic power of transnational political alliances. It considers the ways in which the political and human rights narratives constructed by transnational and supranational agencies work to frame, and to further or limit, the power of the nation-state to wage, as well as to remember, war. Section 4.2 seeks to historicize this argument by identifying what we term the transnational master-narratives of twentieth-century war memory: those of the First and Second World Wars and the Cold War. It examines the frames of war memory constructed by these narratives, and the ways in which memories of pre-existing or related conflicts may be redefined by, or occluded within, these frames. In our current epoch, in which the long-dominant Cold War frame is in process of dissolution, a reawakening and/or re-evaluation of national memories of the two World Wars and other conflicts has been widespread. The transnational relations and processes examined in sections 4.1 and 4.2 not only affect the recognition and remembering of war within the arena of the nation-state, but entail profound consequences for the politics of representation and subjectivity promoted within civil society, and in the private arenas of remembrance explored in earlier sections. Finally, section 4.3 considers the dynamics of change since the end of the Cold War, and argues that globalization, and the growth in importance and power of the narrative of human rights, are transforming the ways in which the politics of war memory and commemoration are now being conducted, and will be conducted, increasingly, in the future.

4.1 The politics of naming

What is 'a war'? The importance of linguistic definitions in the conduct and remembrance of war became apparent in May 1999, when the UK's Prime Minister, Tony Blair, made an illuminating gaffe in the House of Commons. British armed forces had been involved for several weeks in military action in the air and on the ground under the NATO K-*for* command, established to enforce the withdrawal of the Yugoslav army from Kosovo. Responding to Opposition accusations that he had been neglecting Cabinet business, Blair pointed out that, 'with the war on', his time had been pressured. According to *The Times*, this was the first occasion that the government had 'admitted' to the House that Britain was at war, but Blair's office quickly corrected what it called his 'slip of the tongue': 'The Prime Minister has made it absolutely clear that we are in an international armed conflict.' Later the Foreign Office confirmed that 'We are not at war. In order for us to be at war, it would require a formal declaration by at least one side, and probably both. Legally, there have been no wars since 1945.'[174]

While on the surface the Foreign Office's response to Blair's gaffe might appear to contravene both practical understanding about what a 'war' entails and common-sense usage of the word, this was far from being a legalistic quibble. A serious political issue underlay this exchange: namely, whether the military action taken by the UK as part of the NATO alliance could be construed, as NATO claimed, as a humanitarian exercise with a legal basis in United Nations Security Council resolutions on the crisis in Kosovo; or whether, as claimed by the Serbian-dominated government of the rump Yugoslav Federation, it represented an illegal act of aggression by one sovereign state upon the territory of another, and an unwarranted meddling in its internal affairs.

This exchange, about whether or not the UK was 'at war', can serve as a reminder that to use the term 'war memory' at all is to work within implicit assumptions about what a war is. Such assumptions can be opened up for scrutiny by posing two explicit questions: when does the practice of organized violence become something that can properly be described as a 'war'? And who, or what, has the power to institute such a definition? The conflict over language in such cases is always an element in a broader struggle over the representation of the use of armed force: a struggle to make particular meanings effectively dominant, so as to secure legitimacy for 'our' violence whilst rendering that of the enemy illegitimate. Thus, the naming of war is closely bound up with attempts to frame the significance of a conflict in relation to competing constructions of meaning; and, as such, it provides a basis for the contestation of war memory.

At the centre of this 'politics of naming', the nation-state works to institute its preferred definitions of a conflict. The state's centrality and power is closely related to its enjoyment of a privileged legal right to use violence and 'wage war'. Where it is prepared to use the term, 'war', this tends to imply state recognition of the political dimensions of a conflict; that it is indeed 'a continuation of state policy by other means',[175] rather than, say, a humanitarian exercise or an internal 'law and order' problem. Conversely, withholding of state recognition implies a denial of the justice – or indeed the very existence – of competing political claims. As we demonstrated in section 2 of this chapter, such denial of recognition may be directed against those considered to be 'internal' enemies of the state, or even those who have fought for it, such as the French veterans of the Algerian War discussed by Martin Evans.[176] But it is not only an internal matter. The French state's difficulties in shaping a coherent official memory of the Algerian War impacted as well upon its inter-state relations with Algeria, for whom this war – that France was unable to recognize – was the crucial struggle securing independence.

Such withholding of recognition has tended to be the general pattern in conflicts accompanying the process of decolonization. Euphemisms like 'the Troubles' in Ireland, or 'the Emergency' in Malaya and Kenya, helped these British colonial states portray as illegal 'terrorism' the armed actions of those who, in their own terms, were 'freedom fighters' waging a 'war of national liberation', on behalf of an anti-colonial movement intent on establishing a more just form of state.[177] In Northern Ireland, British Labour and Conservative governments from the mid-1970s onwards pursued a policy of 'criminalization', in which recognition of the political aims of the 'armed struggle', waged by the Irish Republican Army to secure an independent all-Ireland state, was withdrawn, and the conflict was redefined as a form of civil disturbance pursued by 'criminals'. 'Prisoners of war', formerly granted 'political status' within the judicial system, had this 'privilege' withdrawn and were forced to wear prison-issue clothing (generating resistance within the prisons which eventually led to the republican Hunger Strikes of 1980–81). Henceforth, representatives of the British state took care to refrain from describing the conflict as a war (at least in public); and the term only returned to common currency in Britain following the announcements of the republican and loyalist ceasefires in 1994 (when an end to 'the war' appeared to be in sight, and its prior existence now could be safely acknowledged).[178]

These conflicts over definition, meaning and legitimacy conducted in the course of an armed conflict inevitably impact upon the ways it can be remembered and commemorated. The UK state has erected no public memorial to soldiers in the British army killed in Northern Ireland and holds no public commemoration, on the grounds that this is a civil conflict. By the same token, certain commemorative practices adopted by the IRA and other republican armed organizations to honour their war-dead – notably the paramilitary symbolism and ritual used at the funerals of many Volunteers killed on 'active service' – were proscribed along with the organizations themselves.[179]

Frequently, in the transition from colonial rule, those who have been criminalized and proscribed by the state they sought liberation from, subsequently win recognition from the newly established state as its revered heroes and leaders. In these circumstances, the securing of state power creates the conditions for the emergence of an alternative frame of memory to that of the colonial state, with the name of 'war' being officially recognized and conferred from this new national perspective. For example, what looked in 1919–21 – to the UK's Prime Minister, David Lloyd George, and to many Unionist observers in Britain and Ireland – to be a squalid vendetta of assassination and ambush conducted by 'murder gangs'[180] in plain clothes, was later dignified in Irish nationalist historiography as the 'War of Irish Independence'. 'You do not declare war against rebels', Lloyd George averred in 1920.[181] Yet the replacement of one dominant narrative with another rapidly occurred once the Irish Free State (later the Republic) had been negotiated into existence in 1921. The new state constructed the IRA campaign in the 'Independence War' as the culmination of its own founding narrative and struck commemorative medals for its veterans.[182] Excluded from public memory within the narratives of the UK state and of the Irish Free State/Republic, is the continuation of the independence campaign that was fought immediately after 1921 by the IRA in the north of Ireland to prevent partition and the establishment of the new Northern Ireland state. This proved awkward to assimilate into the founding narrative of the Free State; it has been described as 'a conflict which appears to have no official name, yet was still effectively a war, even though the Irish government was not officially an "active" participant'.[183]

The power of a state to install a national frame of remembrance, in which the memory of certain named wars is propagated at the expense of other, forgotten wars, has the most far-reaching consequences in contexts where a people has been successfully subordinated and repressed by military occupation, and thus has been unable to secure its own state. The 'stateless' condition of the Palestinian Arabs, for example, stems from their military defeat and dispossession by the Israeli state in 1948, when Palestinians were driven off the land in what had now become 'Israel', and in 1967, when the illegal occupation of the Palestinian West Bank took place. Nevertheless, according to Edward Said, the 'long list of outrages committed against us, the tortures, the killings, the invasions, and above all, the prohibition of our political identity and the systematic dispossession of our people' has been effectively concealed within a hegemonic Israeli frame of memory which foregrounds acts of Palestinian 'terrorism' and its threat to Israel's survival.[184] Yosefa Loshitzky (in Chapter 7, this volume) links this denial of Palestinian oppression to the mobilization of the Shoah within 'the grand narrative of Zionism', as a memory of Jewish victimhood which is used 'to justify not only Jewish settlement in Palestine, but also the Palestinians' expulsion and the negation of their right to freedom and self-determination'.[185]

Among the effects of displacement and repression have been the rupturing and disorganization of Palestinian memory: 'Stateless, dispossessed, de-centred, we are frequently unable either to speak the "truth" of our experience or to make it heard. We do not usually control the images that represent us … Since our history is forbidden, narratives are rare.'[186] Said attributes this 'severance' of memory to what he calls 'the sustained war upon our national peoplehood by Israel'.[187] Yet, paradoxically, where memory is disrupted in this way, use of the term 'war' to name such violence may be problematic. Abu Mohammed, mayor of the West Bank village of Dir Kadis, would only use the term as a simile in his description of the violent appropriation of Palestinians' land, part of the state-endorsed expansion of Jewish settlements which was still under way in 1999:

> When the settlers brought their bulldozers [to destroy the Palestinians' olive-trees] we went down with our papers to prove it was our land. Then the Israeli soldiers arrived and just went mad. *It was like a war here*, they were throwing tear-gas grenades and firing CS-gas cannisters and live ammunition into our houses. (our emphasis)[188]

In the absence of a state arena in which a Palestinian national memory of the protracted war of occupation might be articulated and recognized, the very language used to name the Israeli–Palestinian conflict is undermined and weakened.

Ought we to describe such a case, where state violence is committed against civilians during 'peace-time', as 'a war'; or is it only 'like a war'? Does a war not require there to be at least two sides engaged in armed conflict? Or is it not the case that the existence of 'an enemy within', especially an armed and insurrectionary one, is required by the repressive state to justify its actions; and if one did not exist, it would have to be brought into being, or invented? What other language might be used instead, and with what effects upon the framing of memory? These questions become acute in any society where the full weight of the state's potential for systematic, repressive violence has been unleashed against sections of its 'own' people, as in the military dictatorships of Latin America that flourished during the 1970s and 1980s.

In Guatemala between 1978 and 1985, for example, the army, police and paramilitary forces of the dictatorship committed some 130,000 political killings and 38,000 'disappearances', known and remembered as *La Violencia* (The Violence).[189] In Judith Zur's account, 'nominally [*La Violencia*] was a battle against guerrilla forces … Guatemala's military regime … declared war on "Marxist communism", which it viewed as an armed and dangerous menace within; dissidents and dissidence were to be eliminated once and for all.' In practice, however, the concept of the subversive was expanded to encompass civilians, millions of whom were targeted, especially among the indigenous population, from which the majority of the victims derived; and the victims' widows and other kin, seen as subversive by association.[190] Zur argues that the effects of actual physical violence were compounded by 'a war against memory', encompassing a state prohibition against speaking about atrocities, strategies adopted by the military to obscure their responsibility for killings, and the propagation of 'official truth' that blamed events upon the so-called communist subversives.[191] Thus, according to Zur, the military regime produces an Orwellian contamination of language and memory: 'When they mean war they speak of peace … and when they mean repression, they speak of freedom.'[192] In this environment, it becomes impossible 'to credit any particular explanation of a violent act as definitive', leading to profound disturbances in the personal and common memories of those affected.[193]

Elizabeth Jelin and Susana Kaufman, writing about memories of the systematic use of state violence against the civil population in Argentina (in Chapter 2, this volume), note that a similar justification to that used in Guatemala was advanced by the military junta that seized power and ruled from 1976 to 1983. According to General Videla, one of its leaders, a war had to be fought to destroy 'Godless communists' who threatened the security of the state.[194] As in Guatemala, the definition of those who constituted this threat was inflated by the military: as Videla put it, 'a terrorist is not just someone with a gun or a bomb, but also someone who spreads ideas that are contrary to Western and Christian civilisation'.[195] Jelin and Kaufman argue that, used in these circumstances, the term 'war' is a misnomer. They concede that a leftist guerrilla movement was indeed engaged in 'widespread armed conflict' with right-wing paramilitary death-squads during the early 1970s; and that this conflict, together with the violence unleashed by the state before, during and after the coup, has been 'defined by some as … a "Dirty War" '. However, to use this name implies a particular interpretation – that 'both sides' shared responsibility for the violence – with which they are in disagreement. Argentina, they maintain, did not undergo a war, but a process of 'State Terrorism' undertaken by 'a state apparatus that resorted systematically to terror and repression of citizens and civil institutions without any respect for existing national and international juridical norms'.

For Jelin and Kaufman, this struggle over naming war and framing memory is fundamental in challenging the ability of the nation-state to practise organized violence in pursuit of its political goals. In referring to 'juridicial norms', they identify one of the weapons in this fight: the definitions and precepts of international law, as embodied in agreements such as the Hague, Geneva and Genocide Conventions, and the Universal Declaration of Human Rights. The development during the twentieth century, and particularly after 1945, of international law concerning the conduct of warfare, constituted an attempt to define and regulate the proper conduct of intra-state conflicts and those between the state and its own citizens. It has underpinned efforts to construct a narrative of 'human rights' capable of limiting all forms of violence practised by the state, and the transnational and supranational agencies necessary to institute this narrative: for example, the United Nations Security Council and the International Court of Justice established at the end of the Second World War.[196]

To the extent that individual nation-states are prepared to endorse and adhere to its principles, and acknowledge the authority of its institutions, international law constitutes one of the major ways in which pressure 'from above' may be brought to bear upon the state – with effects both upon its conduct of 'war', and the ways in which this can be remembered. In cases where a state incorporates these principles into its own national law, war memories articulated in terms of the human rights narrative are able to seek recognition within the national arena. Alternatively, the institutions of international law also constitute a supranational arena in which the legitimacy of a state's practice of violence – and thus, its official memory – may be established or challenged. International courts thus become a site of struggle over memory, as when charges of torture practised upon IRA suspects were brought in the 1970s by the Irish government against the British government, and heard by both the European Commission on Human Rights and the European Court.[197] When states wage war against other states, it is usual for them to advance claims for the justice of their cause in terms of international law, as in the diplomatic conflict over rightful territorial sovereignty that accompanied the 1982 Falklands/Malvinas War.[198]

However, the attempt since 1945 to institute an effective, transnational narrative of human rights has had to negotiate the realities of international power relations. These have exposed its relative weakness (although this may now be changing, as we suggest below in section 4.3). The internal state terrorism of Guatemala and Argentina was carried out in spite of 'international juridicial norms', while Israel – whose occupation of Palestine is in breach of the Hague and Geneva Conventions and has been condemned in numerous UN Security Council resolutions[199] – has been able to flout international law with impunity. In these cases, the ability of the nation-state to prosecute its 'war' is underpinned by a different kind of transnational relation: that of inter-state alliances and the balance of power pertaining at the global level of 'world order'. The repression carried out by the Latin American military dictatorships of the 1970s and 1980s took place within the context of the US-sponsored reaction against socialist and communist movements world-wide, particularly under the Reagan regime (1981–89). The dispossession and silencing of the Palestinians was not achieved by the Israeli state alone, but depended upon 'the alliance between Zionism and the US'.[200] Movements engaged in wars of national liberation often embraced the ideology of international revolutionary communism and enjoyed the support of the Soviet Union. In all these cases, local, national and regional conflicts over the waging, naming and remembering of 'war' can be seen to be overdetermined by the Cold War.[201] Indeed, it has been argued that the Cold War polarization that came about in the years immediately following the end of the Second World War and lasted until 1989–90, is directly responsible for the failure fully to implement the human rights narrative as this had been envisaged during the mid-to-late 1940s.[202]

The power exercised by an individual nation-state within its own political jurisdiction is shaped and constrained, not only by the pressure of human rights agencies, but also by these political relations. Hence, the endeavour to construct an official memory of war within the national arena of any particular state is also necessarily circumscribed by external influences operating upon it from this transnational arena, in whose terms national constructions of memory tend to be framed. One abiding image of the way in which transnational alliances may impact directly upon national memory is that of the German Chancellor Kohl and the French President Mitterand, standing hand in hand with heads bowed in the First World War cemetery of Verdun, in a joint commemoration of the seventy-fifth anniversary of the Armistice. This image was constructed for inscription at the centre of a newly fashioned 'European' cultural memory, as a celebration of the reconciliation and alliance between two former enemy states.[203]

Hegemony does not operate solely within the arenas of the nation-state, but also at this transnational level. The uneven distribution of political power between as well as within states defines the dominant terms of legitimacy for their use of violence, installs the wars fought by the most powerful states and alliances at the centre of world history, and establishes the national perspectives and memories of those states and alliances as 'master-narratives'. These place and subordinate other national and local perspectives and memories within the terms of their overarching, transnational, hegemonic frames.

4.2 Hegemonic master-narratives and the shifting frames of war memory

The hegemonic, framing master-narratives of the twentieth century have centred on the European and North American experience of two World Wars, and on the Cold War organized around the military and nuclear stand-off between the American-dominated NATO and the Soviet-dominated Warsaw Pact. Historically, as each earlier conflict was superseded by its successor, the memory of its significance was refocused through the lens of then-current global concerns. Here, too, a politics of naming was at work in the framing and reframing of meaning. For example, in Britain the term which contemporaries used to describe the conflict of 1914–18 was 'the Great War'. The use of the adjective, 'Great', was intentionally ominous, distinguishing the scale of that war from all which had gone before; but the definite article, 'the' Great War, simultaneously carried an unconscious proleptic promise: this would be the apotheosis of war, the 'war to end all wars'. This promise was not fulfilled; and the subsequent terminology – installed after the 'Second World War' had become the sequel to the 'First World War' – marked the failure of this aspiration.

Rupturing and refiguration of previously hegemonic, transnational narrative frames was also a consequence of the reordering of global power relations brought about by each of the World Wars. The marked decline in power of the British and French colonial empires, for example, meant that the impact and legacy of the two World Wars became imbricated with decolonization processes which were also occurring on a world scale. In Kenya, for instance, the involvement of Black volunteers in the British armies that fought Germany in Africa during both World Wars fuelled demands for political reform and African representation after 1918, and for national independence after 1945.[204] After the disintegration of the Tsarist Russian Empire during the First World War, the Baltic states such as Latvia, Lithuania and Estonia achieved national independence, only to be incorporated into the Soviet Union during the course of the Second World War. Here, post-1918 memories of the struggle for national independence, as well as war memories of alliance with Nazi Germany and the destruction of the Jews in 1941–44, became overlaid by the official memory of the Soviet Union's 'Great Patriotic War', and were suppressed along with national culture more generally.[205]

After 1947, memories of the two World Wars became overdetermined by the newly hegemonic transnational frame of the Cold War, which effectively remained in place until 1989–90. As the major Second World War allies became enemies, and the concept of a 'Third World War' began to figure in analyses of nuclear weapons strategy, so dominant memory of the war against Nazism and fascism split and polarized along lines dictated by the global antagonism between 'the West' and 'the Soviet bloc'.[206] In the states of the NATO alliance, particularly after the commencement of rearmament during the 1970s, Second World War memories were reworked around the trope of totalitarianism versus democracy. The Soviet Union was equated with Nazi Germany as an 'expansionist power' that posed a 'strategic threat' to democratic freedoms and ways of life 'graver than at any time since the last war'.[207] To counter this image, the official memory of the Great Patriotic War was mobilized by Warsaw Pact states to reaffirm the 'decisive contribution' made by the Soviet Union to the liberation of Europe from Nazi barbarism, the immense sacrifice paid by the Soviet people and the 'gratitude of the liberated peoples' – all widely acknowledged by the states and peoples of Europe and North America after the war, but now, it was argued, forgotten by Western proponents of the Cold War seeking to deny the 'peace-loving' intentions of the USSR.[208] With the world increasingly polarized into Western-capitalist or Soviet-communist spheres of interest (although many states sought 'non-aligned' status), these competing social systems also overdetermined the bases on which the legitimacy of new wars during that era could be established, thereby turning much of the Third World into a Cold War battlefield.[209]

The effect this had upon war memories more widely can be glimpsed in Bill Nasson's history of the changing role of Afrikaner commemoration of the Anglo-Boer War (1899–1902) in South Africa (in Chapter 3, this volume). Nasson notes how the ruling elite of the new apartheid state began in the early 1950s to commemorate both World Wars. Until then culturally remote and lacking in resonance for Afrikaners, commemoration of the Somme, for example, came to embody 'the national sacrifice and legitimate aspirations for freedom of a European South Africa, inscribed within ... a "timeless" tradition of just wars to preserve Western civilisation'. The Cold War rhetoric here is unmistakable, as is its value in alliance-building at a time when communists were playing a leading role in opposition to the apartheid regime.

The hegemonic master-narratives of the two World Wars and the Cold War established frames within which complex, multi-faceted struggles over war memories, with roots in the histories of specific localities and regions, were reduced to simpler and more abstract battles between communism and fascism, or democracy and communism. The complexity of the conflicts concealed by these hegemonic frames is clearly visible in the case of war memory in Finland, as described by Petri Raivo (in Chapter 5, this volume). Finland's wars of the twentieth century have been local theatres of larger conflicts, first that between Red and White, later that between communist and Nazi. What might be described by outsiders as 'the Finnish experience of the First World War' is remembered by Finns as the Civil War which followed the declaration of an independent Finnish state; while the 'Second World War' is recalled as three distinct though interconnected wars to preserve national independence: the Winter War (1939–40) and the Continuation War (1941–44), both fought against the Soviet Union; and the Lapland War (1944–45), fought against the Finns' erstwhile Nazi allies. The very naming of what was remembered, as well as the forms of commemoration adopted, potentially decentred established paradigms and insisted on the existence and legitimacy of a distinct, Finnish national memory. For this reason, in the face of decades of Soviet hostility during the Cold War epoch, this narrative of national memory remained an entirely 'internal' discourse; while 'externally' the Finns paid lip service to Soviet dominant memory of the Great Patriotic War. As Raivo suggests, comparable complexities, dissolved into a uniform picture, could be found anywhere in the liminal zone of eastern Europe, 'from the Balkans through Poland, the Baltic states, and up to the White Sea'. Here, we can glimpse the ways in which the historic experience of whole regions of Europe came to be marginalized and silenced by hegemonic, transnational memory.

In the post-communist world that has emerged since the dissolution of the Eastern bloc, the Warsaw Pact and the Soviet Union itself in 1989–90, the hegemonic frame of Cold War memory has disintegrated, creating the conditions for widespread reassessment of the war memories of the twentieth century. This has not been an even development and it has taken different forms in different regions; being most acutely experienced in the former-communist states of Eastern Europe, the Russian Federation, and those ex-Soviet states which have won their independence, as well as in Finland, where a relationship of greater dialogue and cooperation has opened up with Russia. But it is a general process whose effects are felt throughout the world. In the UK, for example, the end of the Cold War caused Prime Minister Thatcher to approve in 1990 a public declaration that the British government had 'no selfish strategic or economic interest' in Northern Ireland. This was a statement long sought by the Northern Ireland Office, which Thatcher had refused whilst nuclear patrolling of the Atlantic remained a strategic necessity. It proved crucial to the establishing of the 'peace process' leading to the paramilitary ceasefires of 1994, which in turn have created the conditions for a flowering of new memory-work concerned with the legacy of the Troubles.[210] By contrast, after the withdrawal of Soviet troops in 1989 from the erstwhile Cold War battleground of Afghanistan, the Soviet-Afghan War has been forgotten, and with it the plight of some 1.2 million war refugees stranded since 1979 in camps across the border in Pakistan.[211] While in some circumstances the collapse of the Cold War frame has helped to open up opportunities for conflict resolution and the reassessment of received war memories, in other cases it is not always clear what alternative kind of framing can make sense of the conflicts of the past in its absence.

During the 1990s, the disintegration of the Cold War frame also helped to stimulate renewed interest in the two World Wars and created possibilities for reassessing their continuing legacy in the late twentieth century. The memory of these earlier wars remains 'a stake in the struggle'[212] over the legitimacy and hegemonic potential of present-day political projects in many national contexts. Exploring and opening up these national specificities has tended to involve critical examination of the ways in which these memories were framed during the Cold War epoch, in an attempt to make room for a revised national narrative capable of repositioning the nation in the post-1989 'neo-liberal' global system. This process coincided with the fiftieth anniversary commemorations of the Second World War between 1989 and 1995 which, as we have seen, provided numerous occasions for debate and reassessment. For example, Raivo identifies the approaching anniversary of the outbreak of the Winter War in 1939 as having been an important catalyst in Finland's new relationship with Russia.[213]

Within Russia itself, during the period of *Glasnost* beginning in 1986, 'people ... [began] voicing openly experiences and memories, hitherto only spoken of to family and close friends', and the popular media began to make public 'an outpouring of remarkable testimonies which negated official histories' of the Stalin era and the Great Patriotic War.[214] Since *Glasnost* and the recovery of their independence in 1990–91, the Baltic states have seen a vigorous conflict about the reassessment of their twentieth-century history, including the poisonous legacy of partisan wars, collaboration with the Nazis including participation in the Shoah, and oppression under the Gulag.[215] In the Ukraine, the achievement of independence has opened up a similar process of historical reclamation both internally and in dialogue with the memories of exiled Ukrainian communities, formed in Britain and Canada at the end of the Second World War, largely isolated from their homeland ever since and still deeply divided along the lines that existed in 1944.[216] In attempting to unlock complexities previously hidden within the Cold War frame, alternative designations may be helpful, and might need to be invented where none exist already in subordinated memory: so, for example, Timothy Snyder has proposed 'the Polish-Ukrainian Civil War of 1943–7' as one useful way of redescribing some of the fighting in eastern Europe during the Second World War.[217]

In other regions, reassessment of the legacy of the Second World War has, significantly, *not* occurred. Different kinds of complexity are encountered in the case of war memories in the former European colonies, where the hegemonic

'Western' master-narrative of the Second World War tends to conceal regional specificities within an effectively Eurocentric frame of memory. The fall of Singapore in February 1942, for example, figures in the British story of the war as 'the worst disaster and largest capitulation in British history'; a moment of overwhelming defeat and unprecedented humiliation which marked the lowest ebb in the nation's fortunes before the tide of war turned.[218] From the perspective of Singapore, however, the 'Pacific War' began not in September 1939 but in December 1941, when the Japanese invaded Malaya; and centred on the replacement of the British colonial regime by the Japanese Occupation which lasted until 1945. As Diana Wong has shown,[219] for the Japanese this was a key episode in the '*Daitoa* War': a holy war for the liberation of 'Greater East Asia' from the white colonial yoke; and Japanese rule involved a systematic, institutionalized politics of memory designed to erase all traces of British colonialism and to celebrate its humiliating defeat. Following the defeat of Japan and the reinstatement of British colonialism in 1945, the Japanese Occupation was itself subject to a comparable silencing and erasure of memory, while the British were unable to restore an untarnished memory of their power and benignity due to their defeat in 1942. Wong shows how, in these circumstances, memories of the Second World War came to be 'suppressed' through a reframing of Malayan history: first, in the immediate post-war years, by the narrative of decolonization; and second, during the counter-insurgency war known as 'the Emergency', fought by the British against Malayan communists from 1948 to 1956, by the Cold War. After Malaysia had won independence in 1957, its post-colonial narrative of nationhood emphasized the new state's continuity with pre-war colonialism rather than any fundamental break from it; and consequently the Japanese Occupation became elided as an unfortunate interlude. In the late twentieth century, public commemoration of the Pacific War played no part in national memory in Malayasia (where the fiftieth anniversary of its ending met with official silence).[220]

In the case of the Balkans, the disintegration of Tito's communist Federation of Yugoslavia has led, not to critical reassessment of national memories of the two World Wars, but rather to their reactivation as templates. These have framed the meanings given to a series of new wars – the Serbo-Croatian War of 1991, the Bosnian Civil War of 1992–95, and the Kosovo War of 1998–99 – into which the region has been propelled by the reassertion of Serbian, Croatian, Bosnian Muslim and Albanian ethnic nationalisms, each jostling for state power and position. The re-emergence of Sarajevo as the epicentre of a wider conflagration evoked across Europe and North America a memory of the lessons of 1914. More locally, memories of the Second World War proved to have retained their power to shape current political identities, alignments and antagonisms. The driving force behind the new Balkan Wars was the Milošević regime's pursuit of Serbian hegemony within post-communist Yugoslavia, and its 'Greater Serbia' project. However, a contributing factor leading to the outbreak of hostilities was the ideological play on what Stjepan Meštrović has termed 'displaced emotions from history'.[221] The Serbo-Croatian War in 1991, for example, was fanned by Milošević's warnings to the Serb minority in Croatia about the resurgence of neo-fascism and the 'danger of impending genocide' threatened by the nationalist Tudjman regime, especially after its declaration of independence.[222] Although 'there does not appear to have been any realistic likelihood of anti-Serbian genocide',[223] these fears were fuelled by memories of the genocidal violence unleashed against Serb and other minorities in Croatia and Bosnia-Herzegovina by *Ustase* militias of the Croatian fascist regime that was installed by the Nazis in 1941: memories which Tudjman's own adoption of *Ustase* symbols, and his denial of the existence of Croat-run death-camps such as Jasenovac, did little to dispel.[224]

The complex ways in which the violence of the Balkan Wars of the 1990s has echoed, or sought justification from, the violence of the past, has been bewildering to witness. Second World War memories were re-invoked as Croatian militias drove the Serb communities out of Krajina; while on an altogether larger scale, Serbian 'ethnic cleansing' practised in Croatia, Bosnia-Herzegovina and Kosovo, has caused many opponents to describe the Milošević regime as itself fascistic. Arkan, the Serb 'warlord' linked to the massacre of Croats at Vukovar and other atrocities, argued in justification that: 'During the second world war, hundreds of thousands of Serbs were killed by the Croatian nationalist regime… In my view, the Serbs were entitled to defend themselves against "ethnic cleansing" by Croatia.'[225] The issue of who is or is not 'a fascist' emerged as a central facet of the propaganda offensives conducted from all directions. NATO and the US government tried to make the 'Hitler' label stick to Milošević in 1998–99, while Milošević in turn claimed that the West was 'acting like Hitler' by bombing Belgrade (as did the Luftwaffe).

Appeals of this kind to the memory of the anti-fascist struggle contributed to the Milošević regime's legitimacy at home and abroad. Support for the Serbian-dominated rump Yugoslav state in Russia – popular as well as official – was conditioned by perceptions of a common Slavic cultural heritage buttressed by firmly established popular memories of the Great Patriotic War. As the Russian poet, Yevgeny Yevtushenko, pointed out during NATO's bombing campaign in Kosovo and Serbia in the spring of 1999: 'During the Second World War, the feats of Yugoslav partisans in their struggle against fascism inspired not only our soldiers but also our famous poets.' Despite tensions introduced into the relationship between Russians and Yugoslavs following the collapse of the Yugoslav Federation, Yevtushenko reaffirmed that: 'That war meant so much to us [that] we cannot abandon when they are in trouble those who had to struggle just as much as we did.'[226] A further effect of these templates of the Second World War, which have proved so potent in defining present-day concerns and justifications for violence in the Balkans, has been their eclipsing of other wars of the past, that also have a bearing on the current conflict. By concentrating on the atrocities of the *Ustase*, for example, the massacres and expulsions of Albanians in Kosovo by Serb forces during and after the First Balkan War of 1912–13 (thus consolidating Serbian control of the province it had seized by war from Turkey) may be conveniently overlooked.[227]

4.3 Dynamics of change: the future politics of war memory

Among the most important outcomes of recent work on the memory of past wars, has been an appreciation of the shadows they have cast, stretching far into the future, long after the fighting has ended. Today's wars produce tomorrow's memories of war. So it is salutary, in the light of the well-documented psychic, social and political legacies of the two World Wars, to bear in mind that over 140 'significant' wars have been fought around the world since 1945, while ten international wars and twenty-five civil wars were in progress during the twelve months to 1 August 1999.[228] As we noted in section 1, a further effect of the Cold War has been the emergence of a new pattern in warfare. Since 1990, civil wars have become more common, are fought more intensively and are more likely to escalate into regional conflicts; while effective intervention by the most powerful states to prevent such conflicts has declined significantly since the days when the USA and USSR would routinely flex diplomatic and military muscles to keep subordinate states in order. The casualties of war are increasingly likely to be non-combatants: according to one estimate, as many as 84 per cent of those killed in several recent wars are thought to have been civilians (compared to 15 per cent of total casualties in the First World War, and 65 per cent in the Second).[229] Among the civilians who will be the war survivors of the future, struggling to cope in some way with their own war memories, children feature prominently.[230]

Whilst we worked on this book throughout 1999, the British media were inundated with news stories about the damage inflicted by wars then occurring around the world, and about the ongoing aftershock of recently concluded wars, as societies sought to 'come to terms' with their immediate legacies. These legacies of damage include the exacerbation of political tensions; social problems directly caused by the destruction of lives and infrastructure; cultural damage; and the psychic effects of trauma. By the end of the 'international armed conflict' in 1999 between NATO and Serbia, for example, Kosovo had been reduced to what the UN Secretary-General's Special Envoy for the Balkans, Carl Bildt, described as a 'devastated wasteland that will have to be rebuilt from the bottom up';[231] it had suffered 'the massive destruction of the collective memory' in the form of libraries, archives and legal documents;[232] and it had to undergo the shame-enhanced trauma of the survivors, including the many homeless, the ethnic-Albanian widows whom tradition will prevent from remarrying, and the women rape-victims of the Serb state forces, 'whose life is then essentially over'.[233] During the months immediately following the conclusion of the war, it had to endure fresh waves of 'ethnic cleansing', directed against both the Serb minority (sometimes fuelled by desires to avenge earlier Serb atrocities), and against those Albanians who remained in so-called 'Serb areas'.

To reflect upon the legacies of damage bequeathed by this and other recent and current wars, in the light of the critical debates about war memories that we have addressed in this volume, engenders some chilling reflections. Yevtushenko, responding to the bombing of Serbia, voiced one 'lesson' that might be drawn from the twentieth-century history of the Balkans: 'Like Raskolnikov, history returns to the scene of its crime'; and the wars of the past have proved to be 'a serpent's nest of future wars'.[234] So:

> Be more careful with the Balkans! If into this nest we too arrogantly start to poke either the toe of a soldier's boot or a colonial master's riding crop, or winged rockets, then the awakened serpents may crawl far afield and one of them, like a cobra, puffing out from a local war into a world war, may possibly destroy the entire globe.[235]

The apocalypic fears expressed here can readily be understood in terms of the powerful templates of 'World War' discussed above. However, perhaps with more reason, Yevtushenko also warns of the viperous effects of past conflicts as the breeding ground of 'local wars'. In worn-torn societies, the legacies of damage are liable to be articulated in war memories – of horrors inflicted and sufferings endured – that heighten and intensify binary polarizations of difference formed in relation to historic Others (and the enmities encoded therein); increasing the likelihood of further violence along these faultlines.

The politics of war memory and commemoration in the future will centre on survivor-narratives from these latest wars, as well as on an emerging postmemory politics concerned with the wars of the second half of the twentieth century. The struggle to articulate these war memories of the future will continue to be profoundly affected by the historical shifts in both transnational and national framing, analysed in previous sections. If, as we have argued, the current era is distinguished by the disintegration of the Cold War frame that was long hegemonic in the transnational arena and within the national arenas of many individual states, it is also the case that the nation-state itself is coming under new kinds of pressure from alternative agencies, and its power to construct hegemonic national memories is being contested from above and from below in new ways. These dynamics of change are reshaping the arenas and agencies of articulation, making spaces for new kinds of narrative and generating new kinds of conflict. To conclude this first part of the book, we consider briefly two key aspects of these changes: the increasing importance of the narrative of human rights, and the effects of globalization.

Arguably the most significant development during the post-Cold War decade of the 1990s was the increasing importance internationally of the narrative of human rights. This can be seen in the intensification of efforts to establish effective juridical agencies, operating within the national and transnational arenas, to address human rights issues articulated in memories of past wars. Legislation to facilitate the prosecution of those suspected of 'war crimes' has been passed in a number of national states; former Nazis have been convicted in trials, some of which have attained a high profile in the international media; and there has been widespread public debate (leading to new legislation in a number of states) about the value of making 'Holocaust denial' a criminal offence (as it has been in Germany for some years). During the 1990s, fourteen Truth Commissions – 'official bodies set up to investigate a past period of human rights abuses or violations of international humanitarian law' – were established by national states (double the number set up during the 1970s and 1980s).[236] There have been significant developments also in the transnational arena. The end of the Cold War brought about a dramatic increase in UN peace-keeping operations, from Somalia in 1992 to East Timor in 1999. This has revived aspirations to enhance the role of the UN as an effective supranational agency, and has also stimulated efforts to contain the eruption of violent conflict by the use of international law. The International Criminal Tribunals for the Former Yugoslavia (in 1993) and Rwanda (in 1994) – established by the UN Security Council 'to investigate and prosecute' the perpetrators of war crimes, genocide, and other crimes against humanity – were the first such tribunals since those of Nuremberg and Tokyo after the Second World War.[237] UN efforts to establish a permanent International Criminal Court for these matters, called for in 1948 but 'delayed for decades by the Cold War and refusal of governments to accept an international legal jurisdiction', have been renewed with strong support from member-states in the late 1990s.[238]

The propagation of the human rights narrative can be read as an attempt to construct a new post-Cold War ordering of international relations based on the 'democratic norms' of the Western powers. In the debate about this development, the effective independence of the UN from the control of the USA has been called into question.[239] However, it also entails a reshaping of the transnational frame within which questions of war memory can be articulated and addressed; and this has led to suggestions that, in the post-Cold War era when once-hegemonic political narratives have suffered disintegration or decline, it represents a new 'grand narrative' in its own right.[240] In assessing the political implications of these developments, it is important not to reinvent a new kind of singular model in which international law replaces the nation-state as the ultimate determinant of war memory from above. As we have argued throughout, the inter-relation between the various arenas, agencies and narratives of articulation is complex, always contested and far from harmonious; thus, the effects of the current renewal of interest in human rights are likely to be contradictory.[241]

One significant effect of the growing power of the human rights narrative and its various legal and quasi-legal agencies has been a strengthening of the transnational arena, as a domain where war memories previously unrecognized in the arena of the national state may be articulated. For example, in Indonesia, a state widely condemned internationally for the violence of its armed forces in East Timor, a human rights commission was launched in mid-1999 as a means of securing for the new, post-Suharto, Wahid government the necessary international recognition of its democratic credentials. The commission's report accused the government minister and former head of the Indonesian military, General Wiranto, of crimes against humanity, which led to his removal from his post and demands from the East Timorese independence movement that he be brought to justice.[242] In this way, pressure from above may be exerted upon a state which shifts the internal balance of power in favour of the marginalized and the repressed.

The memory of wars of the past is also being revisited and reframed within the terms of the human rights narrative, as in the repercussions following the 1998 arrest in the UK of the former Chilean dictator, General Pinochet, and his attempted extradition to stand trial in Spain for genocide, torture and other human rights violations committed during the coup and subsequent military rule from 1973 to 1990. Pinochet's arrest was condemned by the democratic Chilean government as an interference in the affairs of a sovereign state, that would 'open old wounds' supposedly being healed through the 'truth and reconciliation' process begun in 1990. But it was welcomed by victims of the junta all around the world and in Chile itself, where developments in the transnational arena made many people aware for the first time of how brutal military rule had been, and reinvigorated national debate about the inadequacies of subsequent 'reconciliation' and democratization.[243] In cases like that of Chile, pressure 'from above' is stimulating renewed pressure from below to articulate war memories previously unrecognized by the state. Where a state is prepared to embrace the human rights agenda in tackling a legacy of war, war memories become publicly recognizable in these terms. This in turn encourages survivors and their successors to use a human rights narrative to frame their own war memories, and to develop new civil agencies to articulate these politically. The human rights narrative is thus proving capable of bestowing a particular form of empowerment on civil groupings that are prepared to couch their own memories in its terms.[244]

Nor are these new possibilities for articulation restricted to societies currently undergoing a 'peace process'. The vicious counter-insurgency war fought by the British colonial state against the Mau Mau guerrillas of Kenya during the Emergency of 1952–60, like other campaigns to preserve imperial interests, was not commemorated and, since the disintegration of imperialism as an ideology during the 1960s, is barely remembered at all within the metropolitan cultural arena. In Kenya itself, the memory of the Mau Mau campaign was suppressed in the neo-colonial state that came into existence after independence.[245] In 1999, research by David Anderson demonstrated the scale of atrocities and of 'appalling abuses of human rights at all stages of the legal process'; and argued that 'a collective blind eye was turned … in the colonial administration'.[246] At the same time, Mau Mau veterans and their families began an attempt to seek legal redress from the British government for alleged human rights abuses committed in its name during the Emergency.[247] Here, we can see the power of the human rights frame to reshape the research agenda of historians (who, as we suggested in section 3.5, have played a key role in mediating survivor memory) and to structure a re-articulation of public memory by a civil agency. Significantly, it was unclear to which legal forum – national or otherwise – the Mau Mau survivors ought to address their case.

The developing power of the human rights narrative is closely bound up with a second dynamic of change, that of 'globalization'.[248] One aspect of this is the way in which the cultural and political arenas of war memory are currently being transformed by the global communications media. The satellite and digital revolution now makes it possible to bring (selected) wars of the world into our living rooms as they unfold moment by moment. Intensive news coverage has played a key role in mobilizing popular support for the so-called human rights interventions of the 1990s – most effectively, perhaps, in relation to Kosovo, where 'media memories' of the 'ethnic cleansing' in Bosnia-Herzegovina were evoked to produce a powerful humanitarian response among viewers in many countries supporting the NATO alliance. In the globalized information world, the news media themselves have become a primary agency for the articulation for war memories.

However, most wars remain unnoticed or barely registered by the global media-eye. This selectivity produces a particular kind of 'forgetting' and a new politics of publicity is emerging to contest it, involving, amongst others, aid organizations and other non-governmental organizations (NGOs). The name, 'forgotten war', is now regularly used by NGOs to draw attention to marginalized war memories. At the start of 2000, for example, Oxfam described itself as helping the 'victims of [a] forgotten war' on the border between Ethiopia and Eritrea, which had created 1 million refugees since 1998, but had gone 'almost unreported' in the British media.[249] In circumstances where a region such as sub-Saharan Africa, one of the poorest in the world, also suffers the highest numbers of wars and war casualties,[250] NGOs increasingly find themselves in the front line of this particular 'politics of naming', which is a fight to win public recognition for the existence and effects of these wars and their connection to poverty. Besides engaging with the news agendas of the global media organizations, NGOs seek a more direct relationship with their public, using mailshots and the Internet. Like other agencies, they make use of the dominant form of personal survivor-testimony, but seek to mobilize it so as to produce new kinds of identifications with others, and 'communities of remembrance' which cut across existing political boundaries. Indeed, under the auspices of the Panos Oral Testimony Programme, new NGOs have emerged with a specific focus on 'testimony collection and communication' as a means to address the legacies of war in developing countries, and 'to reach the Northern public through the media'.[251] In this way, the NGOs are coming to constitute a new kind of transnational agency in the politics of war memory and commemoration.

Finally, the human rights narrative intersects in complex ways with a second aspect of globalization which is also currently reshaping the means available for the articulation and recognition of war memories. This is the continuing expansion of 'heritage tourism', a phenomenon linked to the broader recasting of 'history' into a pleasurable 'heritage' experience.[252] The US Holocaust Memorial Museum in Washington, opened in 1993, is perhaps the best-known case of a site of memory designed expressly to be 'popular' which has attracted criticism, in this case for its appropriation and 'packaging' of the Shoah as an American leisure experience.[253] But this phenomenon is no longer restricted to the West. In Cambodia, as Rachel Hughes has argued, genocide memorialization at key sites such as the Tuol Sleng Genocide Museum and the Cheoung Ek 'Killing Fields' Memorial is now being shaped to address overseas visitors on a South-East Asian tourist trail, who generate one-third of the national income, and since 1993 have outnumbered Cambodian visitors to these sites.[254] In this volume, Nasson notes ironically (in Chapter 3) 'the present shift away from a politically inscribed commemoration' of the Anglo-Boer War towards 'war tourism' in which remembrance is reduced to 'a "heritage" experience of tripping along an African version of the Ho Chi Min Trail'. In contrast, Stanley (in Chapter 10) and Raivo (in Chapter 5) both highlight the potential positive effects of veterans' participation in 'battlefield tourism', which has become an important space of exchange between the veterans and the 'postmemory' generation; while Loshitzky (in Chapter 7) notes how 'Holocaust tourism' to the death camps of Poland has become a 'secular ritual' shared by many second-generation Israelis with their survivor parents. The conflicts over such sites need to be understood as part of the negotiation betwen survivors and their successors, as direct memory passes into cultural memory. In these cases, the argument that sees heritage tourism as displacing or replacing a 'real' politics of memory is being re-posed as a question of the politics *within* historical tourism, concerned with different kinds of commemorative experience and the imaginative relation to the past which they enable.[255]

The clash between the registers of memory encountered in heritage tourism, and those of nationalistic political commemoration or the narrative of human rights, can be acute and painful. The kind of conflict likely to be engendered increasingly in the future is illustrated vividly in the case of 'Leninland', a name bestowed ironically upon a proposed Soviet-era theme-park to be built by millionaire, Viliumas Malinauskas, at Druskininkai in Lithuania.[256] In 1998 Malinauskas won a Ministry of Culture competition with his proposal to relocate in his country park, at no cost to the state, forty-two commemorative statues of Soviet heroes (including various Lenins and a Stalin), torn down from their public sites when Lithuania became independent in 1991. The statues were to be displayed alongside a museum cataloguing the crimes committed by the USSR against the Lithuanian people: some 300,000, including Malinauskas' father and other family members, were deported to Siberia after the Soviet invasions of 1941 and 1944. 'People must know about the past', he argued; 'If it is brushed under the carpet, it will be unhealthy for the next generation.'[257] The park would give Lithuanians 'the chance to look at their history from their point of view, and not from the Russian one, or a foreign one'.[258] The history was to be evoked, however, using heritage strategies: mock electric fences and watchtowers, and a replica of Vilnius train station in 1941, where visitors would be escorted onto cattle trucks by employees dressed as Soviet guards, allowing them to 'experience what it must have been like to be deported';[259] and it would coexist with a range of sports and leisure facilities. With the tourist board predicting it could attract an estimated 2.5 million visitors a year, Malinauskas envisages the park making a major contribution to economic regeneration in a region blighted by unemployment.

The theme-park proposal has met with fierce opposition, especially among those Lithuanians who are pressing for greater international recognition of the crimes of Stalinism. The local MP, Juozas Galdikas, has argued that, rather than providing a memorial to the suffering of the people, the park puts the 'murderers back on their pedestals'; if the statues 'are reconstructed as monuments, the psychological effect for local visitors could be disastrous'.[260] Over thirty organizations, representing former political prisoners and veterans of the partisan war against the Soviet occupation in the 1940s, have denounced the park (located in an area they describe as 'covered with partisans' blood') and the restoration of the sculptures (described as 'monsters from a horror film').[261] Members of one group, Labora, have even gone on hunger strike in protest at the government's decision. What is striking about these veterans' opposition to the theme-park, is their apparent assumption that the statues will retain and evoke their original meaning and power even in this new context, and that everyone who visits the park will read the statues as they do. There is no acknowledgement here of the parodic, even surreal possibilities that the park might hold, as a potentially transformative space where alternative readings of the statues, the regime they commemorated and those who suffered under it, might be made by visitors.

In the case of Druskininkai, the tensions between the imaginative possibilities for reframing memory inherent in this heritage site, and the psychic and political imperatives of survivor memory, are striking. Most poignantly, the proposal has also been criticized by the two sons of Ona Sukackiene, a schoolteacher whose supposed murder by Lithuanian 'bandits' in the 1940s was used in pro-Soviet propaganda and later commemorated in one of the monuments destined for the park. After the KGB archives were opened after independence in 1991, it transpired that she had been killed by the Soviet secret police, not the partisans. Her sons have written to Galdikas asking his help to destroy the statue: 'Nobody asked our permission when they created the monument. Nobody asked us when it was taken down. And now nobody is asking our permission for it to be re-erected.'[262] Here, the survivors of war must confront new forms of power: the cultural entrepreneurs of the global free market, now powerful adversaries in the battle over memory; and the struggle is to ensure that they can be held accountable and made responsive to the agenda of other agencies.

It is within these shifting arenas and frames that the survivors of war and their successor generations must struggle to articulate their own war memories. But the central question in the conflicts of the future will remain that of the politics of agency: who controls the story of war and the way it is told to others?

Notes

1 For the debates on Shoah memory see, for example, P. Friedlander, *Memory, History, and the Extermination of the Jews of Europe* (Bloomington and Indianapolis, 1993); J. Young, *The Texture of Memory: Holocaust Memorials and Meaning* (Yale, 1993); G.H. Hartman (ed.) *Holocaust Remembrance: The Shapes of Memory* (Oxford, 1994); M.C.

Steinlauf, *Bondage to the Dead: Poland and the Memory of the Holocaust* (New York, 1997). See also Yosefa Loshitzky (Chapter 7, this volume).

2 S. Friedlander (ed.) *Probing the Limits of Representation: Nazism and the Final Solution* (Cambridge, MA, 1992).

3 *Veterans: The Last Survivors of the Great War* (dir. S. Humphries, UK, 1998).

4 See section 3.5 of this chapter for further consideration of witnessing and survivor testimony, and section 3.4 for the responses of the successor generations.

5 See, for example, 'Salute to victory 50 years on' and 'Major's message of peace to the Germans', *Daily Telegraph*, 9 May 1995; available on *The Electronic Telegraph*, 9 May 1995, Front Page and Home News, website <http://www.telegraph.co.uk>; *The Battle for Normandy* (dir. C. Wheeler), broadcast on BBC1, 5–6 June 1994; 'D-Day Souvenir Issue', *Daily Mirror*, 7 June 1984; 'VE-Day 40 years on', *Guardian*, 8 May 1985, 15–18; R. Cross, *VE Day, Victory in Europe 1945* (London, 1985); and the programme of the Imperial War Museum's '45–85' Festival (4–15 May 1985).

6 S. Faulks, *Birdsong* (London, 1994); P. Barker, *Regeneration* (London, 1992), *The Eye in the Door* (London, 1994) and *The Ghost Road* (London, 1995). See also the film, *Regeneration* (dir. G. MacKinnon, UK, 1997).

7 *The Boer War* (4 parts), broadcast on Channel 4, 23 Sept.–14 Oct. 1999; *War of the Century* (4 parts), broadcast on BBC2, 5–26 Oct. 1999.

8 'PoW survivors force ban on Japanese at VJ-Day celebrations', *Daily Telegraph*, 12 Jan. 1995; 'Peace ceremony cannot bridge gulf with death railway veterans', *Daily Telegraph*, 14 Aug. 1995; available on *The Electronic Telegraph*, 12 Jan. and 14 Aug. 1995, Home News, website <http://www.telegraph.co.uk>. See also Stanley (Chapter 10, this volume).

9 See section 4.2 of this chapter.

10 J. Vickers, *Women and War* (London, 1993), 6.

11 Vickers, *Women and War*, 4–6; 'Deadly cost of the new warfare', *Guardian*, 22 Oct. 1999, 3.

12 'The miracle children', *Guardian*, 22 Dec. 1999, 14; A. Destexhe, *Rwanda and Genocide in the Twentieth Century* (London, 1995; first pub. in French, 1994).

13 The estimate of deaths is by the Sarajevo Economic Institute, July 1995, cited in the Bosnian Muslim paper, *Slobodna Bosna*, November 1995, and reproduced (with the suggestion that subsequent deaths bring the figure closer to 300,000) on the Serbian Network website <http://www.srpska-mreza.com>, visited 8 Jan. 2000. That of refugees is by the UN High Commissioner for Refugees, found on Institute for Global Communication website <http://www.igc.apc.org>, visited 8 Jan. 2000.

14 See sections 3.1, 4.2 and 4.3 of this chapter.

15 C. Moriarty, 'The material culture of Great War remembrance: review article', *Journal of Contemporary History*, 34.4 (1999), 655.

16 Moriarty, 'The material culture', 655–6. See P. Fussell, *The Great War and Modern Memory* (Oxford, 1975); P. Nora (ed.) *Realms of Memory: Re-thinking the French Past: vol. I, Conflicts and Divisions* (New York, 1996; first pub. as *Les Lieux de mémoire, I: La République* [Paris, 1984]).

17 Moriarty, 'The material culture', 654–5.

18 For the Shoah, see notes 1 and 2 above. For Britain, France and Italy, see T.G. Ashplant (Chapter 11, this volume). For further references to memory and commemoration in these and other national contexts, see Part 1, *passim*.

19 J. Winter, *Sites of Memory, Sites of Mourning: The Great War in European Cultural History* (Cambridge, 1995). Recent conferences have included: 'War and Memory in the Twentieth Century', held at University of Portsmouth, 25–7 March 1994; 'The Re-siting of Memory: Remembering World War Two in the New Europe', held at the Centre for European Studies Research, University of Portsmouth, 12–13 November 1999; 'The Memory of Catastrophe', held at the University of Southampton, 14–17 April 2000.

20 E. Hobsbawm and T. Ranger (eds) *The Invention of Tradition* (Cambridge, 1983); E.J. Hobsbawm, *Nations and Nationalism since 1780* (1990; rev. edn., Cambridge, 1992); B. Anderson, *Imagined Communities: Reflections on the Origins and Spread of Nationalism* (1983; rev. edn., London, 1991); Winter, *Sites of Memory, Sites of Mourning*; J. Winter and E. Sivan (eds) *War and Remembrance in the Twentieth Century* (Cambridge, 1999). These oppositions should be taken as indicative not absolute, especially given the rich interdisciplinary borrowings that characterize the field. This is explicitly recognized by Hobsbawm, for example, who notes in particular the contribution of social anthropologists to the study of 'invented traditions'. See his 'Introduction: inventing traditions', in Hobsbawm and Ranger (eds) *Invention of Tradition*, 14.

21 E. Hobsbawm, 'Introduction: inventing traditions', 9.

22 Ibid., 13.

23 Anderson, *Imagined Communities*, 6–7.

24 Ibid., 9.

25 For a recent study within this paradigm, see J.W. Müller (ed.) *Memory and Power in Post-War Europe: Studies in the Presence of the Past* (forthcoming).

26 Winter, *Sites of Memory, Sites of Mourning*, 11, 79.

27 E. Hobsbawm, 'Mass-producing traditions: Europe, 1870–1914', in Hobsbawm and Ranger (eds) *Invention of Tradition*, 263.

28 Ibid., 264.

29 J. Winter and E. Sivan, 'Setting the framework', in Winter and Sivan (eds) *War and Remembrance*, 9.

30 Ibid., 29. See also 38.

31 J. Winter, 'Forms of kinship and remembrance in the aftermath of the Great War', in Winter and Sivan (eds) *War and Remembrance*, 40.

32 Ibid., 41, 59, 40, 41.

33 Ibid., 41, 51. A rather more complex picture is implied by other references: to 'those who worked through state institutions to help the victims of war' (52), to civic groups who were 'supported by the state' (59) and to the participation of civic groups in Armistice Day ceremonies (59).

34 C. Merridale, 'War, death, and remembrance in Soviet Russia', in Winter and Sivan (eds) *War and Remembrance*, 61–83. Cf. the analysis of funerary practices 'negotiated between the state, various interest groups (such as the medical profession and funeral directors) and the bereaved' in Britain during and after the First World War, in J. Bourke, *Dis-membering the Male: Men's Bodies, Britain and the Great War* (London, 1996), 210–52.

35 Merridale, 'War, death, and remembrance', 71.

36 Ibid., 72.

37 A. King, *Memorials of the Great War in Britain: The Symbolism and Politics of Remembrance* (Oxford, 1998), 6.

38 Hobsbawm, 'Introduction: inventing traditions', 13.

39 D.W. Lloyd, *Battlefield Tourism: Pilgrimage and Commemoration of the Great War in Britain, Australia and Canada, 1919–39* (Oxford, 1998), 50–92.

40 Hobsbawm, *Nations and Nationalism*, 10.

41 Winter, 'Forms of kinship', 55.

42 For further discussion of these points, see section 2.1 of this chapter.

43 Hobsbawm, *Nations and Nationalism*, 11.

44 Winter, *Sites of Memory, Sites of Mourning*, 115.

45 See sections 3.2 and 3.3 in this chapter, for the development of this critique.

46 For an overview of methodological issues, see R. Perks and A. Thomson (eds) *The Oral History Reader* (London, 1998). See also the publications of the International Yearbook of Oral History and Life Stories series (Oxford), and the Routledge Memory and Narrative series (London); for international debates about war mem-

ory, see especially L. Passerini (ed.) *Memory and Totalitarianism* (Oxford, 1992); and K.L. Rogers, S. Leydesdorff and G. Dawson (eds) *Trauma and Life Stories* (London, 1999). Among the wealth of available work on oral history and war memory across Europe, we have been especially stimulated by the contributions of Alessandro Portelli, *The Battle of Valle Giulia: Oral History and the Art of Dialogue* (Wisconsin, 1997) esp. part 2 'Wars'; and the biographical narrative approach of Gabriele Rosenthal, for example 'German war memories: narrability [sic] and the biographical and social functions of remembering', *Oral History*, 19–2, 1991, 34–41; and 'Reconstruction of life stories: principles of selection in generating stories for narrative biographical interviews', in R. Josselson and A. Lieblich (eds) *The Narrative Study of Lives, vol. 1* (Newbury Park, 1993), 59–91.

47 For example, in Winter and Sivan's 'Setting the framework', reference to the international debates in oral history and life history is conspicuous by its absence, especially considering these authors' interest in theorizing the relationship between 'psychological' and 'social' dimensions of personal memory, and the fact that a number of essay contributions in the book (including their own essays) are concerned with life-history material. A similar absence is evident in Jan Müller, 'Introduction: the power of memory, the memory of power, and the power over memory', written from a political-science perspective, in Müller, *Memory and Power.*

48 For further discussion of this point, see T.G. Ashplant (Chapter 11, this volume).

49 Hobsbawm, *Nations and Nationalism*, 11.

50 See Rogers et al. (eds) *Trauma*, which includes four chapters on war memories (in Guatemala, Argentina, Zimbabwe and Northern Ireland).

51 Popular Memory Group, 'Popular memory: theory, politics, method', in Centre for Contemporary Cultural Studies, *Making Histories: Studies in History-Writing and Politics* (London, 1982), 205–52; G. Dawson and B. West, 'Our finest hour? The popular memory of World War II and the struggle over national identity', in G. Hurd (ed.) *National Fictions: World War Two in British Films and Television* (London, 1984), 8–13; A. Thomson, *Anzac Memories: Living with the Legend* (Melbourne, 1994).

52 Popular Memory Group, 'Popular memory', 207.

53 Ibid., 210.

54 Ibid., 211.

55 In addition to the work discussed here, see for example: G. Dawson, *Soldier Heroes: British Adventure, Empire and the Imagining of Masculinities* (London, 1994); L. Noakes, *War and the British: Gender and National Identity, 1939–91* (London, 1998); F. Lorenz, 'The unending war: social myth, individual memory and the Malvinas', in Rogers et al. (eds) *Trauma*, 95–112; M. Evans, *The Memory of Resistance: French Opposition to the Algerian War (1954–1962)* (Oxford, 1997); J. Leonard, 'Facing "the finger of scorn": veterans' memories of Ireland after the Great War', in M. Evans and K. Lunn (eds) *War and Memory in the Twentieth Century* (Oxford, 1997), 59–72.

56 Dawson and West, 'Our finest hour?', 9.

57 Ibid., 10.

58 Ibid., 11.

59 Thomson, *Anzac Memories*, 1.

60 Ibid., 8.

61 Ibid., 216, 220.

62 But see also Peter Sjølyst-Jackson's critical disagreement with Dawson and West (in Chapter 6, this volume) and Stephen Garton's with Thomson (in Chapter 9, this volume).

63 Thomson, *Anzac Memories*, 213, 220–1.

64 By 'nation-state' here we refer to those polities recognized as independent within the international political and legal order. Internally, such states may be multinational in composition, or federal in structure.

65 On templates, see section 3.1 in this chapter.

66 Hence, in part, the significance often accorded to direct, personal memory of war, the claims to unique authenticity which may be built upon it, and their potential to obliterate the validity of other perspectives; discussed in section 3.4 in this chapter.
67 For such social learning, see Winter and Sivan, 'Setting the framework', 16.
68 See further section 3.4 in this chapter on postmemory; and Jo Stanley's account of Secondary Traumatic Stress Disorder (Chapter 10, this volume).
69 Thomson, *Anzac Memories*, 18–20, 170–3, 211–15.
70 A. Grossmann, 'A question of silence: the rape of German women by occupation soldiers', in Robert G. Moeller (ed.) *West Germany under Construction: Politics, Society and Culture in the Adenauer Era* (Ann Arbor, 1997), 38–9, 44–6, 50–2.
71 N. M. Naimark, *The Russians in Germany: A History of the Soviet Zone of Occupation, 1945–1949* (Cambridge, MA, 1995), 132–40.
72 I. Bandhauer-Schöffmann and E. Hornung, 'War and gender identity: the experience of Austrian women, 1945–1950', in D.F. Good, M. Grandner and M.J. Maynes (eds) *Austrian Women in the Nineteenth and Twentieth Centuries: Cross-Disciplinary Perspectives* (Oxford, 1996), 218–19.
73 Bandhauer-Schöffmann and Hornung, 'War and gender identity', 220.
74 Such a process itself is not necessarily a simple lifting of repression, a dispelling of silence. Rather, the (re-)emergent memories may become weapons in a new, contemporary political battle. Grossmann, in 'A question of silence', 33–8, 48–52, critiques one feminist perspective which, she argues, sees the German women raped in 1945 *solely* as victims. This, she suggests, is partially consonant with one of the ways in which the women themselves dealt with their experience during the years of silence, seeing themselves as having both lost their honour, but also preserved it in defiance both of the Russians (cruel but child-like, to whom they were culturally superior) and of their own menfolk (who had failed to protect them). Grossmann's concern is that such a perspective may serve to support the self-perception of many Germans after 1945 of themselves as victims of a barbarous Soviet onslaught, a stance which gained new cultural force during the *Historikerstreit* of the 1980s. Bandhauer-Schöffmann and Hornung, 'War and gender identity', 213–14, express a related concern about how the depiction of Austrian women as victims under the Nazis and then heroines during the harsh years of reconstruction, may mesh with a wider national self-perception of Austria as simply Hitler's victim. In a rather different context, Andrea Peto has noted how the recovery of the experience of Hungarian women raped in 1945 has been assimilated into a nationalist (and masculine) narrative of Hungary as victim of a Soviet onslaught and subsequent silencing, thereby obscuring the role of the Horthy regime as collaborator with the Nazis. 'Rapes in Budapest in 1945: Silencing the past', unpublished paper.
75 B. Archer, ' "A low-key affair": memories of civilian internment in the Far East, 1942–1945', in M. Evans and K. Lunn (eds) *War and Memory in the Twentieth Century* (Oxford, 1997), 45–8, 53–5; cf. Winter and Sivan, 'Setting the framework', 37.
76 Thomson, *Anzac Memories*, 170–3.
77 For the central, heroic figure of the male soldier, see Dawson, *Soldier Heroes*; Thomson, *Anzac Memories.*
78 Anderson, *Imagined Communities*, 9–10.
79 This aspect of war commemoration (and its vital other face, forgetting) – its importance for securing the legitimacy of the nation-state, and underpinning its political claims – is dealt with in detail in Müller (ed.) *Memory and Power.*
80 Dawson, *Soldier Heroes.*
81 D.W. Lloyd, *Battlefield Tourism*, 50–63.
82 Ibid., 76–81.
83 See Ashplant (Chapter 11, this volume).
84 M. Richards, *A Time of Silence: Civil War and the Culture of Repression in Franco's Spain, 1936–1945* (Cambridge, 1999), 7–9, 16–19, 72–3.

85 A. Prost, 'The Algerian War in French collective memory', in Winter and Sivan (eds) *War and Remembrance*, 170–2, suggests that a sense of the lack of legitimacy of the French cause (its methods uncomfortably close to those of the recent German occupiers) lies at the root of the absence of any public (official or sectional) memory of the war.

86 M. Evans, 'Rehabilitating the traumatized war veteran: the case of the French conscripts from the Algerian War, 1954–1962', in Evans and Lunn (eds) *War and Memory in the Twentieth Century*, 76–83. Prost, 'The Algerian War in French collective memory', 170, emphasizes the continuing failure to develop a public memory, laying stress on the division between the more conservative veterans' organization UNC and the leftist FNACA.

87 For the impact of such dominant narratives of memory on post-war Italy, see Ashplant (Chapter 11, this volume); on Germany: Claudia Koonz, 'Between memory and oblivion: concentration camps in German memory', in J.R. Gillis (ed.) *Commemorations: the Politics of National Identity* (Princeton, 1994), 263–6, and G. Knischewski and U. Spittler, 'Memories of the Second World War and national identity in Germany', in Evans and Lunn (eds) *War and Memory in the Twentieth Century*, 239–54.

88 D. Wong, 'Memory suppression and memory production: the Japanese occupation of Singapore', in G. White, T. Fujitani and L. Yoneyama (eds) *Perilous Memories: The Asia-Pacific Wars* (Durham, NC, 2000), forthcoming.

89 A. Gregory, *The Silence of Memory: Armistice Day 1919–1946* (Oxford, 1994).

90 King, *Memorials of the Great War.*

91 P. Nora, 'General introduction: between memory and history', in Nora (ed.) *Realms of Memory*, 19.

92 K.A. Hass, *Carried to the Wall: American Memory and the Vietnam War Veterans Memorial* (Berkeley, 1998), 10–20.

93 Merridale, 'War, death and remembrance', 62, 76–9.

94 Wong, 'Memory suppression and memory production'.

95 Richards, *A Time of Silence*, 7–10, 16–19, 22–30, 47–56, 67–74, 147–69.

96 Winter, *Sites of Memory, Sites of Mourning*; Winter and Sivan, 'Setting the framework', 29.

97 Winter and Sivan, 'Setting the framework', 30. Stanley describes the close relationship between the armed forces medical services, the National Health Service and the voluntary association dealing with veterans affected by PTSD.

98 Evans, 'Rehabilitating the traumatized war veteran', 76–81.

99 Thomson, *Anzac Memories*, 46–72, 142–56.

100 Evans, 'Rehabilitating the traumatized war veteran', 76, notes that in the face of official amnesia about the conflict in Algeria a 'small minority of ex-conscripts … looked towards writing as a way of dramatizing their experience'. He also analyses, 81–3, Bertrand Tavernier's documentary film *The Undeclared War* (1992), which uses interviews with veterans to let them express their suffering and challenge the prevailing taboo about the war; its intent is clearly similar to Marcel Ophuls' *The Sorrow and the Pity* (1969), which helped open up discussion about the Vichy period. See also section 3.2 in this chapter; and de Medeiros's account of novelists' responses to the silence around the Portuguese colonial wars (Chapter 8, this volume).

101 Winter, 'Forms of kinship', 40–1, 47–54.

102 B.M. Doyle, 'Religion, politics and remembrance: a Free Church community and its Great War dead', in Evans and Lunn (eds) *War and Memory in the Twentieth Century*, 223–38.

103 Winter, 'Forms of kinship', 41–7. Garton (Chapter 9, this volume) points out that Lording in his autobiography gives the protagonist a name he himself shared with both his father and his son, thereby affording the text an address across the generations of his family.

104 Richards, *A Time of Silence*; P. Aguilar, 'Agents of memory: Spanish Civil War veterans and disabled soldiers', in Winter and Sivan (eds) *War and Remembrance*, 84–7. M. Achelsberg, 'Mujeres Libres: the preservation of memory under the politics of repression in Spain', in Passerini (ed.) *Memory and Totalitarianism*, 125–6, 135–41, discussing the rupture in the history of Spanish women anarchists, talks of the 'abolition of political memory'.

105 Merridale, 'War, death and remembrance', 63, 81, however, argues that in Russia memories of the Civil War and of Stalin's purges *were* transmitted within families, as secrets rather than as silence.

106 Ibid., 62, 64–6, 76–7.

107 Similarly, Koonz, 'Between memory and oblivion', 264–6, records that, while the official East German memorialization of the concentration camps placed them within the frame of a working-class, socialist, anti-Fascist struggle, which occluded the specific destruction of the Jews, the sites and their memorials nevertheless evoked in some visitors other, suppressed memories of the Holocaust.

108 J. Leonard, 'Facing "the finger of scorn" ', 66–7. See also J. Leonard, 'The twinge of memory: Armistice Day and Remembrance Sunday in Dublin since 1919', in R. English and G. Walker (eds) *Unionism in Modern Ireland: New Perspectives on Politics and Culture* (London, 1996), 99–114.

109 Wong, 'Memory suppression and memory production'.

110 Here, this aesthetic work has connections with another cultural endeavour of civil society, current Israeli revisionist historiography re-examining the origins of the state of Israel in 1947–48.

111 Hobsbawm, 'Mass producing traditions', 307.

112 Winter, *Sites of Memory, Sites of Mourning*, 5.

113 P. Burke, 'History as social memory', in T. Butler (ed.) *Memory. History, Culture and the Mind* (Oxford, 1989), 97–113.

114 A. Portelli, 'The battle of Poggio Bustone', in Portelli, *The Battle of Valle Giulia*, 134.

115 Dawson, *Soldier Heroes*, 23.

116 Ibid., 283.

117 Neil Jarman, *Material Conflicts: Parades and Visual Displays in Northern Ireland* (Oxford and New York, 1997).

118 T.W. Moody, 'Irish history and Irish mythology', in C. Brady (ed.) *Interpreting Irish History: The Debate on Historical Revisionism* (Dublin, 1994); R.D. Edwards, *The Faithful Tribe: An Intimate Portrait of the Loyal Institutions* (London, 1999), 127–276.

119 R. Williams, *Marxism and Literature* (Oxford, 1977), 117.

120 For further discussion see section 4.2 in this chapter.

121 Fussell, *Great War and Modern Memory*.

122 M. Eksteins, *The Rites of Spring. The Modern in Cultural History* (New York, 1989).

123 Samuel Hynes, however, writing before Winter, emphasizes both continuity and innovation. See S. Hynes, *A War Imagined. The Great War and English Culture* (London, 1991).

124 Winter, *Sites of Memory, Sites of Mourning*, 9.

125 Hass, *Carried to the Wall*, 18.

126 Winter, 'Forms of kinship'.

127 Ibid., 43.

128 Winter and Sivan, 'Setting the framework', 10.

129 Winter, 'Forms of kinship', 60.

130 Winter and Sivan, 'Setting the framework', 32.

131 J. Laplanche and J.B. Pontalis, *The Language of Psychoanalysis* (London, 1988), 485–6. See for example Winter, *Sites of Memory, Sites of Mourning*, 113–15.

132 R. Schulte, 'Käthe Kollwitz's sacrifice', *History Workshop Journal*, 41 (Spring 1996), 115.

133 Winter, 'Forms of kinship', 58.

134 M. Jay, 'Against consolation: Walter Benjamin and the refusal to mourn', in Winter and Sivan (eds) *War and Remembrance*, 229.
135 Quotation from Benjamin, 'A Berlin chronicle', in Jay, 'Against consolation', 230.
136 As Portelli notes, 'even individuals are internally divided between the desire to be silent and forget and the need to speak'. 'The massacre at Civitella Val di Chiana', in Portelli, *The Battle of Valle Giulia*, 158.
137 Winter and Sivan, 'Setting the framework', 31.
138 For an example of such an approach, see N. Burchardt, 'Transgenerational transmission in the families of Holocaust survivors in England', in D. Bertaux and P. Thompson (eds) *Between Generations. Family Models, Myths and Memories* (Oxford 1993), 121–39.
139 Hartman, 'Introduction', in Hartman (ed.) *Holocaust Remembrance*, 5–7.
140 D. Porat 'Leoloam lo darchi ha'acharona' (It is Never My Last Road), *Ha'aretz* (15 Jan. 1997), 5, quoted in Loshitsky (Chapter 7, this volume).
141 Quoted in Hartman, 'Introduction', 6.
142 A. Wieviorka, 'On testimony', in Hartman, *Holocaust Remembrance*, 24.
143 Faulks, *Birdsong*.
144 M. Hirsch, 'Past lives: postmemories in exile', *Poetics Today*, 17.4 (winter 1996), 674.
145 Young, *The Texture of Memory*, 5.
146 Hartman, 'Introduction', 8.
147 S. Hynes, 'Personal narratives and commemoration', in Winter and Sivan (eds) *War and Remembrance*, 206.
148 Fussell, *Great War and Modern Memory*, 79.
149 M. Glucksmann, 'German reconcilings', *Soundings*, 6 (summer 1997), 36.
150 The impact of being brought up the child of a badly wounded British First World War veteran is discussed by Doris Lessing (born 1919) in her *Under My Skin: Volume One of My Autobiography, to 1949* (London, 1995), 6–10.
151 Hirsch, 'Past lives', 649.
152 S. Friedlander, *When Memory Comes* (New York, 1979), 155–6.
153 Hirsch, 'Past lives', 664.
154 This is pointed out by Hartman, 'Introduction', 18.
155 Hirsch, 'Past lives', 662.
156 Barker, *Regeneration*, *The Eye in the Door*, *The Ghost Road*. See also Elaine Showalter, *The Female Malady: Women, Madness and English Culture, 1830–1980* (London, 1987).
157 G.H. Hartman, *The Longest Shadow: In the Aftermath of the Holocaust* (Bloomington, 1996), 133.
158 A. Wieviorka, 'From survivor to witness: voices from the Shoah', in Winter and Sivan (eds) *War and Remembrance*, 137–8.
159 Y. Feldman, 'Film as witness: Claude Lanzman's *Shoah*', in Hartman (ed.), *The Longest Shadow*, esp. 91–7.
160 K. Plummer, *Telling Sexual Stories. Power, Change and Social Worlds* (London, 1995).
161 *Yizkor* books are edited memorial books which draw on survivors' accounts of their lives before and during the war, photographs and other memorabilia to construct a collective memory of the dead. They were instigated soon after the war as a means of establishing a historical record of genocide. Wieviorka, 'From survivor to witness', 130.
162 Wieviorka, 'From survivor to witness', 130.
163 Evans, 'Rehabilitating the traumatized war veteran', 81.
164 This is not to claim, however, that the practice of oral history necessarily leads to the endorsement of the narrator's viewpoint. Indeed, such techniques have been used particularly effectively in the critical analysis of testimonies by bystanders and perpetuators of national socialism in Germany and Austria. Gabriele Rosenthal's 'biographical narrative' approach, for example, exposes the ways in which such informants, far from being revealing about their involvement, 'cover it up with

stories'. See Rosenthal, 'German war memories', 34; also 'Reconstruction of life stories'; and R. Sieder, 'A Hitler youth from a respectable family: the narrative composition and deconstruction of a life story', in Bertaux and Thompson (eds) *Between Generations*, 99–121.

165 Koonz, 'Between memory and oblivion', 272–4.

166 Portelli, 'The massacre at Civitella', 143.

167 For fuller description see Ashplant (Chapter 11, this volume).

168 Evans, 'Rehabilitating the traumatized war veteran', 83.

169 Wieviorka, 'From survivor to witness', 140.

170 E. Sivan, 'Private pain and public remembrance in Israel', in Winter and Sivan (eds) *War and Remembrance*, 177–205.

171 Y. Feldman, 'Whose story is it anyway? Ideology and psychology in the representation of the Shoah in Israeli literature', in Friedlander (ed.) *Probing the Limits of Representation*, 223.

172 Sivan, 'Private pain and public remembrance', 179.

173 See section 2.1 in this chapter. A number of contributors to this volume also make reference to this process of 'selective amnesia' by the state, notably de Medeiros on Portugal (Chapter 8, this volume), Curthoys on Australia (Chapter 4, this volume), and Loshitzky on Israel (Chapter 7, this volume).

174 'Blair lets slip the words of war', *The Times*, 20 May 1999, 15.

175 C. von Clausewitz, *On War* (ed. and trans. M. Howard and P. Paret; London, 1993), Book 1, Ch. 1.

176 See the discussion of Evans, 'Rehabilitating the traumatized war veteran', in section 2.1 in this chapter.

177 On memories of the Malayan Emergency, see Wong, 'Memory suppression'; for memories of the Kenyan Emergency, see the novels and essays of Ngũgĩ wa Thiong'o, including *Weep Not, Child* (London, 1964); *A Grain of Wheat* (rev. edn. London, 1986; 1st pub. 1967); *Petals of Blood* (London, 1986; first pub. 1977); *Homecoming: Essays on African and Caribbean Literature, Culture and Politics* (London, 1972); *Writers in Politics: Essays* (London, 1981).

178 T.P. Coogan, *The Troubles: Ireland's Ordeal 1966–1996 and the Search for Peace* (London, 1996), 261–82; T. Collins, *The Irish Hunger Strike* (Dublin, 1986), 31, 41–4.

179 See, for example, the conflict over the funerals in 1981 of republican hunger-strikers, Joe McDonnell and Kevin Lynch, in Collins, *The Irish Hunger Strike*, 395–6, 491–6 (and photograph following 282). Jane Leonard, in *Memorials to the Casualties of Conflict: Northern Ireland 1969–1997* (Belfast, 1997), notes the existence since 1994 of 'an organization of soldiers' families ... established to campaign for the erection in London of a national memorial to service personnel killed in Northern Ireland' (19); and that, '[f]ollowing a ruling by a district council' in 1993, services's casualties may now be named on local cenotaphs (17).

180 Lloyd George, cited in R. Bennett, *The Black and Tans* (New York, 1995), 104.

181 Lloyd George, quoted in D.G. Boyce, *Englishmen and Irish Troubles: British Public Opinion and the Making of Irish Policy, 1916–1922* (London, 1972), 48. This statement led later to a gaffe in the House of Commons similar to that of Tony Blair, when Lloyd George found himself explaining away the killing of a civilian Irish woman by British forces as 'the sort of unfortunate accident that happens in war' (Bennett, *Black and Tans*, 113).

182 R. Foster, 'History and the Irish question', in C. Brady (ed.) *Interpreting Irish History: The Debate on Historical Revisionism* (Dublin, 1994), 139; P. Cooney, 'Unknown soldiers', *The Guardian Weekend*, 24 Oct. 1998, 18–26. Cooney points out that, due to the Civil War between pro- and anti-Free-State wings of the IRA in 1922–24, and the later IRA campaign of 1969–97, memories of the Independence War remain controversial and the veterans are nowadays 'officially ignored' by the Irish state (20). On war commemoration by the Irish Free State and the Republic of Ireland state, see J.

Leonard, *The Culture of Commemoration: The Culture of War Commemoration* (Dublin, 1996), 15–17; Leonard, 'Twinge of memory'; Leonard, 'Facing "the finger of scorn" '.

183 R.K. Campbell, 'The Ulster-Irish War 1922', in *New Ulster*, 17, 1992, 8–13. See also M. Farrell, *Northern Ireland: The Orange State* (London, 1976), 21–65. The awkwardness is exemplified by the National Museum of Ireland's exhibition, *The Road to Freedom: Photographs and Memorabilia from the 1916 Rising and Afterwards*, and the booklet of the same name by Michael Kenny (Dublin, 1993): the story ends abruptly with the transfer of power to the new Free State in January 1922, gestures towards the Civil War of 1922–24 and merely notes 'the exclusion of a portion of Ulster' (46).

184 E. Said, *After the Last Sky: Palestinian Lives* (London, 1986), 133. See also E. Said, *The Politics of Dispossession: The Struggle for Palestinian Self-Determination 1969–1994* (London, 1994), 166–72.

185 Cf. Said, *After the Last Sky:* 'Whereas Nuremberg is invoked as a touchstone for judging war practices in, say, Vietnam, it is quickly withdrawn when Israel is concerned' (139).

186 Ibid., 6, 20; and see also 127–66.

187 Ibid., 149.

188 'Settlers race for land', *Guardian*, 25 May 1999, 17.

189 J. Zur, 'Remembering and forgetting: Guatemalan war widows' forbidden memories', in Rogers et al. (eds) *Trauma*, 45–6.

190 Ibid., 48.

191 Ibid., 48. See 47–9.

192 Ibid., 58 note 9. She is citing a reference to George Orwell's *1984* made by a Swedish representative to the UN, evoked by a visit to Guatemala. Cf. Irina Sherbakova's use of the same passage from *1984* to convey the memory politics of the Soviet regime in her 'The Gulag in memory', in Passerini (ed.) *Memory and Totalitarianism*, 103–4.

193 Zur, 'Remembering and forgetting', 48.

194 General Jorge Raphael Videla, quoted in J.P. McSherry, 'Military power, impunity and state–society change in Latin America', *Canadian Journal of Political Science*, 25 (3), (1992), 467.

195 Ibid.

196 See the website of the International Court of Justice at <http://www.lawschool.cornell.edu/library/>; C.C. Joyner (ed.) *The United Nations and International Law* (Cambridge, 1997); I. Brownlie, *The Rule of Law in International Affairs: International Law at the Fiftieth Anniversary of the United Nations* (The Hague, 1998).

197 Coogan, *The Troubles*, 153.

198 A.R. Coll and A.C. Arend (eds) *The Falklands War: Lessons for Strategy, Diplomacy and International Law* (Boston, MA, 1985).

199 Said, *Politics of Dispossession*, 153, 160, 167.

200 Said, *After the Last Sky*, 133.

201 F. Halliday, *The Making of the Second Cold War* (2nd edn; London, 1986).

202 'Timeline' page at the website of the Coalition for an International Criminal Court at <http://www.igc.org/icc/>.

203 See also the official commemorations of the fiftieth anniversary of VE-Day across Europe on 6–9 May 1995, when 58 world leaders assembled in London and 70 in Paris; further ceremonies took place in Berlin and Moscow. The British Prime Minister, John Major, speaking in Berlin, 'accommodated German sensitivities' by making 'no direct reference to Nazism, the holocaust or war crimes … He referred to the Second World War as part of a "30-years war" waged in Europe from 1914 to 1945 … [and] urged people not to forget the past but to look forward to the future'. 'Major's message of peace to the Germans', *Daily Telegraph*, 9 May 1995; available on *The Electronic Telegraph*, 9 May 1995, Front Page.

204 K. Kirkwood, *Britain and Africa* (London, 1965), 40–86; Ngũgĩ, *Grain of Wheat.*

205 T. Snyder, 'Memory of sovereignty and sovereignty over memory: Poland, Lithuania and Ukraine, 1939–1999', in Müller (ed.) *Memory and Power.*

206 E.P. Thompson, 'Notes on exterminism, the last stage of civilization', in New Left Review (eds) *Exterminism and Cold War* (London, 1982), 2–5.

207 Report of a speech by Margaret Thatcher, January 1976, quoted in Dawson, *Soldier Heroes*, 14. See also G. Dawson, 'Tory defence ideology and the popular memory of World War Two', unpublished paper presented at History Workshop 15, Brighton, 6–8 Nov. 1981, and précised in the summaries of papers, *History Workshop 15* (Brighton, 1981), 48.

208 L. Yeremeyev, *USSR in World War Two: Through the Eyes of Friends and Foes* (Moscow, 1981), 116, 113, 122.

209 F. Halliday, *Cold War, Third World: An Essay on Soviet–American Relations* (London, 1989); Halliday, *Second Cold War.*

210 'Thatcher gave approval to talks with the IRA', *Guardian*, 16 Oct. 1999, 7. See G. Dawson, *Making Peace with the Past? Cultural Memory and the Irish Troubles* (Manchester, forthcoming).

211 'Lessons of peace for children of war', *Guardian*, 6 Dec. 1999, 10; Ockenden International's website: <http://ww.ockenden.org.uk>, visited 7 Jan. 2000.

212 Popular Memory Group, 'Popular Memory', 13.

213 They also provided occasions for retrenchment, as in British right-wing press reports of the commemorations held in Moscow: see 'Ghost of the Red Army rises up for VE Day', *Daily Telegraph*, 10 May 1995; 'May Day Moscow indulges itself in Soviet nostalgia', *Daily Telegraph*, 2 May 1995; both available on *The Electronic Telegraph*, 2 and 10 May 1995, World News.

214 R. Perks, 'Ukraine's forbidden history: memory and nationalism', in *Oral History*, 21.1 (1993), 43–4. See also D. Khubova, A. Ivankiev and T. Sharova, 'After Glasnost: oral history in the Soviet Union', in Passerini, *Memory and Totalitarianism*, 89–102; and I. Sherbakova, 'The Gulag in memory', ibid., 103–15.

215 See R. Hermann, 'History haunts: Lithuania's problems with historical justice' (1998), on the website of *The Electronic New Presence: Central European Weekly* <http://www.enp.cz>; and the following websites, all visited 7 Jan. 2000: Union of Lithuanian Political Prisoners and Deportees <http://www.lpkts.It/en/org_en.html>; Crimes Against Humanity: Latvian Site <http://www.vip.latnet.lv/LPRA>; *Transitions: East-European online news-service* <http://www.ijt.cz>.

216 Perks, 'Ukraine's forbidden history', 43–53.

217 Snyder, 'Memory of sovereignty'.

218 W. Churchill, *The Hinge of Fate* (1951), quoted in A. Calder, *The People's War* (London, 1971), 316, note 26; D. Fraser, *And We Shall Shock Them: The British Army in the Second World War* (Sevenoaks and London, 1988), 196.

219 Wong, 'Memory suppression'. Other recent or forthcoming critical work on war memories in the region of South-East Asia and the Pacific includes: P. Kratoska (ed.) *Malaya and Singapore during the Japanese Occupation* (Singapore, 1995); G. White, 'Remembering Guadalcanal: national identity and transnational memory-making', *Public Culture*, 7 (1995), 529–55; P. Lim Pui Huen and D. Wong (eds) *War and Memory in Malaysia and Singapore* (Singapore, 2000); White, Fujitani and Yoneyama (eds) *Perilous Memories.*

220 For Wong's discussion of the somewhat further-advanced process of reassessment in the independent city-state of Singapore, see section 2.2 in this chapter.

221 S.G. Meštrović (ed.) *Genocide after Emotion: The Postemotional Balkan War* (London, 1996), 2. See his 'Introduction', 1–30, including an analysis of the problems of naming and framing (2–10).

222 N. Cigar, 'The Serbo-Croatian War, 1991', in Meštrović (ed.) *Genocide after Emotion*, 58. See 52–62.

223 Ibid., 58.

224 On the wartime massacres of 1941–45 and their use in the 1990s, see B. Denich, *Ethnic Nationalism: The Tragic Death of Yugoslavia* (Minneapolis, MN, 1994), 30–5.

225 Arkan, letter to *Guardian*, 14 Sept. 1996.

226 Y. Yevtushenko, 'The sleep of reason', *Guardian*, 29 April 1999.

227 N. Malcolm, *Kosovo: A Short History* (London, 1998), xxx, 252–6.

228 J. Ciment (ed.) *Encyclopedia of Conflicts since World War Two* (NY, 1999); *The Military Balance 1999–2000: Global Issues*, International Institute of Strategic Studies (IISS), Oct. 1999. For a summary of the latter, see 'Deadly cost of the new warfare', *Guardian*, 22 Oct. 1999. The figures for 1999 exclude those armed conflicts which the IISS categorizes as 'terrorism'. See also the website of the Military Analysis Network at the Federation of American Scientists, <http://www.fas.org/man/>.

229 *The Military Balance*; 'Deadly cost of the new warfare', *Guardian* 22, Oct. 1999.

230 'The miracle children: Rwandan orphans preach reconciliation', *Guardian*, 22 Dec. 1999, 14.

231 UNESCO Information and Informatics Division, 'Archives and libraries in the Balkans', an e-mail circulated by the British Government's Library and Information Commission, 9 June 1999, referring to the situation in Kosovo.

232 Ibid.

233 Fabienne de Leval, a psychologist in Kosovo with Médecins Sans Frontières, quoted in 'Kosovo's wounded women find no peace', *Guardian*, 19 Oct. 1999, 14.

234 Yevtushenko, 'The sleep of reason'.

235 Ibid.

236 P.B. Hayner, 'Fifteen truth commissions – 1974–1994: a comparative overview', *Human Rights Quarterly*, 16 (1994), 599; 'Truth Commissions' page on the US Institute of Peace Library's website <http://www.usip.org/library/truth.html>.

237 See the official websites for the Tribunals at <http://www.un.org/icty/> and <http://www.un.org/ictr/>; R. Goldstone, 'War crimes: linking peace with justice', *Amnesty*, Sept./Oct. 1997, 12–13; Destexhe, *Rwanda and Genocide.*

238 'Timeline' page at the website of the Coalition for an International Criminal Court at <http://www.igc.org/icc/>.

239 It is significant in this respect that, while the Rome Statute of July 1998 supporting the creation of the proposed International Criminal Court was ratified by ninety-three states in eighteen months, the US government was critical and continued to seek control by veto over its operations <http://www.igc.org/icc/timeline/>. For debate about US/UN relations see, for example, the material collected on the Online Casebook: the Bosnian War Crimes Tribunal, supporting the Introduction to Public International Law course (Ps142K) at Stanford University, USA; website at <http://www.stanford.edu/class/ps142k/casebook/bosnia.htm>, established 5 June 1996.

240 See, for example, the call for papers for a conference on 'Cultures of Political Transition: Memory, Identity, Voice', to be held at the Institute of Commonwealth Studies, London, in September 2000. A further suggestion here is that the 'rhetorical mainstreaming of human rights' has tended 'to entrench a legal/quasi-legal orthodoxy as the preferred way to come to terms with an oppressive past'; and that the mechanisms which institute this 'juridical' narrative, such as truth commissions and war crimes tribunals, 'also structure and restructure memory and identity, and use their own voice to privilege certain voices while suppressing others'.

241 These contradictions can be seen in the case of the NATO alliance's intervention in Kosovo. NATO claimed justification for its bombing campaign as an effort to counter the human rights abuses committed by the Yugoslav state in the province, with the legal sanction of the UN (a claim contested at the time). Subsequently, it has taken pains to secure the memory of its intervention as a just and humanely conducted campaign. However, by subscribing in this way to the human rights narrative,

NATO exposed itself to counter-charges challenging its own war practices on these same grounds; and a case accusing NATO of 'grave violations of international humanitarian law' was taken to the International Criminal Tribunal by Human Rights Watch of New York. ('Nato under fire for choice of targets in Kosovo', *Guardian*, 7 Jan. 2000, 14.)

242 'Indonesia's president says he will punish ex-army chief', *Guardian*, 1 Feb. 2000, 15; 'Wiranto says he will not quit', 'The Editor' section, *Guardian*, 4 Feb. 2000, 7; 'Sacked Wiranto in plea for calm', *Guardian*, 17 Feb. 2000, 19.

243 'People find the confidence to face the truth but fear the general's last laugh', *Guardian* 16 Oct. 1999, 15; D. Campbell, 'Life after death', *Guardian*, 31 Jan. 2000, *G2*, 2–3.

244 See Jelin and Kaufman, Chapter 2, this volume.

245 Ngũgĩ , *Homecoming*; *Writers in Politics*; *Grain of Wheat*; *Petals of Blood*; Kirkwood, *Britain and Africa*, 148–81.

246 David Anderson, quoted in 'Whitehall put in the dock over Kenyan hangings', *Guardian*, 1 Dec. 1999, 19. Anderson's research paper was presented at a workshop at the School of African and Oriental Studies, University of London, Nov. 1999.

247 Ibid.

248 A.D. King, *Culture, Globalization and the World System* (Basingstoke, 1991).

249 'Oxfam helps victims of forgotten war', *Oxfam News*, Millennium Edition (Jan. 2000), 4.

250 'Deadly cost of the new warfare', *Guardian*, 22 Oct. 1999.

251 PANOS Institute's website: <http://www.oneworld.org/panos>. See also O. Bennett, J. Bexley and K. Warnock (eds) *Arms to Fight, Arms to Protect: Women Speak Out about Conflict* (London, 1995).

252 See P. Boniface and P. Boniface, *Heritage and Tourism in the Global Village* (London, 1993); P.J. Larkham and G.J. Ashworth (eds) *Building a New Heritage: Tourism, Culture and Identity in the New Europe* (London, 1994).

253 See E.T. Linenthal, *Preserving Memory: The Struggle to Create America's Holocaust Museum* (London, 1997); P. Gourevitch, 'Nightmare on 15th Street', *Guardian* 'Saturday Review', 4 Dec. 1999, 1–2.

254 R. Hughes, 'Interventionist memory: tourism and memorialisation of the Cambodian genocide sites', a paper presented at the Frontiers of Memory conference, Institute of Education, London, 17–19 Sept. 1999. A collection of papers is being edited for publication by Susannah Radstone and Katharine Hodgkin.

255 See also White, 'Remembering Guadalcanal', a study of Second World War commemoration on the Solomon Islands, which explores the reception of a new American war memorial built to challenge the existing one erected by the Japanese. The new memorial's potential for expanding veterans' tourism led to its acceptance by the Solomon Islands' government and by local people, including some who had previously voiced the view that 'this was *their* [i.e. the outsiders'] war', but who were now prepared to embrace the new commemorative culture even at the expense of their own national narrative of post-war independence. White's essay can be read as an example of how the imperative towards tourism-development may undercut, rather than further, the process of questioning memories constructed from above.

256 H. Jarvis, 'Welcome to Leninland', from the Special Report archive of *Transitions: East-European online news-service* <http://www.ijt.cz/features.html>, visited 7 Jan. 2000; 'Soviet theme park to house fallen idols', *Guardian*, 6 Dec. 1999, 11.

257 Malinauskas, quoted in Jarvis, 'Leninland'.

258 Malinauskas, quoted in 'Soviet theme park'.

259 Malinauskas, quoted in Jarvis, 'Leninland'.

260 Galdikas, quoted in 'Soviet theme park'; Galdikas, quoted in Jarvis, 'Leninland'.

261 Labora, quoted in Jarvis, 'Leninland'; unnamed umbrella-group of ex-prisoners'

and veterans' organizations, quoted in ibid. Malinauskas has dismissed the partisans as 'a group of politically motivated fascists' (ibid.).

262 Letter from two (unnamed) sons of Ona Sukackiene to Juozas Galdikas MP, quoted in Jarvis, 'Leninland'.

Part II

Case studies

2 Layers of memories

Twenty years after in Argentina

Elizabeth Jelin and Susana G. Kaufman

In the midst of deep political conflict and widespread political violence, a military coup took place in Argentina in March 1976. The military government, defining itself as leading a 'Process of National Reorganization', implemented a systematic policy of clandestine repression, including *disappearances* on a massive scale, as the basic policy to handle the political conflict and to wipe out the existing armed political groups. Estimates of disappearances vary, with figures of up to 30,000. Although the military dictatorship lasted from March 1976 until December 1983, repression was harshest during the initial years of the regime.

The purpose of this chapter is to analyse how Argentine society incorporates (or excludes) the memories of its conflicted and painful recent past. It deals not with memories of a war, but with the memories of a period of extreme political violence and of state terrorism. In truth, political and ideological conflict in the early 1970s was extreme in Argentina. There were armed groups on various sides, including a guerrilla movement and paramilitary death-squads. This widespread armed conflict and the military take-over was defined by some as a war, though a 'Dirty War', in which both sides, the leftist guerrilla and the military and paramilitary forces, shared responsibilities. The issue of responsibilities, however, is crucial in alternative interpretations of the period. To talk, as we do, of State Terrorism, i.e. a state apparatus that resorted systematically to terror and repression of citizens and civil institutions without any respect for existing national and international juridical norms, implies a different interpretation of responsibilities and of power relations from the one the military embrace.

Our enquiry is about how memory is produced and constructed. Our goal is to analytically scrutinize processes of reconstructing and rendering meaning to the (diverse, multiple) pasts. It is not an interpretation of what happened in the past, nor the collection of elements (documents, testimonies, 'data') that might help in the societal process of constructing historical memories. Rather than rendering a narrative of the past, we analyse the process of societal remembering (and forgetting), looking at the various *levels* and *layers* in which this takes place. To do this, we first analyse the public/political context in which memory/oblivion of the military dictatorship unfolded, and then present some personal narratives, in which the traces and marks of that past emerge in the development of the life course and in the everyday experiences of people from three generations. Insofar as reality is complex, multiple and contradictory, and that subjective inscriptions of experiences are never a direct mirror-like reflection of public events, one should not expect an 'integration' or 'gelling' of individual and public memories, or the presence of a single memory. There are contradictions, tensions, silences, conflicts, gaps, disjunctions, as well as 'integration'. Social reality is contradictory, full of tensions and conflicts. Memory is not an exception.

For many societal actors in Argentina, the motto *Remember! So as not to repeat!* is a 'given', an unquestionable premise of their lives. In that climate, to carry out critical research is not easy. In the tradition of C. Wright Mills, we attempt to unveil some of the enigmas that emerge at the point of convergence between our personal feelings as involved participants and the public questions and issues that permeate Argentine society. We attempt to do it with analytical depth and rigour, and this leads us to revisit critically our own beliefs and sense of belonging, and at times distance ourselves from those who more forcefully push for memory. Besides the deadening effect of *total memory* – so vividly conveyed in Borges' story of *Funes, el memorioso*[1] – some of the reasons will become clear as we go along.

Argentina: between truth, remembrance and *Nunca más*

Under the circumstances of the political violence in Argentina in the mid-1970s, a local human rights movement gradually emerged.[2] At first almost in hiding and unseen, then becoming steadily more visible, gaining step by step in political relevance and centrality, the human rights movement and the issues it raised gradually occupied the centre of the political arena. Under the military dictatorship (1976–83) the movement developed a range of activities: supporting victims and their relatives, spreading information to break the imposed silence about the nature and scope of the violations, launching open protests, organizing and promoting international solidarity. The most urgent and immediate task was to stop kidnappings, torture and disappearances, and to free those who had been detained. There was a suspicion that key figures in the government were directly involved in what was going on, but the essence of terror is to create confusion, fear, uncertainty and ambiguity. The systematic and planned nature of repression was not immediately apparent to anybody outside the top decision-making apparatus.

A disappearance is a very special type of wound. What disappears is a human being, a body, but also knowledge and information. Those close to the victim and the human rights movement could only say that something has happened, but could not say clearly what. The first demands on the part of the movement were an insistent claim to know, with some certainty, what had happened in each case. There was a clear recognition of the significance of truth in the process of redress of human rights violations. In this process of searching for the truth, the human rights organizations and their international support network were able to shift the interpretation of what was going on from the military definition as an 'anti-subversive war', to a framework of 'human rights violations', a shift that later allowed the introduction of a juridical dimension into political life.

During the last stage of the military regime (1982–83), information became more public, as testimonial reports of victims were published or otherwise made available. As fear receded and popular moral indignation rose, collective actions geared to denounce and demand became widespread, with street rallies that gathered tens of thousands of people. At that time, the reclamation of the public space came hand in hand with first-hand personal narratives, with claims for life, for justice and for memory (also, with an effervescence of artistic activities devoted to the recent past: theatre, fiction, graphics, paintings). Journalism, including a sensationalist 'horror show' in magazines and on TV, devoted itself to disclosing, to telling, to digging up and excavating the pain of those who could not find a grave or a body.

Within this context of generalized visibility, during the electoral campaign and the inauguration of the democratically elected government in 1983, actions of the human rights movements were geared to secure some form of 'justice' for those who were responsible for repression. Human rights activists and organizations felt the urgency of learning about and publicizing the nature of the massive and systematic violations of human rights during the military dictatorship – the demand for 'truth' – and of seeing that the guilty were punished – the demand for 'justice'. The movement was torn between its political and institutional role, expressed in the demand for justice, and its symbolic role in the construction of a historical memory, which implied actively promoting the need not to forget and developing in different ways and in a variety of settings the symbols and events that would foster the preservation of the vivid memory of the lived traumatic experience.

In 1983, there was no clarity as to what kind of punishment *should* or could be applied. Coupled with the fear that political parties or unions would be ready to compromise and accept amnesties for the military, there were some doubts as to whether any wrongdoing could be established in court – in the event that the new democratic government had the political will to prosecute the military. At that time, there was no clear meaning attached to the demand for justice. On the one hand, the issue involved how to act vis-à-vis the victims, whose rights had to be restored and injuries redressed. This included those directly affected and their relatives, including the abducted children of the disappeared. On the other hand, transition to democracy implies the construction of a new institutional order, one that fully protects the human rights of citizens. How to assure these in the future? What kinds of institutional transformations were needed? On the moral and ideological level, what was the content of the rights to be protected and guaranteed?

One of the key issues of the electoral campaign during 1983 and of the initial governmental decisions was the institutional handling of human rights abuses. The problem of how to 'settle accounts with the past' turned into a critical element in establishing the legitimate rule of law. The elected president, Raúl

Alfonsín, took office in December 1983 with an explicit commitment to give an answer to the demands for justice. In his view, however, the political cost of taking the military elite to civilian courts was too great and implied too high a risk for the fragile democratic system. Yet to condone and not to provide any forum for trials was impossible, given the cultural and political climate of the country. The human rights policy of the new government was based on the electoral campaign promise to hold some kind of trials, but also on a desire to limit their scope, as part of the governmental policy towards the military. The proposal was to keep the trials of the military within the realm of military justice.

Human rights activists opposed the decision to prosecute in military courts. The movement wanted civilian courts and a parliamentary commission to investigate repression. The reasoning was that, as elected representatives of the people, legislators would have access to the information in the hands of the military, which would be obliged to provide surviving documents. Such a commission, it was thought, would be able to formulate a political condemnation, a type of punishment that, in spite of it not being in accordance with the juridical procedure, was not irregular. The search was for some sort of legitimate and legal conviction that would serve to reaffirm the basic ethical values of democracy.

The executive acted differently. Instead of a Parliamentary Commission, the investigation was to be conducted by an independent Commission of 'notables', the CONADEP (*Comisión Nacional sobre la Desaparición de Personas*), established immediately after Alfonsín took office, in December 1983. The CONADEP was to receive testimony from relatives of the disappeared and from survivors of the clandestine camps. The Commission set up special offices to collect evidence in several cities. Exiles returned from abroad to testify, and evidence was also taken in embassies abroad. Police and military facilities, as well as clandestine detention centres and cemeteries, were inspected.

The aim of the Commission was not to pass judgement, but to gather detailed information about the fate of the disappeared. At the beginning, there were still some hopes of finding some of the disappeared alive. As it went about its work, the CONADEP became the site of a powerful symbolic indictment of the military dictatorship. Based on the experience and accumulated data of the human rights organizations that had taken it upon themselves to build a data bank on individual cases of abuses and violations, the Commission collected what would become a year later the crucial evidence in the trials of the members of the military juntas.

Its action produced strategic information about repressive techniques and methods. Argentine society learned what had happened. The violations started to become real. Thus, the Commission was an important step towards the moral condemnation of those who were responsible for repression and in the symbolic legitimization of the claims of the victims. Its role was to establish some 'truth'; it was also where the truth could be formally *acknowledged*.

The 50,000 pages of evidence included in the report were presented to the President in September 1984. A special TV programme based on the report, with testimonies of survivors and relatives, was aired at the same time, and the book *Nunca más* (Never Again), with the official report and an annex listing the names of almost 9,000 disappeared, was first published in November 1984. The book soon became a bestseller, and is still being reprinted and distributed widely. As will be seen below when some personal memories are presented, the launching of the book is a benchmark for most people. The book has become part of one's own personal life and recollection.

As could have been expected, it was hard for the general population to realize and believe that these unbelievable stories were part of a very recent and, for most people, silenced past. Many people were 'discovering' what had happened. 'How could I not know what was going on?' was a repeated question. For the victims and their relatives, it was a first step in legitimizing their voice and in having their testimony acknowledged as true. For those who supported the military's 'peace and order', it was a step on their way to perhaps recognizing and condemning the 'excesses'.

The title of the CONADEP report, *Nunca más*, provides a clue to the cultural climate of the country, as well as to the meanings attached to remembrance. That the experience should not be repeated (*Nunca más*) became associated with documenting the 'truth', with collecting a full record of the atrocities the country experienced. In order 'not to repeat', then, one has to keep memory alive: *Remember! So as not to repeat!* (*Recordar para no repetir*) emerged as the message and the cultural imperative.

The trial and the role of testimony

Is 'moral condemnation' enough? If democracy is about functioning institutions, about rules and about participation, how can moral condemnation be transformed into the foundation of democratic rule? The next stage had to be in a different institutional setting. CONADEP was a way to discover what had happened, to find out and acknowledge the *truth*. Now it was the time to ask for *justice*. The trial of the ex-commanders of the military juntas was a moment of utmost impact (both national and international) of the struggle for human rights in Argentina. The nine junta members who ruled Argentina from 1976 to 1982 were brought to trial in the Federal Court of Appeals in Buenos Aires in 1985.[3] The trial was going to show whether the rule of law could impose itself upon the rule of force. It implied a juridical procedure, with all the rituals and formalities that put the judiciary at the centre of the institutional scene: victims were turned into 'witnesses', repressors became the 'accused', and political actors had to remain 'observers' of the action of judges who presented themselves as 'neutral' authorities, defining the situation according to pre-established legitimate rules.

From a juridical vantage point, the task was almost impossible, not least because it had to use penal legislation geared to homicides to try something more akin to genocide. It had to try persons who had not themselves committed murders, who had not given the orders to commit specific acts of violation of rights of specific persons, but who had organized and ordered massive kidnappings, torture and killings of anonymous individuals. The junta members might not have killed any single person; they might not have given the order to kill individual victims; and yet they were responsible for crimes which they had organized. The strategy of the prosecution was to present evidence (there were more than 800 witnesses) that there was a systematic plan, carried out in all parts of the country following the same method of illegal detentions, torture and disappearances. In spite of the difficulties, after five months of moving testimonies, testimonies of persons who had overcome fears of revenge and the reluctance to reveal publicly shameful and humiliating personal experiences, five of the nine commanders were found guilty.[4]

The construction of juridical proof was based on the testimony of the victims, since military records had mostly been destroyed. This implied the juridical recognition of their voices and their right to talk. In that institutional setting, the harm they had suffered had to be presented according to the legal rules of acceptable evidence. Allowing testimony as proof followed the logic that what cannot be shown (the act of aggression) has to be told, but under precise and controlled conditions, so that what is denounced could be verified. In fact, what is acceptable as juridical proof is bodily injury. Feelings and sufferings cannot be measured and included. During testimony, they had to be suspended. When emotion overtook the witnesses, the judges halted testimony until calm and sanity returned. This intermittent pattern had a very special effect: the hidden message was that, in its full details, in its entirety, the experience could not be told; even less could it be heard.

Testimony in court is still the personal narrative of lived experience, but the juridical framework breaks it into components and pieces: the requirement of personal identification, the pledge to tell the truth, the description of circumstances and relationship with the case. The discourse of the witness had to be detached from experience and transformed into evidence. If disappearance is an experience where there is no law and no rule, where the victim ceases to exist as a subject of rights, the testimony in court (of the victims themselves and of those who had been searching for them) was an insistence on recognition. As one witness (who was himself a victim of disappearance and then jail) expresses it, 'The trial eliminated the spectral nature of the testimonies that were moving around in society .... It presented the victims *as human beings*, giving them equal standing with the rest of humanity.'[5]

The outcome of the trial was not only the verdict (in December 1985) condemning the members of the juntas. It was a proof of the systematic nature of repression carried out by the military government, and it implied the collection of a body of information that could become evidence in new indictments of others responsible for repression. Rather than providing a closure to the demand for 'settling accounts with the past' as President Alfonsín had hoped, the verdict of the court actually opened the door for further indictments and trials.[6]

The political temperature rose and, in December 1986, the executive, under mounting pressure from the military, tried to stop further indictments through a

Law of *Punto final* (full stop), which placed a 60-day limit on new criminal summonses. Soon after, the government had to take extreme measures to stop further prosecutions. The *Ley de obediencia debida* (Due Obedience Law, in mid-1987) was the final answer for those cases that had not yet reached a verdict. It was in effect an amnesty for most members of the armed forces. The only offences excepted were rape, theft and falsification of civil status (i.e. irregular adoptions). A few years later, in 1989 and 1990, President Menem used his presidential prerogative to grant pardons to free the members of the military juntas who were in jail, as well as other military and some civilians (prosecuted for their participation in armed guerrillas) under prosecution or sentenced and in jail.

The story did not stop there, however. The human rights movement continued to denounce and demand justice (including presentations in international courts), the Mothers of Plaza de Mayo[7] still carry out their weekly rally in the main plaza of Buenos Aires, cases of illegal adoption and the search of the *Abuelas*[8] still keep the judiciary busy. In this connection, the key military commanders of the Argentine juntas, Videla and Massera, as well as other high military officers, were arrested in 1998, and charged with responsibility for the abduction of children born in clandestine detention centres, in several court cases that the *Abuelas de Plaza de Mayo* have been pushing consistently over the years.

During the whole period, but with much greater force during the 1990s, international and diplomatic pressures have been very active. French and Spanish courts have been investigating the responsibilities of Argentine military in the disappearance of French and Spanish citizens. Other European countries followed. This led to French courts sentencing (March 1990) to life imprisonment (in absentia) ex-Captain Astiz, a famous repressor responsible for the disappearances of two French nuns, who could not be convicted in Argentina because of the indemnity legislation. The Spanish courts have been receiving testimony from numerous witnesses in their investigations of violations of human rights in Argentina. The Spanish courts became the centre of international attention with the Pinochet case,[9] a case that has brought to light the so-called *Operativo Cóndor*, which linked in coordinated joint actions the repressive military apparatus of the six countries in the Southern Cone.[10]

Societal memory, public commemorations

In 1995, the year of the tenth anniversary of the trial, the political and cultural scene of Argentina was shaken by a navy officer telling the story of how repression and disappearances were handled in the navy: aeroplane flights over the ocean, to throw out prisoners who were still alive but injected with tranquillizers. The book that presents his testimony is called *El vuelo* (The Flight). It was the first time that someone who directly participated in repression confessed what was done and how. There was no tone of repentance or remorse – just a story to 'set the record straight'. The confession hit the media:

newspapers, radio, television shows and magazines carried his story. Relatives of the disappeared appeared in the media, in the hope that the repressors who had confessed could give them some clues as to the fate of their disappeared children. Others denounced government policies that prevented the prosecution of confessed killers. In fact, bits and pieces of information, confessed by those who were in charge of repression, became part of everyday news in the country. There was also an institutional response to these confessions. The army Commander-in-Chief recognized (again, for the first time in the history of the previous ten years) that the army had committed crimes, and asked the population for pardon and understanding. His statement was praised in the international community and in the press, while the navy kept silent, not admitting anything.

During 1995, the tenth anniversary of the trial was also honoured in Parliament, and societal memory was revived by the publication of many books, films and videos. It is significant also that during that year a new human rights group came to life: HIJOS, the organization of children of the disappeared, now in their early 20s.

The year 1996 marked the twentieth anniversary of the military coup in Argentina. Throughout the year, and particularly during the month of March, the public sphere was occupied by commemorations. It was an opportunity to talk and to tell what had been silenced or forgotten, an occasion where society felt the emotional impact of the testimonies and personalized narratives, the astonishment of listening to stories unknown before, of recognizing what had been denied or moved away from consciousness. People were faced with the 'reality' of re-enacting fears and disturbing feelings, asking themselves how all that happened was possible, while everyday life seemed to go on, maintaining an appearance of normality.

During such moments, the labours of memory become more inclusive and shared, invading everyday life. It is hard work for everybody, on all sides of the controversies, for all people, of different ages and experiences. Facts are reorganized, existing perspectives and schemes of interpretation are shaken, voices of new and old generations ask questions, tell stories, create spaces for interaction, share clues about what they experienced, what they heard, what they silenced before.

Among the events directly connected with collective memory were the initiatives geared to *mark* the physical sites of torture, death and disappearances, mostly clandestine detention camps. Actually, these were attempts to mark the material and 'real' sites where the real and material bodies of people entered the phantom world of fantasy, disappearing … Commemorations of this sort – as well as attempts to establish memorials, museums and monuments – are part of the political struggles in the country. They are political in at least two senses: first, because their installation is always the result of political conflicts and, second, because their existence is a physical reminder of a conflictual past. Those who call for commemoration often find themselves pitted against those who act 'as if nothing has happened here', and against those who may even demand recognition of the 'heroism' and 'patriotism' displayed by the agents of state terrorism. There are also those who 'did not know', who did not see – the bystanders of horror.

Let us look at one such case. In 1995, a project was introduced in the City Council of Buenos Aires, calling for the creation of a 'Museum of Memory', to be housed in the building that during dictatorship was the clandestine detention centre known as *El Olimpo*. This centre operated during 1978–79, in a block (and building) that now houses an automobile registration office of the Federal Police. In March 1996, the councillors who introduced the project reactivated their proposal, calling for the symbolic inauguration of the museum on 22 March (two days before the twentieth anniversary of the military coup). Human rights organizations included this activity in their commemorations, calling for a gathering on 22 March on that site, to collectively paint a mural on the walls of the building, thus marking the symbolic inauguration of the *Museo de la Memoria NUNCA MÁS* (Museum of Memory NEVER AGAIN).

Action, however, does not follow smoothly according to plans. Negotiations among representatives of different political parties and persuasions in City Council dragged on. As one councillor commented, 'If we do not vote on this project this week, when the coup is commemorated, we will never be able to take it out of the drawer.' Boycotting the project, some council members only agreed to vote on a resolution that would *recommend*, rather than *command*, it to the Mayor. Conflict touched also the high ranks of the Federal Police. The response of the head of the Federal Police to the request made by the human rights organizations was: 'If you paint the Olimpo, I have to resign.'

On 22 March, a barrier of a hundred policemen circling the block of the Olimpo was the 'welcome reception' the marchers got when they approached the area. Nearby, the anti-riot armoured trucks were waiting. Police presence in the area indicated that painting the walls of the building would not be possible. The press reported that 'faced with such a display, the 300 people who had gathered in the neighbourhood of Floresta marched around the block once and, on the wet pavement, they managed to paint "*Museo de la memoria NUNCA MÁS*" '.[11] Next day, newspapers included photographs of the police barrier and of a Mother of Plaza de Mayo, breaking the fence and reaching to the wall to place a bouquet of flowers.

In another attempt to mark a site, in August 1997, human rights organizations called for a *Jornada de la memoria*, on the site where the *Club Atlético*, a clandestine detention centre in the centre of the city, once functioned. In fact, this was the second time such a call had taken place. A year earlier, in July 1996, close to 500 people participated in a gathering that included the construction of a papier maché structure of a tree and a public reminder with the faces of victims of repression. On that first occasion, a firebomb placed at night destroyed the tree and the memorial. On the second *Jornada de la memoria*, a plaque remembering the disappeared was set up, the names of the repressors were engraved, and during the commemoration a monument, a 'totem', was collectively constructed on one of the pillars of the highway. During the following night, the plaque was destroyed, the totem was torn down and the engraved names of the repressors were covered with paint.

The struggle for monuments and reminders is currently unfolding in the Argentine political scene. These are initiatives spearheaded by the human rights organizations, with the support of a wide array of social organizations (labour unions, professional organizations, student and parent school organizations, some churches, some political party groups). There are all kinds of activities: relatives and friends publish remembrance ads in newspapers, dates are commemorated, books are published, and names are proposed for streets and parks. Only rarely are they accepted without controversy or conflict. Societal organizations push, promote and demand. But when these initiatives reach the governmental level, the lack of political will, the lack of an active 'policy for memory', becomes evident.[12] In fact, very few initiatives to commemorate or remember repression and suffering were approved by the government. Yet, societal actors keep insisting.

What happens when the proposals to locate the act of remembering physically, in a memorial or a monument, fail? When the aim is opposed and blocked by other social forces, the subjectivity and the will of those women and men who are struggling for the monument or memorial come out into the public view, renewing their strength to continue struggling. There is no pause, no rest, because it has not been 'deposited' anywhere – it has to remain in the minds and hearts of the people. In a way, the need to turn the unique, personal and non-transferable feelings into public and collective meanings is left open and active. The unsettling question, one that calls for further comparative and interdisciplinary research, is whether it is possible to 'destroy' that which people attempt to remember or perpetuate. Doesn't the oblivion that the opposition/police repression attempt to impose have the paradoxical effect of multiplying memory and making more real the questions and the debate about what the society went through in its recent past?[13]

In sum, in the current socio-cultural Argentine context – and we would claim, in any socio-cultural context – it is impossible to find *one* memory. There is an active political struggle about meaning, about the meaning of what went on and also about the meaning of memory itself.[14] Yet the political arena of memory struggles is not simply a confrontation between memory and oblivion. The 'memory camp' is not a unified and homogeneous front. There are struggles that emerge from the confrontation among different actors within it: struggles over appropriate means and forms of commemoration, about the content of what should be remembered publicly, and also about the legitimacy of different actors to embody memory (the issue of the 'ownership' of memory).[15] Controversies and political conflict about monuments, museums and memorials are plentiful everywhere, from Bariloche to Berlin.[16] They are attempts to make statements and affirmations; they are facts and gestures, a materiality with a political, collective, public meaning.[17] They are public and collective, insofar as they convey and affirm a feeling of belonging to a community, sharing an identity rooted in a tragic and traumatic history. They may also function as a key to the intergenerational transmission of historical memory, though that transmission and its meaning cannot be guaranteed beforehand.

Layers of personal memory

There is no single or easy connecting line between public memory sites and commemorations and personal memory and forgetting. In order to study the way people incorporate in their own lives the experiences of dictatorship in Argentina, we carried out in-depth life-history interviews.[18] Men and women of different social strata and belonging to three birth cohorts were interviewed. There was no claim to statistical representativeness. Rather, the aim was to provide substance, content and meaning to the analytical questions raised. The search for potential interviewees was made through informal networks of colleagues, friends, institutions and NGO activists in working-class neighbourhoods. Within each 'cell' or combination of the three parameters used to define the cases (age, gender and social stratum), the interviewees were selected with a view to increasing as much as possible social diversity (in terms of religion, type of occupational career, family formation patterns, migratory histories and so on). Interviews were recorded, then transcribed and organized so as to allow for multiple readings and approaches.[19]

In the interview narratives, the construction of memory takes on different forms: sometimes a seemingly naive question unleashes anecdotes and remembrances of the times of repression. At other times, memories emerge as major events that were turning points in the life course (experiences of the death of people close to oneself, exile and so on). In some cases the memories of a personal experience with repression are told in such a way that the personal and the historical events of the country are one and the same story. At other times, narratives are much more 'private', with no recognition of the political context.[20]

A first level of personal memory is the 'factual' account of lived experiences:

> J. [a very close relative] was assassinated. He was a political leader; he was kidnapped and then found dead … We had to live for some time away from home, without having been involved in anything … just because our names were in his address book …. We did everything anyone not involved could do …
>
> (Horacio)

> They often found bombs at school … The bombs weren't fake, they were real … My mother bought me earplugs the size of a cigarette filter, and I used to take them with me when I went to the park to play with my friends … Every afternoon at three o'clock a siren, like a factory siren but much louder, would blow, and we knew it was time to put our earplugs in because, according to my mother, that was when they would start knocking down walls … and with the earplugs, we didn't hear the explosions …
>
> (Julia)

> One time, in 1975, I had to make an emergency trip to Tucumán because a sister of mine had died. It was a very cold day. I went by plane. She lived in a village … so I take a bus, and on the way they stop the bus … With a kettle, soldiers drinking *mate*. Christ! was it cold! Then a policeman gets on the bus and asks for IDs, 'IDs, IDs'. I take out my ID and show it to him. He asks me what's the matter with me. 'Please don't make me get off the bus … I'm a wreck! I'm on my way to bury my sister.' The man ordered everyone back on the bus, and we continued on our way. All this happened to me …
>
> (Lucía)

When the events happened, Julia was seven years old; Horacio was a young adult of 25; Lucía, around 45. At the time of the interviews (1995–96), Julia, a young woman of middle-class urban background, was engaged in political activism, devoting considerable time to a working-class community centre geared to help women. Horacio was a 44-year-old businessman, married and with two teenage children, devoted to his family's well-being. Lucía was a 65-year-old woman of working-class rural background. She migrated to Buenos Aires in the late 1940s, at the time when the government of Perón developed many welfare policies for the working classes. On her first arrival she became a factory worker, and then worked for many years as an assistant in a physician's private office. She is retired, and never married.

The three of them remember personal experiences, but in a very different fashion. Besides the influence of class membership and gender, age – or rather the moment in their life courses in which these events took place – is a crucial factor that marks the way in which the events were experienced and the meaning they had at the time of their occurrence:

> When I was a child, it was *normal* for school to be dismissed because of a bomb alarm …. I was 7, and for me, the bad guy was the one who died …. During my entire childhood, all this was perfectly normal. I saw the things going on around me through the eyes of a child.
>
> The anecdote I am going to tell you is not funny, *but seen from a child's point of view*, it might seem amusing … I had a schoolmate, a girl who was a friend of mine. I was 8 or 9. This girl lived in a building belonging to the *Prefectura* [Coast Guard]. The building was fifteen stories high, while the one I lived in was ten … One day this girl invited me to her house. 'Come', she said, 'I want to show you what the soldiers keep in a swimming pool inside the barracks.' [Julia lived close to a military post.] Well, I went one afternoon for *la leche* [the children's equivalent of high tea], and inside the pool we saw mannequins which, according to my friend, the soldiers would gather up and take in at night, and then you might see them again the next day, or not see them again for a couple of days. Can you imagine! Of course I told my parents what I'd seen as soon as I got home. My mother's blood pressure went up to 20, and my father forbade me to ever see this girl or go to her house again. Naturally, I couldn't make any sense of what was going

> on because, as I told you, *things don't seem evil to a child, and neither this girl nor her parents nor anything else seemed evil to me.*
>
> (Julia)

> During that time, you never knew if the guy next to you was after you or not … You couldn't trust anyone because you never knew what side they were on … I would chat, but never give an opinion. I never said that if someone had been picked up, he must have been guilty of something [*si se lo llevaron, por algo será*]. Nor did I say the opposite … I was concerned about what was going on … I tried not to talk much, who knows … I knew too much or I read between the lines … I remember in those days reading the newspaper made me sick. I would get dizzy, my blood pressure would drop, and I would start feeling nervous, almost without realizing it …
>
> (Horacio)

> after Perón, the military already came. How was that called? De facto governments? They came and went all the time … I am a bit forgetful, because after all one's been through, you can't help getting demoralized.
>
> (Lucía)

Twenty years have elapsed since these experiences. Subsequent experiences, with temporal distance and more information, allow for reinterpreting and giving new meanings to lived experience. The stage in the life-cycle when the dictatorship took place is crucial in this regard. The key political marker in Lucía's life is Perón's first government (1946–55). In fact, Perón's first presidency is part of her personal life:

> [I came to Buenos Aires] in February of '46, the time when Peronism was coming in. It was the frenzy of Peronism. Everybody thought that with Perón things would be great, marvellous … I have to recognize that the first Peronist government was truly magnificent. I remember it very clearly, very well.
>
> (Lucía)

She repeatedly expresses her strong emotional involvement with that period of Argentine history. What went on after that is part of her 'forgetting' and her 'demoralization'. She confounds dates and places, and has difficulties differentiating violent acts of the military and those committed by the guerrillas. Violence and violations are not strangers to her, but they are incorporated in her life with disbelief and lack of confidence in politics:

> We lived through some terrible times between '75, '76, up to 1980 … One heard so many things … that a lot of people had been killed, that people disappeared and nobody ever saw them again …. Did you read that book, *Nunca Más*? I've read some chapters, not all … I don't take sides. Not in

> politics, not with governments, not with the military nor anything else for that matter. But there's no doubt that the military did things. You know what politics is like: one person tells you one story, and somebody else tells you another …
>
> (Lucía)

During the dictatorship, Horacio was aware of what was going on. With time, he reinterprets his own feelings and anguish:

> I guess it was the sense of powerlessness … reading things and knowing that much worse things were happening, and somehow one kept on working, living one's life as if it didn't matter … Like it was happening somewhere else, not right here to people who could be friends of mine. After a while people I knew began disappearing. And we kept on hearing more and more … it was impossible to control the amount of information one took in. Inside me, the information was making me sick. It made me dizzy. While reading the newspaper, my eyes would start to cloud.
>
> (Horacio)

For Julia, the search for some meaning in what happened in the country became the leitmotif of her life. She is constantly searching for clues to reinterpret the past, highly aware of her own biography:

> Many years later I came to realize … They weren't knocking down walls; they were executing people at that time! And they used the sirens to cover up the gunfire! *But I didn't figure this out until I was much older* … I was born during a time when '*No te metás*', 'Mind your own business,' 'Keep your mouth shut,' 'See no evil, hear no evil, speak no evil' were the watchwords …

And in another moment of the interview, she locates temporally the stage of 'realizing':

> All these things left their mark on me. It showed up not when I was a child, but after I grew up … When I became active in politics. Getting politically involved was like a game for me, like it always is when you're a kid … The book *Nunca Más* had just been launched, and the second edition would come out a year later. I became active in politics, along with some other kids my age. It was very easy to get started, because there was a Radical Party Comité right across the street from school ….
>
> (Julia)

Actually, most of the interviewees recall the launching of the book *Nunca Más* – whether they read it or not – as a moment when their personal life experience 'touched' the public recognition of human rights violations. The book has become, undoubtedly, a marker for a large part of the population.

When we reach the present, the traces and marks of the experiences that the interviewees went through during the dictatorship are also different. For Lucía, the traces of the military dictatorship of the 1970s seem to fade as one instance in a longer-term perspective, one more case of the political instability and uncertainty that pervaded Argentine history during recent decades. She expresses a sense of living in a threatening environment, insecure and with a lack of protection in relation to her daily life, but also with regard to the political sphere, in sharp contrast with the expectations and hopes that Peronism awakened in her in the 1940s. Horacio suffered the loss of a close relative. The traces are in his fears and in his recurring illnesses. Guilt is a constant mark in his reflective mood, asking himself if he could have done something else or something more. He talks about the fears of the past, of the traces in the present, of the link to his children:

> Times were difficult. There was lots of fear. I remember I used to feel afraid when leaving home. Although one appeared calm and composed, one had the feeling that even though one had no involvement … We were marked. I'll be a marked man for the rest of my life. For example, when walking on Libertador, I can't help looking at the ESMA[21] and turning the corner. If I'm on this side of the street, I turn in this direction, and if I'm on the other sidewalk, I turn in the other direction. I can't take my eyes off that building. And these are the things that make me stop and think: Why? Why? Why do I always look? I can't help it; I have no choice. When I'm with my children, I tell them, 'This is where they used to be … here's where they killed and made people disappear.' I tell them a little so they'll know what went on. Mainly so they'll know what their parents went through.
>
> (Horacio)

Julia is younger. The traces of fear are embodied in her parents:

> The words *guerrilla*, *montonero*, *revolución* were never spoken at home, they were unnameable; not out of rejection of the people, but out of fear … The years go by, and my dad still hasn't lost that fear. Yesterday there was a student demonstration in the Plaza de Mayo, and my husband was one of the security monitors … And his face appeared in the TV news. Well, my father had a fit! He called, and since we were not at home he left a message on the answering machine: 'How can you do these things? You have to be more careful! It won't be long before you find yourself involved.' It's as if he extracts this image and transports it into the past. Or something like that, right? In other words, the fear is still there.

She saw her childhood experiences as 'normal'. She only began to untangle what had happened later on. And at that time, fear invaded her:

> I only began putting together all those stories I told you about what it was

like living under the dictatorship in that context [later on], and *when I realized all the horrible things that had gone on, I was invaded by fear*! I even became afraid of being active in the Radical Party. What would happen to me if the military should come and I belonged to a political party? I would be just another disappeared person! I linked those times with what my parents had conveyed to me, the fear, and that is what paralysed so many people for so many years. But then I began to realize that with Alfonsín in power, the military would not come back.

Julia interprets her present and her future in relation to her fears. Her hopes and her energy are put to many diverse activities, all of them related to political and social activism. Her fears are regarding the future:

> There are times when I am afraid. But you know what I'm afraid of? Not disappearing. I'm afraid of two things: one is exile, because it could happen, and it would be terrible for me because I'm very nationalistic and love my country and my land very much. To have to leave my friends, my people, my customs, just thinking about it makes me sick. That's one fear. And the other is to disappear without having left anything to show I'd ever been alive. I don't mean a monument or plaque. On the contrary, what I'm talking about are concrete deeds.

For Horacio, personal reflection is centred in his relationships to others, on what one does, or doesn't, and on what one could do for others:

> There are many questions one asks about one's life … There are so many things one only realizes later on, and asks, 'How come?' Sometimes I wonder if it is worth looking back. Those times, yes, it was fearful … what's left is that sense of fear … I get this feeling of powerlessness, caught between not having done anything and not knowing what I could have done. It's something like what I felt when my dad died and I was left with a feeling of guilt … One tries to say, OK, I have my family, I'm going to protect them. I'm going to do everything I can so nothing goes wrong. At any rate, I have the peace of mind that comes from knowing that I didn't get my neighbour in trouble in order to protect my family. But who knows? I didn't do anything to help my neighbour either. I think I probably belong to the silent majority of the '*No te metás*' [mind your own business]. Nothing ever happened to me. But I still question myself, and above all I want to be honest with myself: I don't think I would have done much if I had known what was going on next door. And I'm not so sure I wouldn't have said '*por algo será*' ('he or she must be guilty of something') like so many other people did when someone was taken away. That's what I ask myself. It's a dilemma that doesn't go away and will never get resolved, I don't know … I could look for the right answer, I can say, 'No, I wouldn't have … ' and rest satisfied. But

> that's not true. I can't be sure that I wouldn't have … I'm not ready to fool myself …

These are themes that are not talked about in everyday life. The opportunity to talk and reflect upon offered by the interview situation is recognized by Horacio and Julia:

> [The interview] shook everything up inside of me, regarding memories and the past. But it's good to shake things up from time to time because while you reminisce, you remember details you usually forget. The other day I was talking about the interview I had with you with a friend, and in connection with your interview, a lot of memories turned up that you have tucked away without even knowing it …
>
> (Julia)

> What happens is that when I talk to you, it's as if I'm talking to myself.
>
> (Horacio)

Actually, it is in these registers of reminiscence and in the way the past is included in the subjectivity of the present, that the 'labours of memory' display their various layers:

- as narratives of facts and remembrances;
- as feelings that are remembered now, and as feelings generated in the act of remembrance during the interview: feelings about people, personal and group silences, and withdrawals;
- as ways in which memory is conveyed intergenerationally: who tells, who listens, who keeps silent;
- as thoughts and reflection about what one has experienced, considering the moment in the life course when this happened, and one's current thoughts about that past;
- as reflections about one's place in the world, about one's own social responsibility.

Some concluding remarks: levels, layers, strata of memory

Memory is an intersubjective relationship, based on the act of transmission and reinterpretation. Even personal memory requires others to remember: it is group support that makes waking life and memory cohesive and structured. *We are never alone.*[22] When given the opportunity to reminisce, people talk as if their memories were there, waiting to be given the opportunity to be expressed in words. The interview, as an intersubjective occasion, takes advantage of this readiness.

But what is forgetting and reminiscing? Reminiscence implies that there was a previous process of engraving, of fixing something in memory. To reminisce involves the forging of a 'new pact' between past and present, taking the remnants of a memory and incorporating them into the way we deal with the present.[23] To forget, on the other hand, does not imply a void or a vacuum. It is the 'presence of the absence', the representation of what was once there and no longer is, the representation of something that has been erased, silenced or denied. In both cases, we are talking about intersubjective experiences. In the words of Luisa Passerini, we are actually dealing with:

> a memory of a memory, a memory that is possible because it evokes another memory. We can remember only thanks to the fact that somebody has remembered before us, that other people in the past have challenged death and terror on the basis of their memory. Remembering has to be conceived as a highly inter-subjective relationship.[24]

Societal forgetting is also a collective intersubjective affair. It implies a societal cleft, a rupture between individual memory and public and/or collective practices (that may become ritualized and repetitious), or a faulty line in the intergenerational process of transmission. The difference between remembering or forgetting events one has experienced in one's own life, and societal memory, then becomes important. Interpretations and explanations of the past cannot be automatically conveyed from one generation to the next, from one period to another, from those who experienced the events to others who did not. As Yerushalmi notes, the past has to be actively transmitted to the next generation, and that generation has to accept that past as meaningful.[25] Good transmitters, but also open receptors, ready to incorporate what is told to them, are needed in order to make sure that understanding of the past will not die with those who experienced it. An active transmission of memory requires fostering a process of identification that can produce a broadening of the 'we', the active subjects of reminiscing. Yet it requires leaving open the door for the processes of reinterpretation, both on the part of the young and of those who were alive at the time but 'did not know what was happening'. Insofar as rendering meaning is an active subjective task, there is no a priori guarantee that a given meaning will result. There is no way to close off new readings of old stories, because the 'same' story and the 'same' truth gain diverse meanings in different contexts and circumstances. And the succession of generations involves, unavoidably, the creation of new contexts. Memory is, in fact, part of the symbolic and political struggle of each time, of each 'present'.

In this chapter, we have dealt with political confrontations over interpretations of the past and about commemorations, and with the various layers of the most personal and even private memories. In them and in between, a variety of confrontations are being played out. Who is the authority to decide what the 'proper' ways to remember are? Who has the right (and the duty) to embody *true* memory? Insofar as official institutionalized channels to recognize the recent past experience of violence and repression in Argentina are very weak, the struggle over the 'truth' of memories, and what are 'proper' memories, is played out in the societal arena. In this arena, there are competing claims. Personal suffering (especially when suffered in one's own body or among close kin) is for many the most significant criterion, indicating a process of essentializing in 'biology' and in the 'body' the legitimacy of the claim to *truth.* Yet, if only those who had a personal experience of suffering are the legitimate voices to express collective memory, the symbolic authority and power they hold can lead to their 'monopolizing' meanings and contents of memory and truth.[26] This power, in turn, may obstruct the mechanisms of societal involvement (the enlargement of the 'us') and of intergenerational transmission of memory, by not allowing for the reinterpretation – in one's own terms – of the meaning of the passed-on experiences. Thus, there is a double historical danger: oblivion and an institutional void on the one hand; ritualized repetition of the horror and tragic history, with no possibility of the creation of new meanings on the other.

When analysing memory, we are dealing with multiple intersubjectivities, multiple transmissions and receptions of partial memories. Fragmented and contradictory, made up of pieces, shreds and patches, of one layer on top of another, of traces and monuments, slips of the tongue and amnesias. When seen in a collective light, as historical memory or tradition, as the process of searching for the roots of identity, the space of memory becomes a space of political struggle. It alludes to the capacity of preserving the past, a capacity that inevitably implies the participation in the struggle for meaning and for power. Collective remembrances become then politically relevant, as an instrument for legitimizing discourses, as tools for drawing boundaries or for enlarging communities of belonging, and as justifications for the action of social movements.

Notes

1 Published in English as 'Funes the Memorious', in J.L. Borges, *Labyrinths* (Harmondsworth: Penguin, 1970), 87–95.

2 A concise analysis of the political process during the military dictatorship and the transition to democracy is to be found in C. Acuña and C. Smulovitz, 'Militares en la transición argentina: del gobierno a la subordinación constitucional', in *Juicio, castigos y memorias: derechos humanos y justicia en la política argentina* (Buenos Aires: Nueva Visión, 1995), 19–99. For the development and strategies of the human rights movement, see E. Jelin, 'La política de la memoria: el movimiento de derechos humanos y la construcción democrática en Argentina', in *Juicio, castigos y memorias*, 101–46. Initially, the human rights movement was a heterogeneous and loosely connected network of relatives of victims, concerned intellectuals and politicians, church-related activists and a variety of other progressive social actors.

3 The civilian Court of Appeals took over jurisdiction from military tribunals when the latter decided that there was not enough evidence to proceed with military trials. This civilian appeal mechanism was part of the reformed Military Code law adopted by Congress soon after the democratic government took office.

4 General Jorge Rafael Videla and Admiral Emilio Massera were given life sentences; General Viola was sentenced to seventeen years in prison; Admiral Lambruschini was given eight years and Brigadier Agosti was sentenced to three years and nine months. The members of the junta that governed between 1979

and 1982 were acquitted because the court found the evidence against them inconclusive.

5 Norberto Liwski, interview at Centro de Estudios de Estado y Sociedad, Oct. 1990.

6 For an analysis of the effects of the verdict, especially its 'Point 30' on further mandatory indictments, see Acuña and Smulovitz, 'Militares en la transición argentina'.

7 The *Madres de Plaza de Mayo* is one of the best-known organizations of relatives of disappeared people. It is composed of mothers of the disappeared, and it has had (and still has) a highly visible role in the demands for justice and redress.

8 The *Abuelas de Plaza de Mayo* is the organization of women who struggle to denounce the disappearance of, search for, identify and recover their grandchildren, namely children who were born in detention camps or who were abducted with their parents. Many of these children have been appropriated by the repressors or given up for illegal adoption. The abduction and falsification of identity are crimes that are not covered by a statute of limitations. Over the years, the *Abuelas* have participated in the recovery of the identity of more than fifty children. Many cases are still under court investigation, and new cases emerge continuously.

9 The request to extradite Pinochet from the United Kingdom to Spain was front-page news after his arrest in October 1998. This case has raised a major debate about the international handling of 'crimes against humanity'.

10 A. Boccia-Paz, ' "Operativo Cóndor": un ancestro vergonzoso?', *Cuadernos para el Debate* No. 7, Programa de Investigaciones Socioculturales en el Mercosur (Buenos Aires: Instituto de Desarrollo Economico y Social, 1999), 5–21.

11 *Página/12* (Buenos Aires), 23 March 1996, 5.

12 In 1998, however, the legislature of the City of Buenos Aires approved the creation of a Park of Memory (Parque de la Memoria). During 1998 and 1999, several provincial and city governments were discussing creating museums and documentation centres.

13 Claudia Koonz, 'Between memory and oblivion: concentration camps in German memory', in J.R. Gillis (ed.) *Commemorations: The Politics of National Identity* (Princeton, NJ: Princeton University Press, 1994), 258–80, deals with the debates about memorials in concentration camps in Germany and Eastern Europe, showing that the struggle is intensified, rather than diminished, as time passes and political change takes place in the region. One of the controversies about memorials is whether the construction of traditional monuments and memory sites actually discourages engagement, and how to design new ways (not necessarily sites) that would encourage it.

14 Just to give one clue as to the intensity of the struggle, in August 1997, a new book showed up in news-stands all over the country. *La otra campana del 'Nunca más'*, written by a convicted torturer, Miguel O. Etchecolatz, presents and justifies the repressive measures of the military regime. The book generated a heated public debate that showed the extent of the political conflict, the unreality of 'reconciliation' and the depth of historical wounds.

15 Jelin, 'La política de la memoria'.

16 For instance N. Howe, 'Berlin: monuments and memory', *Dissent* (winter 1998), 71–81.

17 That there are historical (and intercultural) transformations in the ways societies deal with history, memory and commemorations, has been amply documented and researched: Y. Yerushalmi, *Zakhor: Jewish History and Jewish Memory* (Seattle: University of Washington Press, 1989); P. Nora (ed.) *Les Lieux de mémoire* (7 vols, Paris: Gallimard, 1984–92); J.R. Gillis, 'Memory and identity: the history of a relationship', in J.R. Gillis (ed.) *Commemorations*, 3–24. Very little of this tradition of research has reached Latin America.

18 The interviews were collected during 1994–96 as part of the project on 'Rights and Responsibilities: life experiences of citizenship in an emerging democracy',

carried out with the support of a Research and Writing Fellowship of the John C. MacArthur and Catherine T. MacArthur Foundation (granted to Elizabeth Jelin in 1994). Marina Elberger participated in the selection of testimonies to be included in this chapter and in their interpretation. As a member of the age-cohort that lived through dictatorship in childhood, the questions she asked and the images she retrieved turned the research process into an intergenerational dialogue. A first interview was designed to elicit a life story, in which each interviewee set the narrative in her/his own parameters and terms. The aim was to get the subject's narrative of her/his family history, including significant links during childhood and adolescence, family involvement in religion, politics and ideology, world-views, educational and political ideals, group belongings. Work experience, and vocational and career issues, were also pursued. A second interview dealt with a more specific set of themes. The guidelines included instances of social participation and views about them (a history of voting behaviour, of encounters with the police and the courts, participation in political and social movements, experiences of social solidarity and so on), and themes of 'historical memory', i.e. experiences and memory of important events in recent Argentine history (including the transition to democracy in 1983, the military coup of 1976, and earlier events).

19 In this chapter, we deal with issues of historical memory of repression. Other published results of the project are to be found in S. Galperin, E. Jelin and S. Kaufman, 'Jóvenes y mundo público', *Investigaciones en psicología: revista del Instituto de Investigaciones de la Facultad de Psicología* (Buenos Aires) 3.1 (1998), 41–53; and in E. Jelin et al., *Vida cotidiana y control institucional en la Argentina de los '90* (Buenos Aires: Nuevo Hacer, 1996).

20 We present here testimonies of a very few interviewees. It was not easy to select the cases to present, since in all interviews there is information that would allow us to show how complex remembering and forgetting is. In subsequent quotations from interviews, all italics are the authors'; the interviewees have been given pseudonyms.

21 ESMA (*Escuela Superior de Mecánica de la Armada*), a large complex of buildings in a very important part of the city of Buenos Aires, was turned into the most important detention camp operated by the navy. It has become the clearest symbol of repression in Argentina. At the beginning of 1998, President Menem proposed tearing down the buildings and constructing a 'Monument for Argentine Unity and Reconciliation' on the site. His suggestion was greeted with the utmost criticism and opposition by almost everybody. There is also an initiative to turn the buildings into a Museum of the Memory of Repression.

22 M. Halbwachs, *On Collective Memory* (1941, 1952; Chicago and London: University of Chicago Press, 1992).

23 'Reminiscence is the result of a psychic process that consists of working with the remnants of a memory, of a dream or a ghost, so as to build a new compromise between what the past that was lived and experienced by the subject and his/her current way of handling that past, what s/he can tolerate ignoring or knowing about it.' M. Enriquez, 'La envoltura de memoria y sus huecos', in D. Anzier (ed.) *Las envolturas psíquicas* (Buenos Aires: Amorrortu, 1990), 121; our translation.

24 L. Passerini, 'Introduction', in L. Passerini (ed.) *Memory and Totalitarianism* (Oxford: Oxford University Press, 1992), 2.

25 Yerushalmi, *Zakhor*.

26 Symbols of personal pain tend to be embodied in women – the Mothers and Grandmothers in the case of Argentina – while institutional mechanisms seem to 'belong' more often to males. The significance of this gender-related dimension of the issues involved, and the difficulties of breaking the gender-typing of roles and power resources require further attention. Further research should also delve into the

significance, in the human rights movement and in society at large, of claims for truth based on suffering, and of the imagery of the family and family relations (J. Filc, *Entre el parentesco y la política: familia y dictadura, 1976–1983* [Buenos Aires: Editorial Biblos, 1997]), in the process of construction of a culture of citizenship and equality.

3 The South African War/Anglo-Boer War 1899–1902 and political memory in South Africa

Bill Nasson

Almost a century ago, Rudyard Kipling, that most ironic bard of the British Empire, despaired of the Boer War as a bad business, plainly and bleakly, 'No End of a Lesson'.[1] Britain had prevailed in its imperialist war to crush the white colonial nationalism of independent Boer republican communities, but it had been a costly job. London's colonial war effort was characterized by humiliating military reverses, an increasing financial burden and a rising moral fuss within domestic Liberal and other anti-war opinion over the rigour of its internment policy and other anti-guerrilla tactics which inflicted great suffering upon Boer civilians, especially women and children. The Transvaal and Orange Free State Boer Republics, products of nineteenth-century migrant settler state-formation in southern Africa, lost a bitter struggle to maintain their national independence from the influence and authority of the Crown. For their radical anti-British-imperialist leadership, the war was a disaster, for it broke what one scholar has nicely termed the Irish or 'Sinn Féin variety of Afrikaner nationalism',[2] as the die-hards or *bittereinders* could not long sustain a will or the means for a fight to the end. The war also signified more than nationalist defeat. Horrendously high mortality rates in British concentration camps, and the loss of perhaps as much as 20 per cent of the tiny Boer republican populations, meant that it represented a form of historical trauma for Afrikaner society, the depth of which imperial and other English-speaking historians have perhaps rarely fully recognized.

Ideologically, the war left an immediate fissure within early twentieth-century Afrikaner society, divisions which began to bite very soon after the signing of the 1902 Treaty of Vereeniging, and which were more or less bound to result from the kind of Boer republican 'people's war' which had been fought. At one ascendant pole, there was an accommodationist, pragmatic Afrikaner war veteran leadership, now reconciled to getting into bed with imperialism, and appreciative of the gains to be made in moving on from the bitter legacy of war to forge an inclusive Afrikaner-English white political nation. At another, there were either *bittereinders* who had jibbed at acceptance of a negotiated peace, or a sulky clump of those who had gone down to defeat, but had in a sense surrendered conditionally, bristling for any fighting opportunity to restore a shattered republican vision by other means. From the start, this restive and militant Afrikaner nationalist constituency defined itself in relation to harrowing memories of the notorious camps and scorched earth of the 'English War', flexing a combative identity against the imagined ethnic and cultural blood sacrifice of the conflict. 'By creating thousands of martyrs for the cause of the volk'[3], and by in one way or another touching every Afrikaner community and family in the country, war and its memory became a prime and most exact register of 'Afrikaner-ness' in the twentieth century.

In social reach and human cost, the South African War was the biggest and most modern of the numerous pre-colonial and colonial wars which raged across the southern African subcontinent. It is the war which today still counts in national memory, however narrow the historical context of that construction; other war memories slumber on through the transmission of nineteenth-century African oral tradition, in popular rural mythologies about Shaka and the Zulu kingdom, and in the odd dreams of regaining some tribal pastoral lost in nineteenth-century settler wars over frontiers or land. There can be no question that these fleeting oral tracings of war valour or war suffering have left little imprint upon the historical consciousness of a modern white-ruled South Africa. British regiments have long had their Zululand monuments; only at the end of the 1990s is a memorial to Zulu warriors to be inscribed on the landscape of official cultural representations of warfare.

In any wide view, political memory of the South African War counts with fairly good reason. Generating a wealth of literature, this early twentieth-century colonial war's literary epitaph remains unrivalled locally, its cultural deposit perhaps making it the modest South African equivalent of an American Civil War, a British Great War or even a Spanish Civil War. In part, that reflects the obvious contrast between the short and easy war that was imagined, and the lengthy and arduous war which was actually fought. It is, even more, an illustration of the degree to which, especially after 1902, the meaning of the war continued to be fought over between new Afrikaner colonial nationalists and unreconciled Afrikaner republicans. For the ruling elite of the new Union of South Africa, which in 1910 united the former Boer Republics and British colonies in an imperial Dominion, Anglo-Afrikaner reconciliation specifically required a moderation of bitter war memories; in benign nation-building rhetoric, the war became a tragedy or a regrettable imperial entanglement in that it had ruptured a natural Boer-British European settler unity. In the cosmology of disinherited and dissident republicans, on the other hand, the war was a grudge to be nursed in concentrated form. Not only had Britain trampled a republican 'nation' underfoot; ex-Boer generals like Louis Botha and Jan Smuts had become London's secondary demons, their silky collusion with imperialism nothing other than a stab in the back.[4] And so, as in post-1939 Spain, a war of arms was now to continue as a memorializing war of words fought by those to whom the Anglo-Boer War became a massive building-block of a patented 'Afrikaner' history, a contagious 'myth of national origin'.[5]

That inflated war of words was certainly of considerable rhetorical importance to the balance of white political forces even into the early post-Second World War era, when the survival of cultivated memories of British concentration camp 'genocide' and cruel conquest through war remained indissolubly connected with the rise and eventual ascendancy after 1948 of a republican nationalist Afrikanerdom. As the emotive expression of a subordinate yet pugnacious national combination of classes, tilting at the political and economic citadels of South Africa's languid English establishment, Afrikaner war commemoration provided a moral legacy of heroic manly struggle and female fortitude and sacrifice. That reflection began in the 1900s, initially through pilgrimages to grave sites, and the disinterring and ritual reburial of the remains of fallen combatants, followed by the later creation of war memorials, such as the 1913 *Vrouemonument* or Women's Monument in Bloemfontein, a male-inspired shrine to female martyrdom, or the 1938 Cottesloe Boer Veterans' Monument in Johannesburg. Other components included the issuing of commemorative veterans' medals, regular wreath-laying ceremonies at camp sites, and more resurrectionary modes of expression, such as mounted parades by veterans under arms. However varied the form of this commemoration, it usually involved an inevitable transition from private or community bereavement to political symbolism through deft Afrikaner nationalist appropriation.

The demand of the political moment was for explicit allegory. For the *bitter-einder* Boer general Christiaan Beyers, therefore, the meaning of the Women's Monument lay in its utility as a condensed repository of 'memories' which could fertilize the future, preparing the soil for the rebirth of a free nation.[6] In similar vein, the annual attendance of coteries of commando veterans or *oudstryders* and camp survivors provided a means of bringing home to an Afrikaner public the enduring feature of a national history: survivors and veterans bobbing along as symbols of an eternal Afrikanerdom, fleshly representatives of a gritty Afrikaner struggle for survival.[7]

It is against this background that myths and patriotic messages declared that the fatal clash between Briton and Boer was not to become just another miserable vestige of the brutalities inflicted by imperial conquest. As prominent nationalistic war poets like Jan Celliers, Eugene Marais and 'Totius' asserted after 1902, cathartic memories of the blood sacrifice of *volkshelde* or armed 'people's heroes' could kindle consciousness of unified national identity and help to renew crucial dignity and purpose. It is perhaps not stretching things too much to see the war enshrined as the Verdun of Afrikaner society, or even as its 1905 mutiny against despotism. In this, Afrikaner nationalist writers and historians beavered away as hard as anyone to keep the war a live and burning issue within the vital social networks of religion, politics, family and friendship.

In the first instance, in the immediate post-war years it was obviously necessary to construct a view of the republican struggle which countered the fairly ludicrous and often offensive portrayal of Boer society peddled by numerous imperialist writers. In the most vulgar of these depictions, the 'Boer' was a degraded rural layabout, an untidy pre-modern with no legitimate place in capitalist modernity, and a warring primitive who displayed cowardly or shifty fighting qualities in pursuing a criminal war against civilizing progress. These stereotypes became refashioned or inverted in comic form by nationalist children's literature in the 1920s and 1930s: the Boer commando, depicted as a bearded toad which had hopped on to General Buller's mess table, would be transformed into handsome and dashing Albert Viljoen, a princely lion-heart who had easily outwitted plodding, slum-born British Tommies. But there was always more to such decorative representations than mere counter-history, or the swapping of mythologies. For influential popular historians such as Gustav Preller, a prominent former war correspondent on the Boer side, accounts of republican doings all served an overriding purpose – to awaken 'the' Afrikaner to the Truth of Their War of Freedom and Their National Mission.

As Albert Grundlingh has emphasized, the impulse was to make a 'war of freedom' inheritance the powerful core of an immanent Afrikaner 'nationalist spirit', with a stream of robust popular histories reminding readers of the Christian Boer crusade 'against the mighty British Empire, and the suffering of women and children'. Memory of war

> had to serve as a constant reminder of the Afrikaners' bitter fight for freedom. Although they had lost ... they were exhorted not to sacrifice a common identity as Afrikaners ... History had to be used in such a way that it enhanced patriotism and national consciousness ... Contemporary Afrikaners had to complete the historic mission of the Boer die-hards – they had to continue the fight for Afrikaner independence in the present.[8]

This meant placing the war in the long-term perspective of Boer-Afrikaner history in general, and of a national martial spirit in particular, filled with images of an old frontier Boer republican manhood.

In practice, the devastating impact of the war upon the fabric of Boer society in the Orange Free State and the Transvaal did not make it easy going for a bookish, middle-class Afrikaner nationalist intelligentsia; it took time to stitch together a meaningful sense of war and a renewable ethnic nationhood. But once the cultural assertion of a standard Afrikaans language became buttressed by official recognition from the 1920s, magic could be worked with war writings. Much of this patriotic alchemy was pursued in popular magazines like *Die Brandwag* (The Sentinel) and *Die Huisgenoot* (The Home Companion). By soliciting rural homes for personal male and female war testimony for regular feature publication, they plumbed a rich seam of earthy folk memory of hardship, privation and brave sacrifice. Homespun, uplifting periodical stories of conduct under conquest circulated through reading circles, public readings and magazine clubs into the 1930s and beyond. Coursing through tiny rural settlements and scattered farms, the influence of this literary flow looks to have been quite considerable, not least through its wide appeal to republican patriots, particularly rural women who had been a fierce repository of wartime resistance in defence of homesteads, as well as to commando veterans and their families. *Die Huisgenoot* and other magazines' correspondence columns filled up with appreciative murmurs from middle-aged readers of personal wartime histories, some urging everyone who had something to say to give a day to the commemorative cause, in which their most raw and 'visionary' 1899–1902 experiences could be documented on paper.[9] In this genre, the war largely ceased to be a disputed and increasingly divisive national struggle against imperial domination; instead, it became resurrected in print as the unity of the pastoral Boer nation at war, a tribal defence of hearth and home by a small and virtuous Protestant people.

While relatively few Afrikaans books on the war appeared between the 1900s and the end of the 1920s, the ensuing two decades witnessed a surge of popular works, ranging from tendentious histories such as Sara Raal, *Met die Boere in die Veld* (With the Boers in the Veld), to animated historical fiction like 'Mikro', *Die Ruiter in die Nag* (The Rider in the Night).[10] Populist writings of this kind helped to lubricate a progressively consolidated collective mentality and public memory by creating an exalted sense of an essentialist national character: hardened, tough, valiant and wily in adversity. And, while harping on war memories of bitterness, anguish and redemptive Christian fortitude, it was equally essential to commemorate superhuman republican bravado, exemplified by the gritty epic of *bittereinder* guerrilla resistance during 1901–2, and the seemingly clairvoyant genius of Boer guerrilla generalship in outfoxing British imperial forces in the field and keeping on the run for so long. In this overdetermined masculine legacy, 'the courage and determination of the die-hard Boer fighters revealed those character traits supposedly typical of the Afrikaner and deemed worthy to emulate'.[11]

One such instinct was embodied in mythical representations of the Boer warrior as *volksheld* or national hero; this image was expressed most powerfully through the presence of rustic *oudstryders* or veterans. Biographies, poems, novels, essays and popular magazines of the 1930s, such as *Die Huisgenoot*, turned generals like Christiaan de Wet and Koos de la Rey into household names.[12] Here were idealized warriors who defined the special quality of heroic Boer masculinity; this lay not in marching to order and in blind self-sacrifice (as with the British enemy), but in swift improvisation, unorthodoxy and mastery of the tactical retreat when the odds against one became too overwhelming. Thanks to such God-given instincts, the superior Boer warrior always knew what to do to live on in order to fight another day. Rousing literature from both non-combatants and veterans reinforced an attitude that 'true' generals, the icons of guerrilla warfare, had never been defeated militarily by British arms. Britain had prevailed solely through superior numbers and resources, and by the barbarism of a levelling war of attrition. For General Jan Kemp, all that tempered the awfulness of defeat in 1902 was that genuinely partisan and therefore 'true' republican warriors had not failed. Through veteran generals, they could continue to speak for national identity, forming a mystical, almost omnipotent brotherhood of the living and the dead.[13]

For nationalist writers in *Die Huisgenoot* in the 1920s and 1930s, the 'fidelity' of heroic military leadership had only been deepened by the 'treachery' of backsliding Boer soldiers who had limply surrendered, or turned traitor by joining enemy forces. In the face of 'unmanly' cowardice and a breaking of ranks, the tenacity with which *bittereinders* stuck to the battle for independence provided a calming solidity for life after war. As warriors, the toughest generals had come back undiminished, still sustaining the honest identity of men 'of the soil', and with the unquestioning conviction of saints.[14]

A striking feature of this manifestation of martial masculinity was its egalitarianism, refracted through a richly layered celebration of a democratic fighting republicanism. *Die Huisgenoot*, *Volksblad* (People's Paper) and other widely circulating organs of nationalist propaganda turned *all* Boer soldiers who had died in combat or held out as die-hards into disciplined, gilded national heroes, placing generals and men on something like an equal footing. Highly emotive examples of valour and sacrifice were easy to come by, as the last months of the war had seen remaining Boer belligerents slowly bleeding to death. By the 1920s, there was an almost insatiable intensity to catalogue everyone who had done their bit as part of a loyal soldiering community; very young boys, the aged, and men who were amputees or otherwise disabled, became a fetish, as writers scrambled to construct various warrior-lineage biographies for those who represented the touching extremities of service in defence of freedom.[15] A minor cult of the less-than-able-bodied provided a moral aura to eclipse almost anything.

As in post-1939 Francoist Civil War historiography, the overlap between popular and more scholarly representations of past war, which turned history into nationalist propaganda, was very marked. Through the 1940s, and for a time even beyond, much academic writing presented the war above all as a militant and emotive moral covenant of Afrikaner nationalist mobilization. Such classic studies of the later 1940s as J.H. Breytenbach, *Die Betekenis van die Tweede Vryheidsoorlog* (The Meaning of the Second War of Freedom),[16] demonstrated a war made up of godless and mendacious British and upright guardians of a freeborn Boer people, men of the elect who listened to the Prophets, and women who would continue to carry the seed of a republican freedom whose time would surely come again.

Integral to the professional promotion of an Afrikaner *volksgeskiedenis* (people's history), this gloss on the war became its historical truth, part of an 'objective-scientific' truthfulness about the Boer past furnished by leading nationalist historians.[17] The psychology of 'scientific' history in the inter-war decades stressed the 'truth' of a Calvinist-ordained, ethnic political unity as one of the eternal verities of an Afrikaner 'being', registered as a mystical 'people's union' or *volkseenheid*. The *volk* was two things. One was an organic body, unified by a common historical culture and a racialized social metaphysics, in which individuals could have no existence outside of nationhood. Another was a core historical identity, defined by a perpetual struggle against predatory foes bent on ungodly business. As more than one historian has pointed out, the ideologies of an Afrikaner nationalist intelligentsia in the 1930s were in various ways linked to those of contemporaneously mobilizing German nationalists.[18] The Afrikaner story became one of seamless ethnic suffering; in the nineteenth century, migrant Boer *voortrekkers* had been bloodied by black African savages, while in the early twentieth century, the 'Boer' (to use the derogatory English sense) had been done in at the hands of the British.

The 1920s, 1930s and 1940s provided a most fruitful environment for a dusting of war memories to be added to other staple ingredients of nationalist feeling. In 1938, the massive centenary celebration of the Great Trek and the Boer-Zulu Battle of Blood River provided a focus for South African War commemorative bonding. Not surprisingly, new monuments and memorials to 1899–1902, along with wreath-laying ceremonies, formed a crucial part of 1938 'people's festivals' or *volksfeeste*, locating the armed struggle and military traditions minted at the end of the nineteenth century in the longer history of a shared nationalist past. Moreover, as the Second World War approached, radical nationalists discovered anew the fleeting late-1914 Afrikaner Rebellion, when several thousand Boer commandos had had enough of the sickly virus of Anglicization, and had launched an armed insurrection against the state, its leaders declaring it an armed protest against the Union's collaboration with the British war effort.

Many prominent rebels had been 1902 *bittereinders*, whose rebellious rhetoric drew heavily on an inherited, dreamy language of republican 'restoration' and 'no surrender'. It spoke of a mystical nationalist past in terms which seemed to relate directly to the latest 'Boer' crisis; and, in due course, later 1930s nationalist discourse ran together the upswings of 1899 and 1914 into an embalmed line of heroic *volkshelde* and selfless armed martyrdom.[19] Like the war, the 'Rebellion' had illustrated a truth about Afrikaner national honour: it was preserved best in adversity. Inculcating and anchoring an irreproachable nationalist 'truth' in these years was the resonance of a *Vryheidsoorlog* (War of Freedom), a pervasive term first coined in the 1880s, after the Boers had defeated the British at the 1881 Battle of Majuba. Retrospectively, therefore, the 1880–81 Transvaal War became represented as the *Eerste Vryheidsoorlog* (First War of Freedom), an epic tribe-of-Israel struggle for justice and independence in the promised land. As a living parable, this imagery went on to season the second major conflict. The confection of a *Tweede Vryheidsoorlog* (Second War of Freedom) served to affirm the key place of this war in the long uphill slog of a republican Afrikaner nationalism. A commemorative grammar of war remembrance fused with the proposition of Afrikaner survival through apartheid, translating into a common offensive front.

This was all, in a way, a doctrinal preface to Afrikaner nationalism's elevation to power through the National Party's electoral victory of 1948. Its achievement of political control changed the picture: there was now a diminishing need to bolster bitter social memory of the 'English War'. Having lent a helping ideological hand, the war's commemorative boom underwent a gradual deflation through the course of the 1950s and 1960s; public memory grew more mellow, and the meanings of the war became less touchy an issue in national white politics. At the most general level, the earlier course of that public memory in the aftermath of Union had always been more about prickly cultural boundaries and partisan allegiances than about any common war heritage. After all, for

English-speakers and Afrikaner advocates of post-1910 white reconciliation, *possessing* and rallying around an Anglo-Boer War memory was bound to prove tricky, for commemoration and reconciliation had become squarely opposed social processes. Any wider commemoration of the fallen could not avoid raking up the more controversial questions of the war's sectional course and consequence. Accordingly, the civic worthies of 'English' cities like Cape Town, Durban, and Johannesburg maintained a mostly muted observance. Nonetheless, in the 1930s and 1940s it was not always possible to avoid becoming embroiled in bitter local quarrels with nationalist Afrikaner political dignitaries and war veterans over their alleged misuse of commemorative processions and gatherings to express anti-English 'anger' and 'hatreds',[20] or at their deriding of civic protocol by excluding mayors from annual memorial invitations. As a pained correspondent to the *Natal Witness* put it in 1939, the emphases in Afrikaner ceremonies were 'neurotic' and 'fixated', when the need was for life to go on, and for the war to be left behind, a 'sad happening' of 'our unlucky past'.[21]

Why was there less of a fixation after the end of the 1940s? Apart from the National Party's political triumph, any answers to this question can probably be little more than conjectural. But a few reasonable suggestions can be made for why exaggerated nationalist reminders of the moral lessons and animosities of the war weakened appreciably in the immediate post-Second World War decades. In part, this was because many Afrikaners, including poorer whites, found themselves enjoying unprecedented prosperity under post-1948 National Party rule. Various Afrikaner interests were specifically secured by the apartheid state, through such measures as a political purge of English-speakers in the Union army and the police, the statutory elevation of the Afrikaans language and the reinstatement to command of Nationalist military officers who had resigned on South Africa's 1939 declaration of war against Germany. In such satisfying circumstances, continuous brooding over the privations and losses of the 'English War' inevitably became a less important element of nationalist propaganda.

There were also other significant factors causing the decline in importance of public memory of the South African War. One, commencing in the early 1950s, was the increasing tendency of the new ruling nationalist elite to extend the frontiers of war commemoration. Both World Wars (previously remote from official republican Afrikaner culture) were now to be commemorated by Afrikaner nationalism; this essentially political dedication even saw the erection of a new monument on the Somme in 1952.[22] More 'universal' than the South African War, these wars encoded a particular commemorative intention for prime ministers like D.F. Malan and J.G. Strydom: to affirm the national sacrifice and legitimate aspirations for freedom of a European South Africa, inscribed within the free and democratic order of what Malan called a 'timeless' tradition of just wars to preserve Western civilization.[23]

A second, later factor was the long-awaited satisfaction at the achievement of a national South African Republic and its withdrawal from the British Commonwealth at the beginning of the 1960s: at least in spirit, the old Boer

Republics could waltz again. Another was the need to shuffle English South Africans into a more companionable and inclusive pan-white supremacist nationalism, in which emergent Afrikaner industrial and financial classes came to share interests with an established English elite. This duly produced a thaw in war sentiment by the 1960s, with a decreasing tendency for staunch Afrikaners to crudely associate English-speakers with the old brutishness of imperial conquest. Whereas in the earlier 1900s the anti-British cause embodied by urban 'Boer' war monuments had caused acrimony and division between Afrikaners and English whites in cities like Johannesburg, by the mid-1960s some larger annual remembrance services had become communal, jointly organized and attended by armed services representatives and members of both English and Afrikaner cultural bodies and war veterans' associations.[24] Still, some residual Afrikaner war sentiment had a knack of breaking through with disconcerting force over the question of monuments to the colonial English war-dead, with English newspapers like Johannesburg's *The Star* and the *Natal Witness* rounding on the Afrikaner press for disagreeable diatribes against any honouring of the old enemy. What they had in mind were individuals like J.J.F. du Toit, chairman in 1964 of an umbrella committee of Afrikaans cultural groupings, and not one for letting bygones be bygones. In August, he came out against the very idea of a combined, transcendent commemoration of the fallen, declaring that to 'commemorate men who died for a British imperialist interest' along with patriots 'who died in defence of the Boer republics' was akin to burying the nineteenth-century Transvaal President, Paul Kruger, and the capitalist imperialist, Cecil Rhodes, 'in one coffin'.[25]

By the early 1970s, such minor-key memory of the war as an ideological totem had largely migrated to more rural and conservative Afrikaner social constituencies, including a khaki-shirt, nationalist ultra-right which continued to maintain a communitarian, '*volk*' ethos through pistol clubs and rifle associations, paramilitary youth camps and regular commando drill. The movement of opinion was gradual, almost imperceptible through the late 1960s, and took wing in response to crucial shifts within the body walls of a nationalist Afrikaner '*volk*'. As Afrikaner capitalism became increasingly less statist and more driven by market economy imperatives, so heavily indebted groupings (like smaller farmers in the Orange Free State and Transvaal) or statutorily protected urban manual workers began to feel the pinch of a capitalist modernization. Their interests no longer fully guaranteed by National Party political patronage and redistributive subsidies, more marginal, insecure and conservative Afrikaner constituencies gravitated towards more extreme and purist Afrikaner political parties and populist movements. Mostly dormant through the fattening 1950s and 1960s, classic 'Boer' traditions of resistance and a warlike identity were now re-invented and re-imagined, given a new lease of life by men and women who, by the late 1970s, had been elbowed out of the core modernizing Afrikaner nationalist constituency.

In a host of small ways, a far-right populism began to course through the 1980s, bearing the old imprint of combative memories of the War of Freedom:

shared by people in organized rural circles, by those attached to fundamentalist Protestant religious tributaries, by scattered clusters of rural army reservists attached to a 'Boer commando' identity, and by the right-wing Conservative Party leadership, this was a hostile and isolationist message. Attempts to reform the economic and political system of apartheid were seen as a 'craven' capitulation to Anglicized and other enemies of the Afrikaner nation: had 'true' memories grown so short that Lord Kitchener had been forgotten? Had Afrikanerdom's national memory forgotten the scarring of 'the Boer nation' by war, and its lesson of unstinting republican vigilance? Oppressive war had 'made' an Afrikaner people, and this was a defining legacy which could not be displaced.[26]

In this sense, the ground so well watered by the Afrikaner nationalist project in earlier decades continued to bear fruit as war memory continued to be invoked and manipulated. Indeed, even when well past its zenith as a generalized mobilizing cult of commemoration, the embers of 1902 could be stoked by moments of crisis, most especially by the combustible terminal turn of apartheid by the 1990s. In a striking 1993 observation, the quixotic Afrikaner historian, Floors van Jaarsveld, concluded that 'Afrikanerdom has suffered two great defeats in its history: the first at the beginning of the twentieth century in the war with Britain, which inflicted a military defeat on it, and the second at the end of the twentieth century – a political defeat at the hands of Africa.'[27] Renewed elements of a *bittereinder* resistance aesthetic took shape around this view, as British imperial domination became African domination, with African National Congress President Nelson Mandela its odious High Commissioner; for its ultra-right, Afrikanerdom once again faced the prospect of complete deracination in a unitary mongrel state. 'Again', intoned Ferdi Hartzenberg of the Conservative Party, 'dark days have come to our people.'[28] In turn, to his right, there was a rekindled yearning for a Transvaal and Orange Free State *Boerestaat* or *Volkstaat* (People's State) to secure the endangered bloodlines of ethnic self-determination. In one memorable burst of unvarnished Anglophobia, Robert van Tonder of the Boerestaat Party called for authentic 'Boers' to consecrate a pure 'Boerestaat' as posthumous revenge upon Queen Victoria, who had seen to it that 'our Boer republics were crushed in 1902 and fourteen other "peoples" were forced to live with us in one state'. This had amounted to a 'declaration of civil war' but then, as now, 'civil war' could not defeat Afrikaners, as, 'after all, it was the Afrikaners who invented it'.[29] In Van Tonder's historical universe the South African War boiled down to an attempt to exterminate a fellow-people, a 'civil war' in which Afrikaners had survived the fire through their imagined creation of a guerrilla 'people's war'.[30]

Equally noteworthy were the pricks felt during the first visit of the Windsors since the 1940s, the 1995 tour of Elizabeth II in the wake of South Africa's recent Commonwealth re-entry, and its transition to majority rule. Reviving old war words and images, an incensed Boerestaat Party formally declared the Queen unwelcome in an imagined 'Boerestaat of Transvaal and the Free State'. From a 'dynasty of conquerors', she was 'the great granddaughter of a cruel queen', whose invading armies had not only 'destroyed our Boers' freedom', but had also committed 'the infamous holocaust in which a sixth of our people were murdered in concentration camps'.[31] Other Afrikaner responses were less apoplectic, if still chastening, even when from non-nationalist quarters. Thus, blustery ANC parliamentarian Carl Niehaus called upon the Queen to observe a more inclusive act of South African war remembrance by commemorating Boer dead. 'If she is going to lay wreaths at World War II and World War I graves here', he declared, 'she ought also to lay wreaths on the graves of the tens of thousands of women and children who died in the camps.'[32]

Through the 1990s, a dissenting Boer War flank engaged on a number of fronts, not so much against the old enemy of rampant Englishness, but against a dawning age of racial equality and majority rule, and against now-despised National Party leaders for yielding to a tide of racial and cultural cosmopolitanism. At one tragicomic pole there was, for instance, the armed seizure of Pretoria's nineteenth-century Schanskop fort and military museum, a symbolic place in thrall to Boer commando heritage and its lineage of highly militarized republicanism. Under the Vierkleur or flag of the old Transvaal Republic, a militant knot of right-wingers protested against multi-racial national political negotiations, their sole purpose seen as a selling-off of the 'sacred' assets of Afrikaner sovereignty. After their arrest and conviction for illegal armed occupation, their khaki-clad leader, Willem Ratte, wrote furiously from gaol to contest his English press depiction as some right-wing anachronism. What outweighed this was a First Anglo-Boer War antecedent. In his torrential and powerful manifesto, Ratte insisted:

> Were the Boers of 1880 called right-wingers, for resisting the imperialist British occupation? Then, as now, you had an alien regime lording it in Pretoria over our people, whose gutless president had betrayed and handed over his sovereign state. Then, as now, the new (neo) colonialist administration pretended to be God's gift to the supposedly 'dirty and dumb Dutchmen', and tried its best to smear the pro-independence party as only a few backward 'Don Quixotes tilting at windmills'. Our struggle has nothing to do with right or left ... this being incidental, like religion in the Irish-British conflict, but everything to do with a nation having an inherent right to be free.[33]

At another eccentric pole, a Kruger Day commemoration in October 1992 saw a sliver of rough-hewn Englishness regain some lustre in right-wing nationalist Afrikaner life. In a cameo of bonding between Afrikaner and English ultra-right interests, indefatigable ex-Rhodesians and new English-speaking adherents of the Conservative Party lumbered into the Vaal River to retrieve rocks from a Boer concentration-camp memorial, reputedly torn down in spite by departing British soldiers in 1903. Under the solemn gaze of a crowd of several thousand, the rock was piled up close to the official Paardekraal camp monument, thereby atoning symbolically for past desecration. Hailing it as 'the greatest conciliatory gesture by English-speaking countrymen since before the Boer Wars', Conservative Party leader Andries Treurnicht announced that 'the time has certainly come for all English-speaking patriots to let bygones be bygones, and to join hands with the Boer to resist the common enemy of black domination'.[34] This, as it happens, was a droll gesture from someone who had spent a good chunk of his political career railing against the perniciousness of English influence in South African life.

In innumerable other ways, fragments of the old imperial connection entered the grain of radical nationalist war remembering in the later 1990s. In April 1998, for example, there was gleeful Afrikaner press appreciation of news from Ireland of opposition in Listowel to the raising of a plaque to commemorate Kitchener's birthplace, because of his ruthless prosecution of the closing stages of the war. One typical correspondent noted that it would be received with 'great satisfaction by Afrikaners in general, but particularly by descendants of the Boers who fought against Kitchener's barbarism'. Several others invoked the obvious analogy, declaring that Kitchener's concentration-camp policy had been the genocidal work of a British 'Hitler' in South Africa; Paardekraal was an Auschwitz, a war crime never to be forgotten. J.A. Marais, 'son of a Boer father' exiled and imprisoned with other captured soldiers on St Helena island, and 'a mother who was interned in the Klerksdorp concentration camp', urged on the campaigning idea of a republican Irish-Afrikaner 'war crimes tribunal for Kitchener' in 1999, as the most meaningful 'commemoration of the outbreak of the Boer War'.[35] It was all caught sublimely by a *Sunday Times* call for Britain's Prime Minister, Tony Blair, to apologize for his country's 'appalling concentration camp record', in the light of public contrition before the Irish for the potato famine, and Indians for the Amritsar massacre. In response, a correspondent cautioned readers against expecting much sentiment, for while there were 'many Irish and Indian voters in the UK', there 'is no Afrikaner vote'.[36]

Such flickerings can probably be seen as being among the last thorny episodes of an antique Afrikaner past, dredging up for a self-conscious and ailing 'Boer' minority the nostalgias, iconography and resentments of a world in steep decline. For isolationist politico-military bodies like the *Wit Wolve* (White Wolves), there was a knowing sense of things sliding away, as some displaced farmers and alienated blue-collar workers clung doggedly to a *volkisch* memory of the war. In a sulphurous style which was on the way to becoming almost formulaic, *Afrikanerweerstandsbeweging* (AWB, Afrikaner Resistance Movement) leader, Eugene Terreblanche, likened black political demands to those of the English *Uitlanders* (foreigners) who had been so unwelcome a blot upon Paul Kruger's Transvaal Republic. Terreblanche growled that not to grant Afrikaner 'freedom' in an autonomous people's state would be 'to play with fire', as British statesmen had discovered in the 1890s. Sceptics were referred to Thomas Pakenham's major 1979 history, *The Boer War*. A redemptive English work of 'truthful' war scholarship which, in his view, gave the Boers their due and more, Terreblanche recommended it as being as admirable as Shakespeare, therefore 'not part of any English conspiracy'.[37]

Within the AWB on the extreme white right, the coming of black majority rule represented 'a second invasion', requiring the invoking of a national Boer state defence. This produced overstatement: a heightening of the inclination to clip about under arms on horseback in a steely display of the Boer commando tradition of bearded masculinity, and a gingering-up of the key tribal bonds of kinship and language. Because of the danger to 'authentic' Afrikaners from untrustworthy news editors, slippery National Party politicians and meddling American imperialists (Washington having now replaced London), the call was for a 'return' of 'the generals', for nationalist Afrikaners had historically always looked to their toughest and sharpest military men, the de la Reys and de Wets, for salvation. Something of that vindicated old martial spirit was now again urgently required.[38]

Just how much things turned on emotive memory of the ashes of 1899–1902 could be glimpsed in puffy Conservative Party talk of an imminent *Derde Vryheidsoorlog* (Third War of Freedom), in the cocky use by the AWB and Boerestaat Party of the old Transvaal republican flag and its admonitory anthem, *Ken u die Volk* (Do You know the People), in sniping gunshots at the British Embassy in Pretoria in 1990, and in a bomb attack on Melrose House, where the Boer leadership had signed its surrender in May 1902. Back where it all started, radical ideologues of national purity demanded the reclamation and renewal of the old Boer Republics. This was not something to be recreated, for historically they had always been in full, legitimate existence, based on a Boer occupation which had 'enjoyed internationally recognized independence until 1902'.[39]

In a further poignant echo from the early 1900s, there was denunciation of reformist National Party leadership as *verraiers* (traitors) to the cause of Afrikanerdom for having submitted to the ANC. Fringe politicians to their right went for the government as 'a lot of traitors', guilty of appalling 'treachery' and 'acts of treason against its own people'.[40] The base appeal of the notion of 'national treachery' was to a form of atavism, a bitter replay of the closing moments of the Anglo-Boer War which had produced a full-blown split between Boer *hensoppers* ('hands-ups'), opting for peace rather than going on to the wire, and *bittereinders*, some of them stiff-necked women, who wanted to hold out against imposed peace terms. In other words, here again were some leading Boer 'generals' turning traitor through premature capitulation or spineless surrender. The chief modern incarnation of this yellow behaviour was, of course, President F.W. de Klerk, who in 1997 tellingly let slip in London that he had bowed to the inevitable necessity to 'surrender the right to national sovereignty'.[41] In terms of the conventional nationalist content and discourse of Afrikaner history, the very term 'surrender' has long been peculiarly pregnant in partisan meaning; through his treacherous submission, de Klerk had in effect become guilty of complicity in the monstrous crime of war against the freedom of his own nation.[42] Equally, this seizure of the didactic precedent of 1899–1902 war memory may well be seen as perhaps the final historical moment when the fires of traditionalist nationalist bitterness and self-righteousness could still be fanned by the historical stimulus of the moral relevance of the South African War.

The essential way in which the South African War has been remembered is sectional – Afrikaner nationalist, political and polarizing; and, over the course of the present century, that war commemoration as political vision has passed through a number of phases, variously inflected by changing material and psychological contexts. This present consideration of war remembrance arises in an increasingly contradictory climate of commemoration. On the one hand, the current centenary of the war is providing for an obvious surge of war remembrance and deliberation over its meaning. On the other, the centenary coincides with the obvious displacement of Afrikaner nationalist power by a post-apartheid political order. At the time of writing, it is perhaps never more obvious that shifting historical contexts affect rooted ways in which war is popularly remembered and commemorated. As external repositories of public memory, war memorials and monuments now face a future in which they are divested of symbolic power accrued in the nationalist apartheid era; they may, as some scholars have recently argued, be transmuting into open 'texts', mnemonic reminders of how a particular South African community (no longer special, nor privileged) viewed the place of a defining war in its history.[43]

Nor is this the only issue worth consideration. In a changing society directed by the liberation aristocracy of a black African nationalism, busily creating its own national mythologies of martyrs and liberators, what will happen to national memory of a historical Anglo-Boer War, in a context in which the stance of the country's black majority has long been to dissociate from a perceived 'white man's war'? One possibility is that the war may come to be remembered as an increasingly remote episode from a vanished European imperial age. As the Johannesburg *Sunday Independent* put it on the centenary of the 1895 Jameson Raid on the Transvaal, that madcap moment of imperial buccaneering which Winston Churchill considered to be the real start of the Anglo-Boer War: 'with this tumultuous century drawing to a close, hindsight puts the do-or-die battle between Afrikaners and English-speakers into its proper, smaller context. White men were never going to win indefinite control of this African continent.'[44]

There can be little doubt that perpetuation of the war in the public spaces of historical memory will wane, a revivalist centenary notwithstanding. It is even conceivable that with the disappearance of a distinctively constituted Afrikaner 'political nation', the war may come to have little meaning at all as core national political heritage. Even the key arenas of war are already not what they were, unlike say, the Somme or Flanders. The Orange Free State, seminal to freehold Boer republican identity since the mid-nineteenth century, has had to fall in with African nationalist times, losing its revered Dutch 'Orange' stamp in 1994 to become the Free State, and shedding Afrikaans in favour of English as its sole official language three years later. Its fighting sister Transvaal has also been administratively dismembered, retitled and demarcated as new territorial entities with chunky non-colonial names like Gauteng and Mpumalanga. Sanitized of their Boer republican identity, on a simple map-reading the old warring territories have become the provinces-from-nowhere a century later.

At the same time, other impulses seek not so much to deflate war remembrance as to recompose it, through detaching its Boer commando and concentration camp past from conservative nationalist Afrikaner history, and imbuing it with new national patriotic meanings. In this, the purpose of imagining the war in a new way is to construct consensus. Accordingly, an ANC judge of the South African Constitutional Court has reclaimed idealism and the universal story of human freedom as the shared meaning of the South African War. Sliding together the language of the present with that of the past, Judge Albie Sachs asserted his personal 'pride in the heroic struggle of the Boer fighters in the history of the world and in our history'. Any history of a liberal human rights culture had to 'take into account the fate of the women and children in the concentration camps. So much of Afrikaans history is part of the struggle for freedom. *Vryheid* (Freedom) has real resonance and meaning.'[45] Assimilation of the war to a more inclusive new South Africanism is a model illustration of how those pursuing a present political agenda seek instinctively to re-invent or re-imagine the past, so that the Anglo-Boer struggle is to be remembered through a new lens of understanding as one of a series of South African Wars, in which 'virtually all ethnic groups' played a shared role, thereby forging 'the common historic destiny of all South Africans'.[46]

Yet, outside elite cultural and academic missionary circles, the South African Anglo-Boer conflict remains what it has always been, a sectional war, and one not *felt* to be a shared legacy, let alone a shared tragedy. In any final analysis, it is hard to see how it could ever be otherwise in a colonial country of such limited commonalities. The South African War is not, and has never been, an American Civil War, a 'never-to-be-forgotten moment' in 'the collective consciousness that makes Americans American'.[47] It produced no Stephen Crane to mythologize a tragic war as a national rite of passage to 'manhood' for both the Blue and the Gray. Instead, it produced an ideological memory as far removed from a single all-encompassing statehood as could be imagined. In our own time, it is this political heritage, sentimentalized or institutionalized, which is now fast trickling away. In the longer run, the war may continue its present shift away from a politically inscribed commemoration of a sense of self, to more anodyne remembrance as historical heritage, or as a site of pilgrimage for British and Commonwealth regiments to honour the sacrifice of their fallen. Already, able-bodied devotees of General Christiaan de Wet are taking guided hikes along his legendary flying escape route in the Western Transvaal, and there are centenary plans for the construction of concentration-camp trails for overseas visitors. As war tourism, South Africa's Anglo-Boer War may end up providing a 'heritage' experience of tripping along an African version of the Ho Chi Minh Trail. That would be a long way from the struggle of Afrikaner generals to resolve the fate of a white nation, but it would be another turn in the cultural negotiation of South African War commemoration.

Notes

1 Quoted in Jonah Raskin, *The Mythology of Imperialism* (New York: Delta, 1971), 71.

2 Deryck Schreuder, 'The power of colonial nationalism in shaping post-colonial South Africa', in John Eddy and Deryck Schreuder (eds) *The Rise of Colonial Nationalism, 1880–1914* (Sydney: Unwin, 1988), 55.

3 Alan Lester, *From Colonization to Democracy: A New Historical Geography of South Africa* (London: I.B. Tauris, 1998), 68.

4 G.D. Scholtz, *Die Ontwikkeling van die Politieke Denke van die Afrikaner, Deel vi, 1910–1924* (The Development of Afrikaner Political Thought, Pt vi, 1910–1924) (Pretoria: Perskor, 1979), 215–36.

5 Isabel Hofmeyr, 'Building a nation from words: Afrikaans language, literature and ethnic identity, 1902–1924', in Shula Marks and Stanley Trapido (eds) *The Politics of Race, Class and Nationalism in Twentieth Century South Africa* (London: Longman, 1987), 109.

6 Albert Grundlingh, 'The National Women's Monument: the making and mutation of meaning in Afrikaner memory of the South African War', unpublished conference paper presented at 'Rethinking the South African War 1899–1902', University of South Africa Library Conference, Pretoria, August 1998.

7 Elsabé Brink and Sue Krige, 'Remapping and remembering the South African War in Johannesburg and Pretoria', unpublished conference paper presented at 'Rethinking the South African War', Pretoria, 1998.

8 Albert Grundlingh, 'War, wordsmiths and the "volk": Afrikaans historical writing on the Anglo-Boer War of 1899–1902 and the war in Afrikaner historical consciousness, 1902–1990', in E. Lehmann and E. Reckwitz (eds) *Mfecane to Boer War* (Essen, Germany, 1992), 52.

9 *Die Huisgenoot* (Pretoria, Cape Town), 19 Aug. 1924; 30 Mar. 1932.

10 Sara Raal, *Met die Boere in die Veld* (With the Boers in the Veld) (Kaapstad: Nasionale Pers, 1938); 'Mikro', *Die Ruiter in Die Nag* (The Rider in the Night) (Kaapstad: Nasionale Pers, 1941).

11 Grundlingh, 'War, wordsmiths', 45–6.

12 Kobus du Pisani and Louis Grundlingh, ' "Volkshelde": Afrikaner nationalist mobilization and representations of the Boer warrior', unpublished conference paper 'Rethinking the South African War', Pretoria, 1998.

13 J.C.G. Kemp, *Vir Vryheid en vir Reg* (For Freedom and Justice) (Kaapstad: Nasionale Pers, 1941), 86.

14 *Die Huisgenoot*, 18 Aug. 1923; 25 Mar. 1935.

15 Ibid., 14 July 1937; 6 Dec. 1940.

16 J.H. Breytenbach, *Die Betekenis van die Tweede Vryheidsoorlog* (The Significance of the Second War of Freedom) (Johannesburg: F.A.K., 1949).

17 Albert Grundlingh, 'Politics, principles and problems of a profession: Afrikaner historians and their discipline, c.1920–c.1965', *Perspectives in Education* (Johannesburg: University of the Witwatersrand) 12.1 (1990/1), 6–14.

18 T. Dunbar Moodie, *The Rise of Afrikanerdom: Power, Apartheid, and the Afrikaner Civil Religion* (Berkeley: University of California Press, 1975); Lester, *From Colonization to Democracy*, 98–100.

19 Sandra Scott Swart, 'The rebels of 1914: masculinity, republicanism and the social forces that shaped the Boer Rebellion' (MA dissertation, University of Natal, 1997), 143–6.

20 *Die Burger* (Cape Town), 21 Feb. 1937, 19 Aug. 1940. Signs of such fractiousness continued into the 1960s: *Die Transvaler* (Pretoria), 14 July 1963; *The Star* (Johannesburg), 19 Aug. 1964; Brink and Krige, 'South African War'.

21 *Natal Witness* (Durban), 14 Aug. 1939.

22 *Delville Wood Commemorative Brochure* (Johannesburg), 1952; this presented South Africa as a European nation, with a war record of willing sacrifice for liberty through all its modern conflicts.

23 *Die Burger*, 17 Nov. 1952.

24 *The Star*, 27 July 1965; *Rand Daily Mail* (Johannesburg), 2 Sept. 1966.
25 *Rand Daily Mail*, 20 Aug. 1964.
26 *Die Burger*, 22 June 1989.
27 *Beeld* (Johannesburg), 12 Feb. 1993.
28 *Die Burger*, 17 Dec. 1993; *Rapport* (Johannesburg), 20 Dec. 1993.
29 *Beeld*, 18 Jan. 1993; *Aida Parker Newsletter* no. 163 (1993), 2. This is a far-right, Transvaal-based opinion news-sheet of the 1980s and early 1990s, rumoured to have been established with conservative US foundation funds.
30 *Die Burger*, 21 Mar. 1995.
31 *Cape Times*, 21 Mar. 1995.
32 *Cape Argus*, 23 Mar. 1995.
33 *Sunday Independent* (Johannesburg), 31 Dec. 1995.
34 *Weekly Mail* (Johannesburg), 16–22 Oct. 1992.
35 *Rapport*, 12 Apr. 1998; *Sunday Times* (Johannesburg), 5, 19 Apr. 1998.
36 *Sunday Times*, 26 Apr. 1998.
37 *Sunday Independent*, 21 Dec. 1995; the reference is to Thomas Pakenham, *The Boer War* (London: Weidenfeld, 1979).
38 *Die Volksblad* (Pretoria), 6 Oct. 1994.
39 Johann van Rooyen, *Hard Right: The New White Power in South Africa* (London: I.B. Tauris, 1995), 43.
40 *Patriot* (Pretoria), 5 Apr. 1991; *House of Assembly Debates*, col. 106, 6 Feb. 1990.
41 *Cape Times*, 19 Feb. 1997.
42 *Patriot*, 7 June 1991.
43 Brink and Krige, 'South African War'.
44 *Sunday Independent*, 31 Dec. 1995.
45 Cited in Grundlingh, 'War, wordsmiths', 54.
46 *Cape Argus*, 1 Oct. 1996.
47 Noah Andre Trudeau, *Out of the Storm: The End of the Civil War, April–June 1865* (Baton Rouge: University of Louisiana Press, 1995), 422.

4 National narratives, war commemoration and racial exclusion in a settler society

The Australian case

Ann Curthoys

On 15 February 1965, a ragged line of some thirty students from the University of Sydney found themselves outside the Returned Servicemen's League (RSL) club in the tiny dusty town of Walgett, in far north-west New South Wales (I was one of them – oh, so long ago!). They were protesting against the refusal of the club to allow membership or entry to Aboriginal people (locally the Murris), including ex-servicemen. In the hot sun, from noon to 7 p.m., the students, only one of them – Charles Perkins – Aboriginal though not from that part of the country, held up placards saying 'Acceptance, Not Segregation', 'End Colour Bar', 'Bullets did not Discriminate', 'Walgett – Australia's Disgrace', 'Why Whites Only?', 'Educate the Whites', and 'Good Enough for Tobruk – why not Walgett RSL?' Local townspeople, black and white, soon gathered, remarking on this strange event, some jeering, many just watching. Never before, almost certainly, had people held up placards of protest in Walgett, and definitely not outside the RSL Club. From time to time heated three-cornered discussions broke out, between Murris, white townsfolk and students. The atmosphere was very excited, the levels of hostility high. Later that evening, when the students left town, the bus was forced off the road by a truck driven by some angry young white men. This violence brought the student protest to national media attention the next day, attention which stayed focused on it for the rest of its two-week tour against racial discrimination in country towns.[1]

This incident is a rich one for historical ethnography and detailed textual and cultural analysis. Here we see articulated several conflicts, first between indigenous Australians and a segregationist white settler community, and then between that community and a relatively new post-war urban internationalist desire for racial equality. The site of protest is significant, taking place as it did outside a RSL club, one of hundreds around the country providing a social centre for ex-servicemen, a space for a drink, a meal and entertainment. Later protests would be held outside swimming pools and picture theatres, also significant sites of segregation in these small country towns where black and white lived close to, but rarely with, one another. But this protest was the first and it raises many questions. Why was it outside an ex-servicemen's club? Why were indigenous ex-servicemen excluded? Why did the excluded men care?

In this chapter, taking Australia as an example of wider phenomena in the histories of colonial-settler societies, I use this incident to examine the ways in which racial difference, segregation and conflict continually threaten the nation. Like so many others, white Australians yearn for an inclusive national narrative, while at the same time practising exclusions on the basis of race. Such racial exclusions shadow and undermine national narratives of harmony, unity and common purpose, revealing their fragility and provisionality. Such narratives emerge not from a secure centre but from a majority formation and collectivity (to which I myself belong) far from its European origins, whose own sense of relation to place, nation and history is deeply fractured and constantly contested. Racial exclusion, that is to say, illuminates the contradictions and otherness *within* white Anglo-Celtic Australian culture, a restless, anxious, complex culture of white nomads who claim to be settlers in and on an antipodean land not their own.[2]

* * *

This otherness, this unsettlement over questions of race and identity, is revealed most clearly in war commemoration. For it is in the commemoration of war that Australian popular culture finds its most profound sense of nationhood, as Stephen Garton also notes in this volume. War is commemorated in a distinctive way, through the observance of a national holiday, Anzac Day, on 25 April each year. Anzac stands for Australian and New Zealand Army Corps, and Anzac Day remembers neither the end of war nor a major victory, but rather the unsuccessful landing in 1915, quite early in the First World War, of Australian, along with British, New Zealand, French and other troops on the Dardanelles peninsula, in an ill-fated attempt to open the way for an attack on Constantinople. The Australian and New Zealand soldiers came to be known as the Anzacs, though in Australian usages of the term the New Zealand part tends to be elided and half-forgotten. This landing began a military engagement which lasted for months, with much loss of life, before the troops withdrew. It is now held, in Australian official and popular discourse alike, to symbolize courage and sacrifice, and the emergence of the Australian nation, formed only fourteen years earlier, as an independent entity on the world stage. The narrative of Anzac is powerful as a story of innocence betrayed, a story in which the fittest young men of the young nation gave their all for their country and Empire, only to be shot down cruelly, the fault not so much of their Turkish enemy as of the brutal idiocy, the criminal foolishness, of the British command who sent them there. The innocent young white men of the new nation proved themselves in heroism in the face of forces outside their control, in courage amidst pain, death and suffering.[3]

The importance of this story to the Australian imaginary cannot be overestimated, resting as it does on the collective memory of the trauma of the Australian experience of the First World War, marked by long absences of soldiers far away in Europe and North Africa, and extremely high casualty rates directly affecting the majority of families. What is interesting about this

Australian narrative of nation, as some have noted, is that it is based on the commemoration of a massive defeat, though that defeat, to be sure, resides within a larger narrative of ultimate allied victory.[4] The Anzac archive of story and legend, allegory and myth, started with Charles Bean, a wartime journalist whose eyewitness reports provided the basis, the original story, on which all subsequent and multiplying narratives have been based. Anzac stood, he wrote, 'for reckless valour in a good cause, for enterprise, resourcefulness, fidelity, comradeship, and endurance that will never own defeat'.[5] Since Bean wrote his first newspaper reports from the front, the Anzac story has been endlessly retold, changing its inflections over the years. One of the best-known versions is historian Bill Gammage's eloquent 1974 evocation, *The Broken Years: Australian Soldiers in the Great War*. Gammage makes the important point that the Australians who landed at Gallipoli felt both a traditional loyalty to the British Empire and a newfound Australian nationalism, seeing the two as quite compatible: 'They felt themselves to be not merely sons of the Empire, but Australians, owning no masters.'[6] Another influential narration is Peter Weir's film *Gallipoli* (1981), starring Mel Gibson, in which the tale of the pathos of innocence betrayed and bravery in defeat, is inscribed again. Public memorialization of the Gallipoli story multiplied in the wake of the service held at Gallipoli itself on its seventy-fifth anniversary in 1990, and in recent years there has been heightened interest in the last survivors of the Gallipoli campaign.

In the story of Anzac lies the emotional locus of Australian narratives of nation. Except for the Bicentennial celebrations in 1988, Australia Day – held on 26 January to mark the anniversary of the first convict settlement at Sydney Cove – has far less meaning to most Australians than Anzac Day. Interestingly, there is no holiday commemorating the actual political formation of the Australian nation, through the federation on 1 January 1901 of the separate Australian colonies and thereby the acquisition of a national constitution. Where federation, an event that might at first glance seem a more appropriate occasion for the celebration of nationhood, is seen as a rather boring and technical event, the carnage at Gallipoli is still remembered. Nationally, the process of memorialization is fostered both officially, by prime ministers, and unofficially, by the many thousands who march, pray and watch on Anzac Day, and by the thousands of young backpackers who now visit the site of Australian graves at Gallipoli each year. Since the 1920s the innumerable war memorials found in practically every country town, and the major war memorials in the capital cities, have become the focus of Anzac Day ceremonies all over the country.[7] In Walgett, where my story began and will shortly return, the war memorial is at the centre of the town, at the intersection of its two main streets. Location of practically everything in the town is marked by its relation to 'the Monument'.

In drawing on a sense of victimhood, of courage in defeat, the Gallipoli legend draws upon a more general Australian popular historical mythology. Beginning with their origins in a British colony of punishment and exile, and continuing with present-day economic restructuring and globalization, white Australians have traditionally regarded themselves as victims of natural and political forces outside their control, whether it be a hostile nature and ecology, rapid economic change and crisis, or powerful nations overseas.[8] This rhetoric of victimhood has an almost biblical quality, recalling the story of the Exodus. The Israelites went to Egypt to escape famine, subsequently grew to a great multitude and were enslaved by a later, unsympathetic, Pharaoh. Under the leadership of Moses they fled from their oppressors; after wandering through the desert, they came to the Promised Land, the land of Canaan, which they conquered and occupied. Settler societies like Australia, it seems, all develop their own versions of the Exodus story, in order to reposition themselves as toilers and victims, rather than invaders or oppressors. Several scholars have already noted the way reworkings of the story of Exodus provide the foundations for both American and Israeli national historical narratives. In the American version, the pilgrims left Britain for America, a new Promised Land reserved by God for his new chosen people, liberating themselves from the tyranny of the British Pharaoh. Identifying New England as the New Canaan, the Puritans saw the worldly enterprise of colonization as a mission to restore mankind, an idea which became the foundation of an enduring American dream. [9]

In the Australian version, there is also a story of leaving the Old World to make good in the new, the migrant's chance to find salvation and redemption, and to start anew. But here it is accompanied and shadowed by another biblical story, the story of the Fall and expulsion from Eden, the leaving of Britain at Britain's will, and the finding in the colonies of a hell on earth.[10] Australian victimological narratives and sensibility rest on the one hand on notions of expulsion and exile from mother England, and on the other on the optimism of a new society free from the oppressions of the Old World. We can see this uneasy mixture in the Australian *Catholic Leader* of 5 June 1988 to which Andrew Lattas draws our attention:

> For some of us Australians, our forefathers forsook the green fields and teeming cities of Europe and beyond, setting out like Abraham and Moses to find a promised land. They knew Exodus and Exile, condemnation and chains, desert wanderings and struggle, inequality and injustice, the crucible of tragedy and suffering, the childbirth of a new people.[11]

These national narratives begin with the convicts, the men and women expelled from their homeland, and forced to live and labour in a hostile distant place, and continue in narratives of subsequent settlement, exploration and pioneering, of white settlers locked in a harsh and bitter struggle to farm the often dry and hostile land. They reach their pinnacle in the narrative of Gallipoli, the sending of young men as expendable fodder for the Empire.

These stories of adversity and struggle have profound allegorical consequences for Australian race relations, both past and present. In the biblical story, the original Exodus, the necessity to flee bondage under the Pharaoh becomes the moral justification for the later conquest of the Canaanites. The earlier victimhood warrants the later aggression. And so it is in white Australian historical mythology. The convicts who were forced and the immigrants who chose to flee their economically disrupted home in Europe deserve our admiration for their struggles and suffering, not our criticism for their acts of colonization. Indeed, in earlier forms of white historical mythology, violent conflict between settlers and indigenes scarcely happened at all – the latter simply faded away, 'doomed to extinction' as the result of some mysterious force of colonization.[12] Typical of nineteenth- and twentieth-century views was that of writer George French Angas, in 1865:

> As British colonisation is gradually spreading over the Australian continent, so the primitive inhabitants of the soil are fast dying out, and, in many places, they will, ere long, have totally disappeared. Wherever the white man locates himself, so surely do the inferior races give way: as the Red Indian and the Bushman have vanished before the colonists of North America and the Cape, so will the degraded nomads of Australia perish in like manner.[13]

Since a rapid decline in indigenous populations could be observed in all colonies, it followed that the settlers in Australia could hardly be held particularly or specifically responsible – it was simply a sad consequence of settlement itself.

These ideas of an impersonal colonizing process have been replaced in recent years by some acknowledgement of settler agency and the brutality of the settlement process. Aboriginal activists and spokespeople have consistently drawn attention to a history of land-taking and brutality, and have insisted on describing 'settlement', suggesting something relatively peaceful and benign, as 'invasion', something unwanted, violent and destructive. In addition, the work of historians has drawn to public attention a history of violent dispossession. Influenced by such activists and historians, Prime Minister Paul Keating, in his now famous 'Redfern speech' in December 1992, spoke openly of a history of dispossession, murder and brutality.[14] Several years later, in 1996, the television series *Frontier* popularized the findings of the historians among a large audience.[15] Such acknowledgements, however, while sincerely made by some, continue to be denied by others. Talk of invasion, dispossession, massacres, brutality and the like is regarded as abhorrent, and as a distortion of Australian history. Most notable among the many who question the notion that Australia has a brutal colonial past is Prime Minister John Howard, who said in October 1996, after six months in office, that he sympathized fundamentally with Australians (meaning non-indigenous Australians) who are insulted by such negative characterizations of their past.[16]

* * *

Australian colonial and later national identity was thus in its deepest sense built on the exclusion of indigenous peoples from foundational historical narratives. In a political sense, the Australian nation was founded on their exclusion from the national sphere of government. Even with the processes of inclusion that have accelerated since the Referendum of 1967, which gave the federal government powers to legislate in relation to Aboriginal people and also meant they were included for the first time in the national population census, popular white historical mythology remains largely a story of white settlement, in which the white settlers rather than the indigenous peoples they displaced are both the heroes and the victims.

These white national narratives are, however, not altogether straightforward. Like such narratives everywhere, they have their shadow, their dangerous supplement, which lies inside the story, threatening to undo them. The danger comes from the existence of Aboriginal servicemen in both World Wars. For the association of white Australia with the Anzac legend to retain its narrative and symbolic power, those who fought in the First World War ought to be white Australians, and this is in fact how they are largely now remembered. Aboriginal people at this time were considered to be outside the nation, and were governed by the various states, with a range of techniques of surveillance, 'protection', economic exploitation and control. There were restrictions on voting, marriage, drinking, movement, employment, education and freedom of religion. Segregation was commonplace in many parts of the country. Yet, despite their lack of the most basic legal, civic, political and human rights, at least 300 Aboriginal people enlisted.[17] There was probably a mixture of motives: the promise of a steady income; a desire to better themselves through contribution to the war effort; a pride in their masculinity (most were men) and skills; perhaps, a belief shared with white Australians that the country's future depended on British victory and a German defeat.

If Aboriginal servicemen hoped for a better deal after the war, they were soon disappointed. When they returned in 1919, they found they were excluded from the Soldier Settlement scheme, designed to help returned soldiers settle on the land, and, indeed, the scheme accelerated the revocation of existing Aboriginal reserve lands for soldier settlers.[18] Aboriginal returned servicemen in northern New South Wales protested against the conditions they faced, such as their continued exclusion from the facilities of the town, and petitioned for civic rights, supported by the local branches of the Returned Sailors' and Soldiers' Imperial League of Australia, the forerunner of the RSL.[19] Ex-servicemen might be granted exemption certificates from the various restrictive Acts then governing Aboriginal people in the various states, but this was not automatic.[20]

This same pattern was repeated, but on a much larger scale, during and after the Second World War. Governments again wanted an all-European fighting force, the Defence Act specifically exempting persons who were 'not substantially of European origin or descent' from call-up for war service and from compulsory training, and Service Regulations also barred them from serving voluntarily.[21] Yet this policy was soon reviewed, and recruitment officers were allowed discretion in dealing with Aboriginal applicants. As the threat from Japan and therefore the need for servicemen and women increased after December 1941, the regulations were interpreted ever more liberally to allow Aboriginal recruitment. The approach to Aboriginal recruitment remained for the rest of the war inconsistent and changeable, indicating a deep uncertainty as to whether

Aboriginal people were properly Australian or were akin to enemy aliens. There were occasional fears that Aboriginal people might side with the Japanese, given Japanese appeals to anti-colonial sentiment. And indeed white Australians had very good reason to fear that they might not hold or deserve the allegiance of Aboriginal people.[22] Amongst Aboriginal people, there was some hesitancy about enlisting in defence of a country which had served them so ill. When the question of Aboriginal eligibility to enlist arose in the early months of the war, some organizations such as the Australian Aborigines' League wanted, as a precondition, the removal of all legal and other disabilities. Yet Aboriginal people were not sufficiently numerous and therefore critical to the war effort for them to be able to bargain for such benefits. Such political activism did not inhibit recruitment, and some leaders such as Jack Patten enlisted in the army to prove their patriotism and to help prevent the Aboriginal political movement from being discredited.[23] Many simply hoped that war service itself would strengthen their case, individually and collectively, for full citizenship after the war.

Despite the discriminatory laws and regulations, almost 3,000 Aboriginal and Torres Strait Islanders did serve, both in units based in northern Australia, and in ordinary army units on a variety of fronts overseas.[24] Though Aboriginal soldiers were mainly men, some Aboriginal women also served in the armed forces and in the Women's Land Army.[25] Furthermore, the strategic location of many mission stations in the north led to their Aboriginal residents, perhaps as many as 3,000, supporting the war effort in various ways.[26]

The war did destabilize the existing pattern of race relations to some extent. The high demand for labour during the war, for example, meant many Aboriginal people could walk away from the protectionist and regulatory Protection and Welfare Boards and live back in their own country where work was available.[27] It also brought white soldiers from the urban south and Aboriginal people in the isolated north into contact for the first time. Robert Hall, the historian of Aboriginal involvement in the Second World War, suggests that these white soldiers, having no long-term stake in their own economic survival in the Northern Territory, and therefore little direct concern with land issues or longer-term labour exploitation, had no particular reason to fear and control Aborigines in the way their predecessors in the Territory had, and were thus more egalitarian in their approach. The war also brought to North Queensland numbers of African-American troops. The intermixing between Aboriginal women and African-American men, who were well educated and enjoyed equal pay with white US soldiers, had the potential to unsettle the existing patterns of segregation in the Australian countryside.[28]

* * *

After the war, Aboriginal war service was to pose awkward challenges to the established racial order. Aboriginal soldiers returned with an expectation of reward in the form of full citizenship, not only for themselves but also for their communities. They wanted and expected recognition in the form of an end to well-established practices of 'protection', surveillance and segregation, and increased employment and other opportunities.[29]

These expectations, however, were not to be realized, at least not in the decades immediately after the war. Insofar as any recognition was given, it was to ex-servicemen individually, and not to their families and communities. In March 1949, for example, the Labor government granted Aboriginal ex-servicemen the right to vote in the federal elections but this right continued to be denied the vast majority of Aboriginal people, who were still declared wards of the state, that is, without adult legal rights, under the Acts operating in the various states.[30] In 1961, sixteen years after the war's end, Aboriginal rights remained extremely limited everywhere, with restrictions on freedom of movement, control of property and freedom of association with Europeans. Only if one had an exemption certificate could equal rights be granted. In Queensland and Western Australia there were government controls over marriage and Aboriginal people did not have the right to vote. There were restrictions on the right to drink alcohol in all states other than Victoria. The existence of Aboriginal ex-servicemen highlighted this particular form of discrimination very clearly. On Thursday Island, it was reported by a trade union delegation in the late 1950s:

> An Islander, in uniform, may drink beer, and is bound by the same laws regarding his actions as a free man. Should he take his uniform off however, he can be arrested and put 'under the Act.' In uniform, he may enter a hotel, but his own brother may not. Islanders who served in the last war were free to enter hotels then. Now they are not.[31]

In New South Wales, the focus of my story, though there were fewer controls over marriage, and a formal right to vote, surveillance and control of Aboriginal people were sustained through the post-war years. Aboriginal people were barred by law from drinking alcohol, unless they had an exemption certificate; this did not actually prevent drinking, but it did make it more clandestine and expensive, and it especially served to exclude Aboriginal people from white public bars, clubs and other forms of sociality. As mobility rose, with Aboriginal employment increasing on road construction and the railways, and men travelling on vast seasonal agricultural harvesting and pastoral circuits, the Aborigines Welfare Board created a network of district welfare officers to supervise and maintain surveillance over Aboriginal people wherever they went. In the name of the new policy of assimilation, it also encouraged the destruction of Aboriginal communities by tempting individuals to cut themselves off from family, culture and land: those with exemption certificates could gain better access for children to schools and eligibility for a house – one without Aboriginal neighbours – in town.[32]

Or at least that was the Welfare Board's plan. It was undermined partly by the decline in rural employment with the effects of mechanization, but especially by the fact that white townsfolk consistently refused to accept Aboriginal people into the towns. They opposed Aboriginal access to housing, schools and town facilities including picture theatres and swimming pools.[33] In Moree, another north-western New South Wales town with a significant Aboriginal population, they were excluded not only from the municipal thermal baths but also from the Memorial Hall, built in honour of those who served in war.[34]

None of this went uncontested. There had been for some years an active Aboriginal protest movement and white support movement, and a variety of local and international political pressures pushing Australian governments closer towards granting formal equal rights. These protest groups usually had strong connections with either the Churches or the political left. The RSL, known for its political conservatism, sought full citizenship rights for Aboriginal ex-servicemen, but remained aloof from campaigns for full citizenship rights for Aboriginal people generally. In RSL thinking, the returned soldier could acquire rights as an individual that could not be acquired by his or her racial group as a whole; as Stephen Garton points out, for the RSL the status of returned soldier transcended all others.[35] In the late 1940s it pressed for ex-servicemen to be given the right to vote, and later pressed for the exemption of ex-servicemen from restrictions on the right to drink alcohol.[36] From time to time local RSL clubs assisted Aboriginal ex-servicemen and their families, for example building a cottage on the Brungle Aboriginal Reserve in southern New South Wales for the widow of an Aboriginal ex-serviceman.[37] Yet these calls for particular rights for ex-servicemen did not lead to an opposition to racial policies generally.[38] In a letter to the Prime Minister in December 1961, the National Secretary put the RSL position clearly:

> It is the firmly expressed opinion of the Returned Servicemen's League that service abroad in the armed forces of the Commonwealth in time of war should entitle any Aborigine to special consideration. It is quite clear that the Aboriginal problem is not nearly as simple as many people would have us believe, having regard for the fact that many of them are ill-equipped to cope with conditions of modern day society. However, our application here is made on the basis of a very small number who have served their country to an outstanding degree. Our simple belief is that as they were advanced enough to share the dangers of active service they are, by the same token, sufficiently advanced to cope with the privileges and responsibilities of full citizenship.[39]

The internal and external pressures for change in Aboriginal policy were, however, beginning to have some effect. Racial exclusion had become an embarrassment for many within white Anglo-Celtic Australian culture, a sign of provincial backwardness rather than progressive cosmopolitanism. Under pressure at home and abroad, governments in the early 1960s removed their racially discriminatory legislation. The Commonwealth government in 1962 extended the franchise to Aboriginal people in the Northern Territory and both Queensland and Western Australia followed suit. The restriction on alcohol was finally withdrawn in New South Wales in March 1963, removing the final function of the Exemption certificate, and the hated Exemption certificate was itself removed soon afterwards.[40]

By the mid-1960s, then, racial discrimination was the result not so much of state or Commonwealth legislation as of continuing practice and convention, sometimes embodied in local council ordinances. The right to drink in hotels and to belong to clubs fell into the category of rights now allowed in legislation but still denied in practice. It was to be another decade before laws were in place *preventing* racial discrimination in public life. In this situation, this interregnum between legislation embodying racial discrimination and legislation against it, there was considerable social tension over the issue. Aboriginal people continued to be excluded from bars and clubs around the country, as a matter of convention. The RSL appears to have had no policy specifically preventing its clubs from excluding Aboriginal ex-servicemen from membership, leaving it to local clubs to determine their own policies.[41] Some of them, like the Walgett club, did practice racial discrimination, and excluded Aboriginal ex-servicemen from membership.

* * *

And so I have returned to the story of the protest in Walgett. Local activists, people like Harry Hall and George Rose and Ted Fields, had been protesting about the club's exclusionary politics (and much else) for some years. When two Aboriginal children were gaoled for several days in July 1964 for stealing crayons, the Walgett activists invited the New South Wales Labor Council to visit the town to investigate. The gaoling, and the visit of the delegation from the Labor Council, were widely publicized in the Sydney press. The delegation's report went far beyond the original gaoling incident which had prompted the visit: it detailed many other instances of racial segregation. It drew attention to the segregation of drinkers at the Oasis Hotel Motel, and to the exclusion of returned Aboriginal diggers from the RSL club, except sometimes on Anzac Day. In the words of the report, the RSL officials 'denied that this was racial discrimination and said that they were only following past practices'.[42]

Charles Perkins and his Freedom Riders took up the case of the Walgett RSL six months later. They knew it was an excellent focus for protest against racial discrimination. In a society where ex-servicemen were held in high esteem, and seen as embodying nationhood and national identity, racial discrimination against an ex-serviceman, even an Aboriginal one, was easily ridiculed and attacked. Perkins and the Freedom Riders quickly realized that the exclusion of Aboriginal people from membership and regular admittance to a Returned Servicemen's League club had major symbolic significance for both white and black Australians, and was thus an excellent focus for an attack on racism. For Aboriginal people, their exclusion, frequently even on Anzac Day, symbolized social discrimination and denial of political rights in a particularly stark and offensive way. Instead of the inclusion they had expected on their return,

Aboriginal ex-servicemen and their communities had been met with a deep symbolic exclusion. When retracing the Freedom Ride in 1991, I spoke to Lyall Munro (senior), a respected Aboriginal leader in Moree. He spoke extensively of the feelings of Aboriginal returned servicemen in the north west, of whom there were hundreds. Lyall said he sometimes used to join the Aboriginal returned soldiers down on the river bank, and listen to their stories. They felt their exclusion from the RSLs very keenly, he said. The first Aboriginal pilot in the Second World War, he said, lived at Toomelah, near Boggabilla, a town to the east of Walgett. There was Private Gunner Suey living in Walgett, who had been a prisoner of war after the fall of Singapore. Some had been Rats of Tobruk, like Herb Leslie, and then were refused service in the pub at Moree. After the war, Munro said, they found they couldn't get returned servicemen's loans at the bank. Many of them had quite conservative politics, voted Liberal or Country Party.[43]

Not all the ex-servicemen, however, were politically conservative, by any means. For many their war service had raised their expectations and broadened their knowledge of the world. Many of the leaders in the Aboriginal political movement of the 1950s and 1960s had served in the war – Bert Groves, Joe McGinness, Kath Walker (later known as Oodgeroo Noonuccal), Ted Loban and, from the Women's Land Army, Faith Bandler.[44] Their war experiences had given them, it seems, both the desire and the confidence to seek lasting political change.

And so it was that the Freedom Riders chose the RSL at Walgett as their first target for a demonstration. It was actually very quiet for most of the day. As one of the Freedom Riders, Aidan Foy, remembers it:

> I think the locals were just absolutely shocked. They were stunned. There was no opposition really. There wasn't the sort of scene that occurred later. After we'd been there for a little while, the employees of the RSL came out and offered us soft drinks, which of course we all refused to take. There were some speeches made. Charlie made a speech, I'm sure Harry Hall made a speech and George Rose made a speech, I'm fairly sure, just talking about the segregation of the club.[45]

Pat Healy remembers that 'Charlie spoke, very well I thought, contrasting the treatment of the black servicemen who weren't allowed to use or be members of the Walgett RSL. I guess there's just a sort of feeling, it's an emotional thing about remembering these white, highly emotional males, really'. Charlie Perkins remembers it this way:

> Walgett was the start of it all – that was the beginning. That was the jewel in the crown in terms of where things really began in recreating and re-establishing relationships in the context of race in this country I think. That's where it all began. The demonstration in the street outside the RSL, the mass of Aboriginal people that were watching and were involved, young

people that were participating in our demonstrations, who are now grown up men and women around NSW and throughout Australia they've got prominent roles. And that sort of crystallized our thinking on what we should do and we set about it then. The first one, of course, was to demonstrate outside the RSL, the great sacred cow at that time, you know, nobody criticizes the RSL, they can do whatever they like. They thought they were 'God's gift' to not only Australia but everybody else. And so we decided that that would be the prime objective … to fly in the face of the most cherished institution in Australia, because they were very racist, and they were not only that, they were arrogant, and they were self-centred and very discriminating and opinionated.

The demonstration outside the RSL club did have an effect, though it took some time. The immediate response of local clubs was to place guards on the premises to prevent trouble from the Freedom Riders, and to insist that there was no discrimination in their own club.[46] The National Secretary of the RSL said it was a matter for the local clubs, but also denied there was any colour discrimination in the clubs.[47] Four years later, three ex-servicemen – Tom Lake, Sye Morgan and Eric Thorne – were still trying to join the Walgett RSL.[48] The Federal Council for the Advancement of Aboriginal and Torres Strait Islanders, a highly active body coordinating a multitude of local pro-Aboriginal organizations which had been at the forefront of the successful Referendum campaign of 1967, took up their case, resolving at its annual conference in April 1969 to condemn RSL branches refusing membership, and urging that this matter be discussed and made public at the RSL congress.[49] A group of law students from the University of Sydney reported that their survey of the issue in May 1969 found no Aboriginal people had yet been admitted as members of Walgett RSL.[50] Reginald Saunders, an Aboriginal ex-serviceman who had become the first Aboriginal commissioned officer in the army, and who was now acting as the Liaison Officer for the newly formed government advisory body, the Council for Aboriginal Affairs, also put pressure on the RSL on the men's behalf.[51] (Although a relatively conservative man, Saunders had been appalled when the National President, Brigadier Alf Garland, called for blood-testing of Aboriginal people as a way to determine an individual's entitlement to benefits.)[52] Pressure mounted when the National Aborigines Day Organising Committee arranged a march and wreath-laying ceremony by Aboriginal ex-servicemen at the Cenotaph in Sydney on National Aborigines Day on 11 July 1969. Support for the event was especially strong from the northern towns – Dubbo, Armidale, Kempsey, Lismore and Newcastle. Unable to do anything else, the RSL supported the event, asking its sub-branches to support the ceremony, its state president participating by laying a wreath.

In the face of such renewed attention to the fact that Aboriginal people had served in war, the Walgett branch was finally brought into line in 1971.[53] Tom Lake, who had served with the army in the 39/52nd unit in New Guinea, was finally admitted to the RSL in that year, an event covered by ABC television in a

Four Corners programme.[54] Mrs Gladys Lake, Tom's wife, was, she told me in 1991, the second member. Her father, she said, was also a returned soldier, who had fought in the First World War in France. He used to march on Anzac Day, but had never tried to join the RSL.[55]

* * *

A great deal has changed since the demonstration outside the Walgett RSL in 1965. Government policies have resulted in the formation of the Aboriginal and Torres Strait Islander Commission (ATSIC) to institute Aboriginal management of government funding for Aboriginal people. Land rights of varying kinds have been acknowledged, and there has been substantial return of land in the Northern Territory. Aboriginal access to education has increased significantly, though is still far less than that of non-Aboriginal people. While complaints of discrimination and segregation in pubs and clubs continue, it is against the law, and there are strong bureaucratic means in place to redress such grievances.[56] And there are signs of inclusion of Aboriginal people into national narratives in a multitude of ways. Notably, Aboriginal people have been increasingly incorporated into the Anzac legend, in ceremony and exhibitions. As part of the International Year of the World's Indigenous People in 1995, the Australian War Memorial developed a travelling exhibition called 'Too Dark for the Light Horse', on the role Aboriginal and Islander peoples played in the defence of the nation. A gallery version toured regional centres during 1998 and 1999, and a modified version went to remote Aboriginal communities in the Northern Territory in July 1997. This version has since toured through south-western Queensland and western New South Wales. Everywhere it has attracted attention; along with many others, I saw it in a shopping centre in Canberra in mid-1998. This is a truly popular exhibition, with an emotional impact deriving from its contrast between Aboriginal war service on the one hand and practices of discrimination and exclusion on the other.

Yet if there have been some profound changes in policy and practice, white settler hostility to Aboriginal people remains strong. Now, instead of exclusion from schools, hospitals, swimming pools, picture theatres and RSL clubs, we see in the rapid growth in support for the conservative populist party, One Nation, virulent opposition to native title, Aboriginal management of welfare provision through ATSIC and any notion of a national apology for the child removal policies of the past. For its part, the RSL has become one of the most vigorous opponents of land rights and recognition of native title, agreeing at its National Congress in 1993 that 'Thousands of Australians, both Aboriginal and white, died in the field of service to preserve this nation for all Australians', by which it meant that there should be no recognition of special indigenous rights to land.[57] Instead of an advocacy of racial separation and segregation, we see now a language of equal rights, taken to mean treating everyone the same without reference to relevant differences, and especially to claims to land on the basis of prior occupation. In this way, the discourse of equality so strongly fought for in the two decades after the Second World War has transmogrified into a discourse of refusing to acknowledge the past and its effects, and in refusing strategies to redress those effects. In this refusal of a history of colonization, we see some continuities between then and now. We see the same difficulty in acknowledging the legacy of brutal land-taking and subjugation which underlies our present polity, which provides the foundation of the nation itself. Because of the powerful Exodus-related narrative of victimhood which underlies so much white Australian identity, both historically and in the present – the focus on white hardship, suffering and victimhood – popular acknowledgement of the violence and exploitation on which the nation is built is still extremely difficult, still in process, a project incomplete and unfinished. Just as a combination of local and international pressures forced changes to formal legal and civil rights in the 1960s, so, we might hope, may they do so in our current crisis, our current discontents.

Notes

1 For a fuller description, see P. Read, *Charles Perkins: A Biography* (Ringwood, 1990). I am currently completing a book on the Freedom Ride, and this chapter arises from that research. For assistance with research for this chapter I thank Lani Russell, Inara Walden and Ann Genovese, and the Australian Research Council for its financial support; Mr Derek Robson, National Secretary of the Returned and Services League of Australia Limited, for permission to peruse the RSL papers in the National Library of Australia (NLA); and John Docker, with whom I have extensively discussed the ideas in this chapter.

2 Meaghan Morris, *Too Soon Too Late* (Bloomington, 1998), p. 25.

3 See K.S. Inglis, *Sacred Places: War Memorials in the Australian Landscape* (Melbourne, 1998) and *Anzac Remembered: Selected Writings of K.S. Inglis*, edited by John Lack (Melbourne, 1998); Stephen Garton, *The Cost of War: Australians Return* (Melbourne, 1996).

4 Inglis, *Anzac Remembered*. The commemoration of the unsuccessful Gallipoli campaign can be contrasted with the emerging commemoration of Australian military involvement in Vietnam, which focuses increasingly on the battle of Long Tan, a qualified victory within a larger military defeat.

5 C.E.W. Bean, *Anzac to Amiens* (Canberra, 1946), 181.

6 B. Gammage, *The Broken Years* (Canberra, 1974), 94.

7 Inglis, *Sacred Places*.

8 This attraction to a history of suffering, sacrifice and defiance in defeat has been noted by several commentators: J. Brett, 'John Howard, Pauline Hanson and the politics of grievance', in G. Gray and C. Winter (eds) *The Resurgence of Racism: Howard, Hanson and the Race Debate* (Melbourne, 1997), 7–28; R. Gibson, *South of the West: Postcolonialism and the Narrative Construction of Australia* (Bloomington, 1992), 173–4; A. Lattas, 'Aborigines and contemporary Australian nationalism: primordiality and the cultural politics of otherness', in G. Cowlishaw and B. Morris (eds) *Race Matters: Indigenous Australians and 'Our' Society* (Canberra, 1997), 234–5; S. Dermody and E. Jacka, *The Screening of Australia: Anatomy of a Film Industry*, vols 1 and 2 (Sydney, 1987, 1988).

9 S. Bercovitch, *The American Jeremiad* (Madison, 1978); E.W. Said, 'Michael Walzer's Exodus and Revolution: A Canaanite Reading', in E.W. Said and C. Hitchens (eds), *Blaming the Victims: Spurious Scholarship and the Palestinian Question* (London, 1988), 161–78; E. Shohat, 'Antinomies of exile: Said at the frontiers of national narratives', in M. Sprinker (ed.) *Edward Said: A Critical Reader* (London, 1992), 140–1; R.M. Schwartz,

The Curse of Cain: The Violent Legacy of Monotheism (Chicago and London, 1997). There is as yet relatively little comparable critical biblical scholarship in discussions of Australian frontier narratives. Biblical scholar Roland Boer is investigating explorers' journals in these terms. Bill Bell, as part of his research for 'The History of the Book', has found extensive references in the shipboard diaries of Scottish migrants to the Bible, particularly Exodus.

10 D.B. Rose, 'Dark times and excluded bodies in the colonisation of Australia', in Gray and Winter (eds) *The Resurgence of Racism*, 205.

11 Andrew Lattas, 'Aborigines and contemporary Australian nationalism', 235.

12 Russell McGregor, *Imagined Destinies: Aboriginal Australians and the Doomed Race Theory, 1880–1939* (Melbourne, 1997).

13 G.F. Angas, *Australia: A Popular Account* (London, 1865), 49–50.

14 Keating's speech was particularly indebted to Bruce Elder, *Blood on the Wattle: Massacres and Maltreatment of Australian Aborigines since 1788* (French's Forest, 1988), as discussed in Klaus Neumann, 'Remembering victims and perpetrators', *UTS Review* (published by the University of Technology, Sydney) 4.1 (May 1998), 1–17.

15 The most prolific and well-known of these historians is Henry Reynolds, especially *The Other Side of the Frontier* (Ringwood, 1981), and *Frontier* (Sydney, 1987). His most recent book is *Why Weren't We Told? A Personal Search for the Truth about our History* (Ringwood, 1999).

16 *The Australian*, 25 October, 26–27 October 1996.

17 R.A. Hall, 'Aborigines and Torres Strait Islanders in the Second World War', in D. Ball (ed.) *Aborigines in the Defence of Australia* (Canberra, 1991), 32. Note that it is very difficult to be sure of the actual number enlisted, given that 'race' was not recorded.

18 H. Goodall, *Invasion to Embassy: Land in Aboriginal Politics in New South Wales, 1770–1972* (Sydney, 1996), 124. Only one Aboriginal soldier received a soldier settler's block in New South Wales: George Kennedy drew a 17,000 acre block outside the town of Ivanhoe, but like many others left it after five years. D. Huggonson, 'Aborigines and the aftermath of the Great War', *Australian Aboriginal Studies* 1 (1993), 7.

19 Huggonson, 'Aborigines and the aftermath of the Great War', 7; Goodall, *Invasion to Embassy*, 149.

20 Huggonson, 'Aborigines and the aftermath of the Great War', 3.

21 Army recruiting regulations on this matter were not officially changed until 1949. See Hall, 'Aborigines and Torres Strait Islanders in the Second World War', 34–5.

22 Ibid., 57.

23 Ibid., 34; D. Huggonson, 'Struggle for equality', *North West Star*, 7 April 1995, 14.

24 Hall, 'Aborigines and Torres Strait Islanders in the Second World War', 36–46.

25 D. Huggonson, 'Our women and the war effort', *Koori Mail*, 12 July 1995.

26 Hall, 'Aborigines and Torres Strait Islanders in the Second World War', 46.

27 Goodall, *Invasion to Embassy*, 262.

28 D. Huggonson, ' "Jim Crow" in Queensland: US spells it out in black and white!', *The Chronicle*, 10 January 1995, 11.

29 In Western Australia for example, a 1944 Act gave the vote to servicemen who had obtained a certificate of citizenship. Since such certificates required abandonment of all association with their own communities for two years, only 25 of the 400 Aboriginal servicemen in WA gained the right to vote through this Act. Hall, 'Aborigines and Torres Strait Islanders in the Second World War', 34.

30 Ibid.

31 Building Workers Industrial Union, *Report on Cape York Peninsula and Torres Strait Islands*, undated, apparently late 1950s, 20.

32 Goodall, *Invasion to Embassy*, 263–7.

33 H. Goodall, 'New South Wales', in A. McGrath (ed.), *Contested Ground: Australian Aborigines under the British Crown* (Sydney, 1995), 86–7.

34 This decision was the subject of considerable dispute in Moree at the time. See *North West Champion*, June–August 1955.
35 Garton, *The Cost of War*, 55.
36 R. Hall, *The Black Diggers: Aborigines and Torres Strait Islanders in the Second World War* (Canberra, 1997), 191; Garton, *The Cost of War*, 55.
37 'Cottage for widow', *Dawn*,14.5 (May 1965), 12.
38 In response to a request from a parliamentary inquiry into Aboriginal Voting Rights in 1961, the RSL declined to express a view: Letter A.G.W. Keys, RSL National Secretary, to A.R. Browning, 29 May 1961, RSL Collection, MS 6609, box 21, series 1, item 4797, NLA.
39 A.G.W. Keys, RSL National Secretary, to R.G. Menzies, PM, 21 December 1961, RSL Collection, MS 6609, Box 21, series 1, item 4797, NLA.
40 Goodall, 'New South Wales', 106.
41 Letter from A.G.W. Keys, RSL National Secretary, to Miss S.H. Mahomet, Auckland, New Zealand, 23 February 1965, responding to her letter of 17 February, RSL papers, MS 6609, Series 2, Box 314 , NLA.
42 NSW Labor Council, *Report of Delegation to Walgett, NSW, July 1964*, in Records of the Communist Party of Australia, MSS 5021 and 691/70, Box 73 (159): 1952–1975: Aborigines, item 39, p. 2.
43 Ann Curthoys, Personal Diary, 6 April 1991, Moree.
44 Joe McGinness, *Son of Alyandabu: My Fight for Aboriginal Rights* (Brisbane, 1991); D. Huggonson, 'World War the start of political battle for indigenous Australians', *ATSIC News* (published by the Aboriginal and Torres Strait Islander Commission), Spring 1994, 11.
45 Aidan Foy, interview in Newcastle, 7 April 1994, interviewed by Inara Walden. George Rose and Harry Hall were both leading Aboriginal activists in Walgett.
46 *Newcastle Morning Herald*, 16 February 1965.
47 Letter from A.G.W. Keys, RSL National Secretary, to Miss S.H. Mahomet, Auckland, New Zealand, 23 February 1965, responding to her letter of 17 February, RSL papers, MS 6609, Series 2, Box 314 , NLA.
48 Letter A.J. Russell, Acting National Secretary, to Mr R. Saunders, Liaison Officer, Council for Aboriginal Affairs, 21 May 1969, RSL papers, MS 6609, Outward Correspondence, Box 7, NLA.
49 *Rights and Advancement* (FCAATSI Newsletter), no. 19 (May 1969), 15.
50 P. Tobin, 'Fringe dwelling rural Aborigines and the law in NSW', *Smoke Signals* (March 1970), 43.
51 Letter A.J. Russell, Acting National Secretary, to Mr R. Saunders, Liaison Officer, Council for Aboriginal Affairs, 21 May 1969, RSL papers, MS 6609, Outward Correspondence, Box 7, NLA. Reg Saunders later participated in the campaign for Aboriginal land rights, and tried to change the RSL's hardline stance against such rights, *The Age* (Melbourne), 2 August 1984.
52 R.M. Hall, *Fighters from the Fringe: Aborigines and Torres Strait Islanders Recall the Second World War* (Canberra, 1995), 86.
53 *Reveille*, 1 June, 1 July, 1 August 1969. A report entitled 'Walgett – then and now' in *New Dawn*, August 1971, mentioned that there were now Aboriginal members in the Walgett RSL.
54 Thank you to Frances Peters-Little for drawing my attention to this programme, a copy of which is held at the Australian Institute for Aboriginal and Torres Strait Islander Studies, Canberra. And a special thank you to Mrs Gladys Lake, now deceased, who, when I visited Walgett in 1991, told me about the admission of Tom Lake and herself to the RSL.
55 Ann Curthoys, Personal Diary, 3 April 1991, Walgett.
56 The Brewarrina RSL in particular continued to be accused of prejudicial treatment of applications for membership from Aboriginal people. See John Austin and Michael

Dodson, 'What it means to be black in country towns', *Identity*, 2.2 (October 1974), 37; and 'Aboriginal actress denied membership of RSL club', *Sydney Morning Herald*, 2 April 1991, 2, concerning exclusion of Essie Coffey, film-maker and actress, from the Brewarrina RSL club.

57 Reported in *Queensland Times* (Ipswich), 23 June 1993.

5 'This is where they fought'

Finnish war landscapes as a national heritage

Petri J. Raivo

The historical landscapes of war are fundamental parts of the national iconography of modern states. They stand there not just for their own sake but to evince a nationalistic narrative of a unified history and culture from the past up to the present day. Externally, these landscapes and the monuments erected upon them (usually little more than engraved or cut stones) may not be striking to look at, but through state rituals they are invested with national soul and memory.[1] This is very true in the case of Finland, the contemporary territory of which is literally covered with battle sites, monuments, remains of defence structures, boundary-stones, trenches, pits, graves and landmarks of previous wars. Especially important are the sites in memory of the Second World War. It has been said that the story of this war – a story of the defence of Finland's independence, to which the nation was driven against its own will and which it survived only by its own will-power – was, and remains, the great national narrative of the Finns.[2]

This chapter discusses Finnish national memories of the Second World War and the ways in which these have been embodied in the making of commemorative landscapes and sites, including cemeteries, monuments, battlefields and contemporary heritage sites. It explores the changing functions, forms and meanings of these sites during the post-war decades, and considers how and to whom they tell their stories of the national past on the threshold of the new millennium, sixty years after the outbreak of the war.

Behind the memory and memorials: geography and Finnish war history

In common with previous episodes in Finnish war history, Finland's specific national experience of the Second World War was shaped by the country's geography. 'There is nothing we can do about geography', is the famous phrase uttered by the Finnish statesman J.K. Paasikivi (1870–1956) to describe the political problems that Finland has always had because of its location between the cultural, political and military realms of East and West (see Figure 1 on p. 146). The area of present-day Finland has been a contested territory throughout its history, and the borders have moved time after time either eastward or westward.

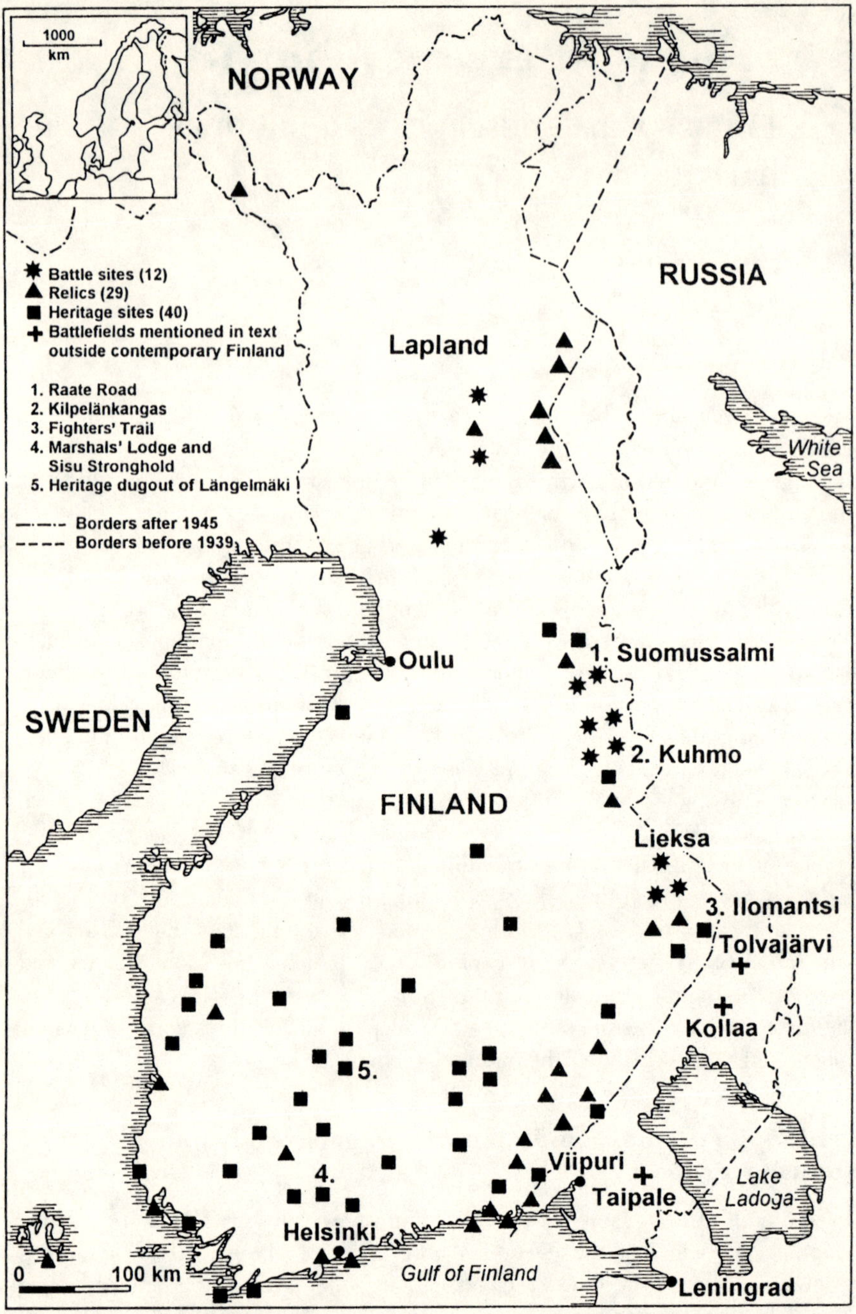

Figure 1 Map of battle sites and tourist landscapes of the Second World War in Finland (Petri Raivo). Most of the tourist sights are former battlefields and relics of war - old strongholds or defence lines - or heritage sites like museums or replica dugouts built in the late 1980s or 1990s.

Consequently Finland, and especially its eastern area, Karelia, has often been seen as a cultural buffer, part of the marginal zone that stretches from the Balkans through Poland, the Baltic states and up to the White Sea. For centuries this zone has been both a battlefield and a bridge for cultural transformation.[3]

Until the middle of the twelfth century, the area between the Gulf of Finland and Lake Ladoga settled by Finnish and Karelian tribes was a political vacuum, and therefore an object of interest to both its western neighbour, Sweden, and the Roman Catholic Church there; and its eastern neighbour, Novgorod (Russia), and the Russian Orthodox Church. Sweden came out on top, as under the peace treaty of 1323 between Sweden and Novgorod the western and southern parts of Finland were tied to Sweden and the Western European cultural sphere, while only eastern Finland (Karelia) became part of the Russo-Byzantine world. The border was not clear or secured, however, and both sides launched military campaigns across it to gain influence over the other. During Sweden's period as a great power (1617–1721), it extended its realm around the Baltic and managed to exploit the weakness of Russia by pushing its borders still further east. When Sweden lost its position as a great power in the early eighteenth century, Russian pressure on Finland increased, and after temporary occupations and wars that shifted Russia's borders westwards, it finally conquered the whole of Finland in the 1808–09 war with Sweden. Finland was joined to Russia in 1809 as a Grand Duchy enjoying extensive autonomy, and gradually came to form a state within a state, with its own administration, legislation, monetary economy and official borders to separate it from the rest of the Russian Empire.

This arrangement continued until 6 December 1917, when, in the middle of the disorder created by the Russian Revolution, the Finnish Parliament approved a declaration of independence drawn up by the Senate. This was recognized by Soviet Russia on 4 January 1918. At the same time, the breach between the parties of the left and the right had become irreconcilable. At the end of January 1918, the left-wing parties staged a coup and the government was forced to flee Helsinki. The ensuing Civil War – in which the government and its supporters, known as the Whites, received help from the Kaiser's Germany, and its opponents, the Reds, from Bolshevik Russia – ended in May with victory for the government's troops. The Civil War left 15,000 victims, both soldiers and civilians, and in its aftermath more than 80,000 prisoners of war from among the Reds were sent to prison camps, where 12,000 of them died of hunger and disease, or through arbitrary execution. However, subsequently it was commemorated as a war of independence, because in the course of these hostilities the Russian troops stationed in Finland were expelled from the country. Finland was taking its first steps as a sovereign and independent, but divided republic.[4]

The troublesome geopolitical location of Finland was again apparent in the late 1930s, when it became embroiled in another European-wide conflict. In August 1939, Germany and the Soviet Union signed a non-aggression pact which included a secret protocol placing Finland in the Soviet sphere of interest. When Finland refused to allow the Soviet Union to build military bases on its territory, or to cede certain areas in order to strengthen the strategic defence of

Leningrad, the latter revoked the non-aggression pact of 1932 and attacked Finland on 30 November 1939.

In the so-called Winter War (1939–40), which was the first of three separate wars that the Finns had to fight during the Second World War, Finland stood alone, receiving only sympathy and modest assistance from other countries. Despite the size and weakness of the Finnish army – its sixteen divisions (compared with the Red Army's fifty-five divisions) suffered shortages of artillery, heavy ammunition, transport and signal equipment – and the almost total lack of a defence structure, Finnish military successes exceeded all expectations.[5] At the beginning of the war the hardest battles were fought in the areas of Tolvajärvi and Suomussalmi (see Figure 1), where Finnish troops beat back the attacking Red Army. The enemy was held up in Kuhmo and Kollaa, for example, and engaged in a tough defensive battle until the end of the war. To the north-east of Lake Ladoga large detachments of Russian troops (six divisions of the Russian Eighth Army) found themselves cut off from communications and supplies when Finnish ski troops isolated them in pockets during a counter-offensive. Eventually, in spite of the stunning victories, courage and fighting ability of its troops, this small nation had to capitulate in a war of attrition against overwhelming Russian superiority in terms of both manpower and armaments. The decisive battles along the Mannerheim Line on the Karelian Isthmus (the main Finnish defences) started in February 1940. Heavy fighting supported by tank troops and continuous artillery fire enabled the Russians to pierce the defensive line and the Finnish troops were forced to fall back towards new positions near the city of Viipuri, where fighting continued up to the signing of a peace treaty in Moscow on 13 March 1940.

As a consequence of the Winter War, Finland was forced to cede 10 per cent of its territory to the Soviet Union, including most of the area of Karelia. Some 25,000 Finnish soldiers were killed and 43,000 wounded. The losses on the Russian side were of course much heavier. It has been estimated that the dead alone amounted to something like 100,000 men.[6] When the Winter War ended in March 1940 the Second World War was in its early stages. The Soviet Union continued to exert pressure on its tiny neighbour, whose fears grew when the Baltic states were forcibly incorporated into the USSR in August 1940. Isolated from the West and with Sweden staunchly neutral, Finland turned towards Germany. This former silent supporter of the Soviet Union during the Winter War became interested in Finland as a support area and useful ally in a possible war with Russia. Germany was also interested in the nickel resources of the Finnish mines in Petsamo on the Arctic Ocean, as this was an essential raw material for arms production in the Third Reich. For the Finns, whose country's independence had lately been saved, alliance with Germany seemed to offer possibilities of regaining the ceded areas and gaining revenge over Russia for all the bitter defeats of the Winter War.

When Germany invaded the Soviet Union in the summer of 1941, Finland, now better prepared and equipped than during the Winter War, entered the fray as a co-belligerent with Germany. This second phase of Finland's involvement in the Second World War is called the Continuation War (1941–44). In July 1941 Finnish troops started an offensive to liberate the ceded areas of Karelia, and by autumn 1941 they had penetrated deep into Eastern Karelia, beyond the old border. In December 1941 they ceased their offensive on all fronts. This was the beginning of a period of trench warfare that lasted until July 1944, when, at almost the same time as Operation Overlord in the west, the Russians began a massive counter-offensive on the Karelian Isthmus, forcing the Finns to leave their battered positions. The fighting continued around the new defensive positions near Viipuri, where the largest battle ever fought in the Nordic Countries took place (see Figure 1). During this month-long campaign the Finnish troops were able to halt the attacking Red Army both on the Karelian Isthmus and in the Karelian borderlands north of Lake Ladoga. In Ilomantsi in North Karelia, Finnish troops even surrounded and incapacitated two attacking Russian divisions as late as August 1944. The Continuation War ended in an armistice in September 1944 followed by the Treaty of Moscow, the terms of which were confirmed in the Paris Peace Treaty of 1947.[7]

Under the armistice Finland was ordered to drive from the north of the country German forces that had been stationed there since the beginning of the Continuation War, partly in order to release Finnish troops for service in Karelia. During this third war, known as the Lapland War (1944–45), Finnish forces pushed 200,000 German soldiers out of northern Finland, thereby hastening their redeployment to Norway which was already in progress. When the Second World War in Finland ended finally on 27 April 1945, a total of 85,000 Finns had been killed and some 200,000 permanently wounded out of a population of fewer than 5 million. Finland had had to cede territories to the Soviet Union amounting to about 10 per cent of its total area, including those parts of Karelia lost after the Winter War and reclaimed after the Continuation War, and the Petsamo area. In addition, the area of Porkkala on the coast near Helsinki had to be leased to the Soviet Union (not to be returned to Finland until 1956) and a heavy burden of war reparations had to be paid. At the same time 423,000 refugees evacuated from the ceded areas had to be resettled in the remaining parts of the country. Even so, despite the harsh effects of the war, Finland survived as an independent and sovereign state; and Helsinki was one of only three capitals of European states involved in the Second World War which were never occupied by foreign forces (the others being London and Moscow).

Commemorative landscapes: inscribed narratives of the nation

Finland's difficult geopolitical situation has also shaped Finnish remembrance and commemoration of the fight for national survival during 1939–45. In the era of the Cold War, Finland had to be constantly on its toes regarding its relations with the Soviet Union. This meant, among other things, the development of two different discourses – an external and an internal – concerning the memory of the Second World. The external discourse was tied to Finland's foreign policy and the special rhetoric needed to deal with the Soviet Union. In terms of this dogmatic policy it was not possible to remember or remind people that the war was an act of national self-defence into which Finland was forced by others, at least not without remembering the problematic alliance with Nazi Germany during the Continuation War. The official (and thus correct and objective) interpretation of the Great Patriotic War in the eyes of the Soviet Union was that it started in June 1941, when Germany and its allies launched Operation Barbarossa.[8] Finland and its political representatives never remembered it in that way, but they did not protest against the Soviet version in public. Officially the country kept silent.

Because of this policy of official silence, there has always been a strong, visible internal discourse in which the war and its memory have been acknowledged as important and elementary parts of Finland's history. This memory and memorializing has not been tied to particular political parties or ideologies, even though the extreme left (being pro-Soviet Union) did not participate in it. The war touched all layers of society. Almost every family had members who died or were wounded, and one family in every ten lost its home or property because of the unfavourable outcome of the war. As a result, war books, for example, are still the most avidly borrowed and read of all those in public libraries. The most popular of them all is *The Unknown Soldier* by Vainö Linna, a realistic story, based on personal experience, of an infantryman trying to survive in the hell of war; first published in 1954, it sold 615,000 copies during the 1970s, a figure that would be equivalent to 6 million copies in Great Britain.[9] Together with literature and coverage in the media, national service in the Finnish Defence Forces, and later, voluntary participation in various defence and reservist organizations, have been key elements in the process of initiating each new generation into the legacy of the Second World War.

It is in the context of this process of remembering that the Finnish commemorative landscapes are significant. Finnish collective memory of the Second World War is most evidently inscribed on the landscape in the form of cemeteries for 'the fallen'. There is one in almost every town and village, acting as an icon and symbol of the war. The burying in the cemeteries of their home parishes of soldiers killed in action was a particularly Finnish phenomenon and institution, and about 83,000 soldiers (the vast majority) are buried in 622 such graveyards all over the country.[10] One example is the cemetery of 'fallen heroes' in the village of Sippola in Anjalankoski, south-east Finland, some 50 kilometres from the Russian border. There are 214 war dead buried in this cemetery, and Anjalankoski as a whole has four such cemeteries, where 530 war dead are buried. The cemetery of Sippola is part of the larger civilian cemetery in the local churchyard, from which a section has been marked off for war graves. It is typical for Finnish war cemeteries to be situated in conjunction with local civilian graveyards in this way. Only in the main cities such as Tampere or Helsinki are there separate war cemeteries outside the civilian burial grounds.

The previous war experience that provided the Finns with a model for commemorating the Second World War was not the First World War (as in many

Figure 2 The memorial to soldiers who fell during the Second World War, in the cemetery of Polvijärvi, North Karelia, in 1997 (photo: Juha Kokko). The most typical features of Finnish commemorative landscapes are the cemetery and local war memorial. About 83,000 soldiers are buried in 622 comparable graveyards all over the country.

other European countries) but the Finnish Civil War, fought in 1918.[11] The background to the tradition of burying dead soldiers in the cemeteries of their home area has its roots in that war, in which the bodies of the victorious Whites (unlike those of the defeated Reds) were sent back home, where special areas for 'fallen heroes' were set aside in their parish cemeteries to honour them. This custom continued after the outbreak of the Winter War, but in that context it was applied inclusively to the dead of the whole nation, not only to one exclusive part of it.

During the Second World War the responsibility for returning soldiers killed in battle to the cemeteries of their home parishes lay with either the chaplains of the units or special centres for the evacuation of the fallen. Their bodies were brought back home partly for religious reasons, but partly to raise spirits on the home front as well. In the 1930s Finland was still a highly agrarian country, with 54 per cent of the Finnish labour force gaining its livelihood from farming and an even higher percentage living in the countryside.[12] The ideological base of the young nation lay firmly in the rustic values and strong Lutheran faith of the peasants and small-holders. It was therefore important for the families and local communities to have their own boys buried in the sacred ground of their home churchyard. On the other hand, it was also a message to the troops on both sides that the Finns, as defenders of Western values and Christianity, were not going to leave their fallen heroes in the hands of their atheistic Bolshevik enemy. 'Never abandon your companion in arms' is a well-known slogan that is still quoted in the Finnish defence forces today, meaning that those serving beside you will take care of you even if they have to bring your body back home. In Finnish war literature or war movies the leaving behind of the bodies of soldiers or their burial in temporary field graves is a strong metaphor to describe desperate and hectic fighting.[13]

In Sippola, as in all the other 621 Finnish cemeteries for those killed in the war, the graves are arranged in straight lines and regular rows. Each carries a small, white-painted wooden cross, with the name of the person, date of birth, date of death and place of death marked. This accurate marking was made possible by the identity discs that the men carried and by the careful treatment given to the dead even under difficult conditions at the front. But even so, some of the graves are empty. Some of the bodies could never be brought back home but had to be left in temporary field cemeteries or even behind enemy lines. This is especially the case with those killed in summer 1944, when Finnish troops were retreating rapidly to new defensive positions. But even without a body, a cross with the name and dates stands there among the others. In some of the cemeteries the white crosses have been replaced with small headstones. The important thing is that the cemetery has a military uniformity about it and that all those buried there are equal. It is a graveyard of brotherhood, without differences between farmers and workers, or Whites and Reds, or private soldiers and officers. This idea of brotherhood-in-arms, and of equality, is an essential part of the memory of the Second World War in Finland; and the rows of crosses in cemeteries are its physical manifestation.

Among the small crosses or headstones there is often, as at Sippola, a monument to preserve and honour the memory of over 7,600 soldiers as well as civilians who were buried in cemeteries outside the present borders of Finland, in the ceded areas of Karelia.[14] After the Treaty of Moscow in 1944, fifty-seven cemeteries were left on the other side of the closed border and the deceased had to be commemorated through new monuments erected all over Finland and through books, registers and rolls of honour dedicated to them. The new owners of the ceded areas did not respect the cemeteries (or anything else Finnish that still existed in the landscape) and they were mistreated or abandoned, in many cases becoming targets of the systematic demolition and denigration that was part of Stalinist policy in the occupied territories. The area was also closed to foreign visitors until the late 1980s.

In the main part of Finnish war cemeteries like Sippola, there is usually a large stone memorial decorated with a relief or statue, as a symbol of the holiness of the site. The practice of decorating cemeteries with sculptures was also inherited from the commemorative tradition of the Civil War. The first sculptures were erected in Second World War cemeteries as early as 1940. Three years later, the National Commission for Commemoration of those Fallen in the War was established to organize and assist with the planning of cemeteries and monuments. With the help of the Commission some 600 burial grounds and numerous monuments were planned. Its work came to an end in 1959, when almost all the war cemeteries on Finnish territory were in an appropriate condition.[15]

Although the cemeteries of those killed in the war are the most symbolic national landscapes of commemoration in Finland, the majority of the war memorials and monuments are found elsewhere. Most monuments outside cemeteries are erected to commemorate famous battle sites, or certain military detachments, or civil casualties suffered during the war. Many monuments have been erected, and are still being erected, by war veterans' organizations, but because of the age and natural thinning of the ranks of veterans, others such as local history and heritage societies, communal administrations and various reservist defence organizations have taken a much more visible role in erecting monuments during recent decades. Such monuments are most likely to be located in the areas where battles took place, where civilian casualties were high, or where particular units were formed. But they may also be situated in the home areas of war veterans. This is because during the Second World War the Finnish army used a conscription system in which local men served together in the same detachments, and thus took part in the same battles. Therefore some of the battlefields, like the famous Winter War sites of Kollaa and Taipale, are not only national symbols of war but are also bound up strongly with certain regional and even congregational veterans' traditions and monuments.

In Finland, as elsewhere in Europe, war monuments are probably the most widespread mode of public statuary. Even the smallest and poorest of local authorities insisted on erecting a monument to its local war dead as a matter of honour, and monuments and memorials of this kind employed Finnish sculptors almost entirely for the first two post-war decades.[16] To erect monuments and build sites of commemoration was an especially important part of the local mourning process. Finnish war cemeteries were places of personal and collective grief and veneration, where local people could mourn their loved ones. The spirit behind the monuments erected in them may be seen as a mood of longing in which mourning, regret and farewells are the dominant themes. In the cemetery at Sippola, for example, the relief on the main monument (erected in 1953) shows mourning civilians: a mother with a child. In spite of this sad motif, the name of the monument, 'Belief in the Future', is full of optimism. Thanks to the sacrifices of the dead, it suggests, there is still a hope for a better future, even for war widows and orphans.

War cemeteries are not the only sites, and monuments are not the only forms, in which the dead may be mourned and commemorated. In Sippola the local parish has honoured the memory of its war dead both by publishing a book and by hanging nameplates on the wall of the local parish centre. The names and dates of birth and death of those who died in the war can be found on the wall of most school halls. Every Finnish church has a framed order of the day issued by Marshal Mannerheim, the wartime Commander-in-Chief of the Armed Forces and a national hero, dedicated to the mourning mothers of the nation.

The primary function of all these commemorative sites and landscapes has been to contribute to the mourning process. At the same time, the war cemeteries in particular have also been places of invented traditions and rituals, for example on national anniversary days. Remembrance Day for the Fallen (also known as Veterans' Day) and the Day of the Finnish Defence Forces are both anniversaries originating in the Second World War. The former was celebrated for the first time just after the Winter War, since when it has been celebrated on the third Sunday in May every year by flying the Finnish flag and visiting the war cemeteries; while the latter is celebrated annually on 4 June, the birthday of Marshal Mannerheim. Finland's Independence Day, celebrated on 6 December, has its origins in the pre-war traditions of the nation, but since the Second World War the memory of the fallen heroes has become a visible part of the rituals and solemnities of the day. On all these days, as also on All Saints Day and at Christmas, thousands of candles are lit in the war cemeteries all over the country by relatives and families, while veterans' organizations and defence organizations pay tribute with wreaths and guards of honour. Thus, these places of private mourning also serve as places of collective recollection of the war, places where the nation at large, as well as the local community, can worship itself through the cult of the fallen.

Thus, if the memory of the Second World War in Finland has been, and remains, a socialization process for the nation, where citizens are introduced to the national narrative of the Finnish state and its history, a visible part of this process has been the construction and use of memorial sites and monuments. A war monument is a material embodiment of remembrance; a social and physical arrangement of space and artefacts intended to keep alive the memories and ideas of the generations which participated in the war, and to last for generations to come.[17] It is invested with a national soul and memory and can be regarded as a national landscape in its own right; that is, a representation of Finland's national narrative inscribed on the physical environment.[18] The epic narrative of the defence of Finland's independence is made visible in war monuments by means of metaphors and symbols. These represent the collective memory in terms of a number of themes derived from Finnish national war experience and heritage. The major themes have been those of sacrifice, victory and unity.

The Second World War was a great tragedy for Finland. National independence was secured by the sacrifice of many lives and a substantial part of the national territory. The cemeteries of the fallen are of course the most visible of the monument sites dedicated to the lessons of sacrifice, but there are others. The idea of sacrifice is tangible in most of the monuments to the memory of special detachments or battles, such as the main monument to the battles of Suomussalmi and the Raate Road, where Finnish troops encircled and annihilated two Red Army divisions in December 1939 (see Figure 1). This monument, designed by the famous Finnish architect Alvar Aalto and unveiled on the twentieth anniversary of the battles, in 1959, does not celebrate overtly the glorious victory achieved by the Finnish troops over the enemy so much as remind us of the spirit and sacrifice of the men who fought there, as reflected in the following words inscribed on the stone: 'Many brave warriors paid for the victories of Suomussalmi with their blood. They pointed the way of glory to a nation that was resolute but alone.'

The theme of sacrifice is also evident in monuments dedicated to some 2,700 civilian victims of the war and to the evacuated civil population of the ceded areas who met with death, sufferings and misery during their wanderings. The majority of civilian dead were killed in air raids. Others were victims of Soviet partisan activities. Most of the latter were old men, women and children living in remote villages in northern Finland, such as the village of Seitajärvi in Lapland, where partisans murdered eleven people, mostly children, in July 1944. A monument erected to them in 1971 carries the words: 'From later generations: all victims are precious to the fatherland.'

The second major theme of Finnish war memory is that of victory. According to the Finns' national narrative, they did not lose the war, although they were on the side of the losers and had to cede territories. Rather, the outcome was a successful defence, in that the Finnish troops were able to frustrate the strategic aims of the Red Army to conquer and occupy the territory of Finland (in 1940 and 1944), and Soviet attempts to overthrow the Scandinavian way of life and traditions of democracy. In Finnish war memory there is no real difference between the three wars the Finns had to fight in 1939–45, because the image of the enemy was always the same – the Soviet Union and the Red Army; and the threat was always the same – the loss of independence and sovereignty. The 1941–44 war is known in Finland as the 'Continuation War' because it is understood as an extension of the Winter War and an attempt to make up the losses suffered in it. Even the Lapland War against the Germans can be remembered in this same way because it was fought against a background of the harsh terms of peace dictated by the Soviet Union and the continual fear of occupation if Finland did not fulfil its obligations.

In one sense, every Finnish war monument and cemetery is a silent but visible symbol of victory. These monuments stand there only because of the sovereignty and independence of the country (whereas a defeat would have led to the construction of monuments and memorials to honour the Red Army, as in so many Eastern European countries after the war). Although some of the monuments manifest the theme of victory more loudly (especially since the

1980s), in most cases a plain, modest style and an almost stoic statement of facts is found, rather than any provocative attempts to declare national triumphs and heroic deeds. This is most clearly visible in the monuments dedicated to the final fronts on the battlefields. The most typical of these point out quite concisely what happened. For example, the wording on the monument to the battle of Salla (erected in 1971 by the local reserve officers' society) says simply, 'at the end of the Winter War the front in the Salla area was situated in this terrain'. The theme of successful defence is expressed even more tersely in a series of monuments to the Winter War in the Lieksa area in North Karelia (see Figure 1), which carry only the text, 'This is where they fought', followed by the date of the battle. These monuments are nonetheless powerful symbols of a victorious defence of the nation's independence. Their simple inscriptions – like, 'Here the enemy advance was halted on 9.12.1939' (at Haukiperä on the Raate Road, erected in 1959); or, 'Here an enemy battalion was destroyed on 20.12.1939' (at Saarijärvi in Lieksa) – tell visitors an unpretentious but straightforward story of victory. 'Here we fought', 'Here we halted', 'Here we destroyed', and here we are still on the soil of an independent nation.

Probably the most important theme in Finnish war memory and memorials is that of national unity. For the Finns, the Second World War was a war that they never wanted but were forced to fight in defence of the nation. But it was also a turning point that showed that Finland could stand as a united, sovereign nation against an overwhelming enemy. When the Finns talk of the spirit or the miracle of the Winter War, for example, they are referring not only to the decisive victories that they scored over the Red Army, but, more importantly, to the unity of the nation that stood as one against the enemy when war broke out, and made those victories possible. According to this narrative the spirit of the Winter War brought about the unification of the nation and the beginning of a healing process, twenty years after independence and a bloodthirsty, cruel Civil War that had divided society and filled people with mistrust and bitterness.

There are reasonable arguments to support this narrative of unification. However, the Winter War was not solely responsible for bringing about unity. There had been serious and successful attempts at national integration in Finnish society during the late 1930s, with the first majority government incorporating the Social Democrats and the Agrarian Party being formed in 1937. The Social Democrats had been the biggest party in Finland ever since the Civil War, so that no war efforts were possible without the support of the working class and the trades unions. The practical consensus politics between parties and ideologies that was typical of post-war Finland has its roots in both the political cooperation of the late 1930s and the experience of wartime unity. The Molotov-Ribbentrop non-aggression pact signed by the Soviet Union and Germany and the reckless bullying of tiny Finland by Stalin were events that shocked most of the leftist Finns into forgetting the previous disputes and taking up arms in 1939 to defend their country.

The symbolic role of war memory and memorials in creating and maintaining this unity is clearly visible in the case of the Remembrance Day for the

Fallen. Celebrated for the first time on 19 May 1940, when it was called the 'Day of the Fallen Heroes and Unity', this was dedicated to the memory of all those killed in the Winter War and all the victims of the Civil War in 1918.[19] Thus the defeated side in the Civil War, that had earlier been excluded or discriminated against, was now included in the commemoration. This meant that new memorials were needed. Ever since the 1940s, as a part of the process of national unification and integration, monuments to the Civil War Reds, usually called monuments 'to those who fell because of their convictions', have been built alongside the monuments to the Whites erected in the 1920s and 1930s. Hence it is possible to argue that the monuments to the losing side in the Civil War not only commemorate the victims and the suffering of 1918, but also support the national narrative and politics of unity related to the Second World War generally, and the Winter War in particular.

On the other hand, the Continuation War of 1941–44 has never carried the same unambiguous message of unity as has the Winter War. Although it, too, is seen as an unavoidable conflict and the Soviet Union is named as the aggressor on account of its bombing of Finnish cities, the second war raises more problematic issues. Arguments related to the role of Finland as an ally of Nazi Germany and as an occupying force in Russian Eastern Karelia existed during the campaign, and arguments against the memory of unity during the Continuation War have continued after the event. Nevertheless, the majority of the Finns supported the country's war effort, and national unity remains the central message of the memorials to the war of 1941–44 as well as to the earlier conflict.

Some memorials offer a controversial interpretation of the memory of the Second World War because they fail to fit into the framework and themes of the nationalist narrative. This is especially so in the case of memorials dedicated to the hardline communists (few in number) who were imprisoned or executed because they actively resisted the national war effort on grounds of their political convictions; by refusing to do military service, by deserting, or by acting as fifth columnists or saboteurs. One of these controversial monuments is the Memorial to the Opponents of Fascism in the Malmi cemetery in Helsinki, erected in 1963 by former political prisoners and dedicated to those who were sentenced to death by the Finnish government because of their political activities. A similar communist monument, erected in 1948 and bearing the inscription, 'My memorial is oblivion. We fought on behalf of freedom', can be found in the city of Turku. And oblivion would seem to be very much the case with these monuments, which are not mentioned in any tourist guides or in the pages of local history books. However, the controversial interpretation of war continues to exist within the commemorative landscape, even though it is tightly excluded from the popular memory and the national narrative.

Battlefield tourism and national heritage

Although the setting-up of memorial sites and monuments has been a continuous process since the war, the forms, contexts and intensity of commemoration has varied throughout this time. The period from just after the war until the mid-1960s was a time for coping with loss and sorrow, both individual and collective, and for establishing cemeteries for the dead. The need for more precise and detailed remembering and commemorating of the war, and the resources to do so, came later. From the 1950s to the late 1970s there developed the tendency to commemorate specific battles and detachments, and civilian and other victims of the war. From the mid-1970s, and especially during the 1980s, there occurred a monument boom. In fact, half the existing monuments that stand outside cemeteries have been erected since 1975, and more than 30 per cent of them between 1985 and 1995.[20]

A visible change, not necessarily in the forms of these monuments but in their context and the attitude adopted to commemorating the war, took place in the mid-1980s. One important reason for this was the approaching fiftieth anniversary of the outbreak of the Winter War in 1989. This was celebrated nation-wide through the publishing of new books, the production of documentaries, and the holding of civic ceremonies, as well as the construction of monuments. It was a time for wide discussion and reappraisal of the reasons, events, outcome and legacy of the wars in which Finland was involved in 1939–45.[21] Not long after the anniversary, the Soviet Union ceased to exist and it became possible to begin discussing and evaluating Finland's twentieth-century history and its relationship with its eastern neighbour quite openly. Since then, attitudes in Russia have begun to change. In 1992 Finland and Russia signed a state agreement of cooperation to perpetuate the memory of the soldiers who fell in the Second World War, and the Finnish heritage societies dedicated to the lost areas of Karelia have repaired and enclosed the old cemeteries and erected new memorials to replace those that had been broken, stolen or removed.[22]

The internal re-evaluation prompted by the fiftieth anniversary of the Winter War, together with the collapse of the Soviet Union and thus the external reassessment of the meaning and outcome of the war (especially as far as the Continuation War was concerned) are two reasons for the recent changes in Finnish war memory. A third factor is the younger generation's frank interest in and recognition of the sacrifice made by its grandparents, and the will of the wartime veterans to tell their stories to the post-war generation. This has led some veterans to become involved in the making of new kinds of commemorative site.

One interesting example of this is the 'Fighters' Trail' ('*Taistelijan taival*') on the battlefield of Ilomantsi, North Karelia (see Figure 1, site 3). The country's fate was decided in places like Ilomantsi, where in summer 1944 Finnish troops halted (and even humiliated) the attacking Red Army. There has been growing public interest in these events, which are seen as decisive, the last stand of the nation. The site at Ilomantsi is also important because most of the battlefields of summer 1944 are in the ceded territories. The Fighters' Trail is a footpath with signs telling of what happened there one night in August 1944. The trail was made in the mid-1990s by the veterans of the Troopers' Guild (*Ratsumieskilta*), men who themselves participated in the battle fifty years ago. Their task – to destroy the Red Army regiment threatening the rear of the Finnish troops by means of a surprise attack at night – was very dangerous and large numbers of casualties were expected. For this reason no family men were enlisted for the operation. The signposts were made by the veterans themselves and recount very private stories about their fears and hopes on that night. Most of these narrators were no more than 19-year-old boys when the battle took place. The signs are written in the first person, and the writers use their own names and also name the wounded and dead. Walking along the path as it winds through the forest, and reading the signs, visitors to the Trail may experience the special aura or sense of place that battlefield sites are often felt to possess. For example, when the writer of a sign tells how he was wounded and paralysed for the rest of his life in this very place, the cruelty of war, death, fear, pain and hopelessness is made very powerfully present. It is not a question of troops and their movements but of human beings, young men, almost children, and the personal horror that they felt.

The development of the Fighters' Trail is linked to a fourth reason for changes in the commemoration of war that have taken place in Finland since the mid-1980s; that is, the emergence of processes of preservation, re-creation and representation that belong to contemporary heritage-making and its associated tourism. Museums, sites and milieux connected with the Second World War have become more visible tourist attractions. Former battlefield sites, in particular, have undergone very obvious transformations on account of the new demands of the heritage industry and heritage tourism.

There are 170 marked battlefields or sites of minor skirmishes in previous wars on the Finnish soil, fifty of them from the wars of 1939–44. Some of the most famous battlefields have been tourist sites for thirty years or so. They have been marked on maps and in travel brochures, and equipped with the monuments, signposts and other information symbols which mark their special meanings and tell us what we should pay attention to and what we should see. Without such marking, historical sites would not exist in our present-day landscapes.[23] Through the commemorative function of the 'marker', a battlefield – an otherwise undifferentiated area of land – becomes an ideologically encoded landscape. When a marker inscribes the meaning of war on material soil, it becomes an historical 'sight' which we stand in front of as we do displays in museums or exhibitions.[24] A former battlefield, then, is 'both material and meaning'.[25] It is a landscape of memory composed of physical relics left behind by history, such as remnants of fortifications, pits, graves or ruins; as well as names, stories and interpretations related to them. These historical signs and stories, inscribed on the surface of the landscape, strengthen, direct and validate both personal and collective memory.

As a part of the national narrative and memory, battlefields are often described as mystical places where it is still possible to experience imagined visions and sounds of the past, as at the 'Fighters' Trail' at Ilomantsi. However, the style of narrative marking of this particular landscape is exceptional, partly because there are hardly any visible marks left from the time of the battle. Most of the former battlefields of eastern Finland are marked in the more traditional way, with signposts and notice boards that tell us the course of events, monuments that point to and commemorate the deeds of the soldiers, and repaired or totally reconstructed relics such as trenches or pillboxes, that highlight the authenticity of the place.

This sense of the authenticity of an historical site does not exist fortuitously, but has to be maintained by tending the physical landscape. As Edward Tabor Linenthal argues in the case of famous battlefields in the United States, physical preservation is designed to preserve the sanctity of the site itself, to separate the sacred space from the surrounding secular space. Thus the landscape of the battlefield is 'frozen' as it was at the time of the battle, so that visitors can reflect on the meaning of the epic event in authentic surroundings. One consequence, of course, may be that a previous landscape of horror and death is transformed into a park of beauty and tranquillity.[26] In Finland, the battlefield of Kilpelänkangas in Kuhmo, where extremely heavy artillery fire took place during the Winter War, is one of the sites where the scars of war – grey, broken trunks of trees – are still visible in the landscape (see Figure 1, site 2). A footpath has been made for visitors, whose highlight is a scene in which there are three dead trunks standing in the middle of the open landscape. Presented in this way, the dead trees become symbolic. They evoke images of the three crosses on Calvary giving a visible testimony to the scale and destructiveness of the shelling. The sanctity of the site is thus created by associating its material landscape of tended, and thus artificially frozen, scenery, with ideological codes such as events and sites from the New Testament, and the ethos of Finnish war history.

What has been particularly a phenomenon of the 1980s and 1990s is the active development of these sites by the heritage industry, as exemplified by the new heritage centres constructed in the old battlefield areas. A good example of this transformation is the famous battle site of the Raate Road (see Figure 1, site 1). Previously there had been a Raate Winter War exhibition in one classroom of what had once been the local primary school, but in the mid-1980s a new visitors' centre, 'The Gateway to Raate' ('*Raatteen Portti*'), was built to cater for 25,000 visitors annually, mostly during the summer season. This is one of twelve major battlefield sites of the Second World War in contemporary Finland that are now marked as regionally and nationally important tourist sites. In addition to these, a number of other war relic sites have become local tourist attractions; the most important of these being the Salpa Line of forts, built across eastern Finland from south to north in 1940–41, and intended to be the last line of defence in summer 1944. No actual battles ever took place in the trenches, pillboxes, bunkers or tank obstacles of the Salpa Line. But as a visible, and quite imposing, part of war heritage, it is now regarded as significant, and twenty-nine major tourist sites have been developed along its length.

One question that arises in transforming battlefields and other military installations into heritage sites, is whether visitors should be shown what is most authentic or what is most spectacular. This is because the original relics, which are most authentic, can often be a disappointment for the tourist, something less than was expected. Many heritage developers believe that, for most tourists, it is all the same whether the trenches still visible in the landscape of famous Winter War battlefields in Suomussalmi and Kuhmo are the original ones that have been repaired, or reconstructions that look original. Their authenticity is not so important provided that the timber does not look too new. The traces must be recognizable as part of an imagined past and thus appear old, whatever their real age.

This perspective has been important in addressing a problem faced by war heritage-makers in Finland: that there are relatively few battle sites left inside Finnish territory. Most of the Winter War sites and almost all those of the Continuation War are situated on the Russian side of the border. Until recently these areas were beyond the reach of tourists. To compensate for this lack of access, and to honour the generation who fought in those particular battles, new types of heritage site based on replica landscapes and simulacrum milieux have been built in Finland itself. Since the late-1980s, more than fifteen sites of this kind have been built all over Finland, but particularly in the south and west, far away from the historical front lines. There are both replica copies of structures such as dugouts and trenches, whose originals are situated somewhere else; and simulacra that accurately resemble such structures. In most cases they have been constructed by veterans themselves, thus they are meant to serve not merely as tourist attractions but as reminders of the war efforts of local veterans for the post-war generation (see figure 3, below).

Figure 3 'This way to the trench experience'. The dugout of Veterans' Park on Mäntyharju, built in 1955, is an example of a replica or simulacrum landscape commemorating the Second World War (photo: Petri Raivo).

An example of these replica landscapes can be found in the area surrounding the former hunting lodge of the Finnish war hero, Marshal Mannerheim, on the shore of Lake Puneliajärvi (see Figure 1, site 4). The lodge was built originally in 1942 as a birthday gift from the Finnish soldiers to their leader and situated in the wilderness of Eastern Karelia occupied by Finnish troops. Before the end of the war it was dismantled and reconstructed in Loppi in the southern part of Finland, where Mannerheim continued to use it as a hunting lodge for several years. Later it was opened to the public, and it now attracts 14,000–15,000 visitors a year. Although the hunting lodge and sauna are situated in southern Finland, the local landscape of a lake, hills and dense forest can easily evoke images of the remote, romantic wilderness of Eastern Karelia. To strengthen this illusion, additional replicas have been built in the area during the last few years. In 1992 a new trench construction, '*Kenttävartio Sisu*' (the 'Gutsy-ness' Stronghold),[27] was built on top of the hill close to the lodge, to represent the nature of war in the wilderness of Eastern Karelia. To enter the stronghold, visitors have to pass through a barbed-wire entanglement and a fake minefield. Inside the site all the constructions – trenches, shelters, lookout and even the latrine – have been given names like Repola, Tiiksi, Ruka, Tsirkka and Outa: those of the mystical places and battles of eastern Karelia.

The 'Heritage Dugout' of Längelmäki in Central Finland, built by local war veterans in 1982, is an example of a simulacrum (see Figure 1, site 5). It does not represent any particular place but mimics the typical accommodation dugout built somewhere near the front line. Like most of these sites, it carries the status of a local war museum because of its collection of uniforms, weapons and other objects related to the war. In their leaflet, the builders of the site hope that 'with good imagination it is possible to get a feel of the wartime front conditions where the life of our nation was really decided'.[28]

This experiential and tangible dimension is the key to the memory-making function of these replica and simulacrum sites. In one way they act like the traditional memorials and marked battlefield sites, both commemorating the wartime generation and supporting the national narrative with its themes of sacrifice, unity and victory. What is different in the case of replica landscapes, simulacra and heritage centres as compared with traditional sites of commemoration, is that now these functions are all wrapped in a cloak of educative experience and entertainment. The new heritage sites are made to manifest, maintain and support the memory of war, and thus the national narrative, for the generations that do not have any personal experience of the wars concerned.

* * *

In many ways the experience and heritage of the Second World War is still a part of the Finns' everyday lives, and thus a part of their everyday cultural, social, temporal and spatial experience. One comes across signs of those who participated in the war everywhere: in their ideas and values, in their literary works and books, and in their memories and memorials. The Second World War has also left its mark on the Finnish landscape. The remains of old battlefields in eastern and northern Finland and hundreds of war monuments and cemeteries all over the country are an integral part of local and national memory and heritage, whereby the past history and culture of the nation is inscribed on historical landscapes and sites. These marked sites represent a special 'map' on which the spatial dimensions of the national memory are made visible. This map has been created by means of 622 cemeteries, 930 monuments, and forty heritage centres, museums, or replica landscapes and simulacra. These war memorial sites overwhelmingly confirm the nationalistic narrative of a unified Finnish history and culture maintained from the past up to the present day.

Although the process of setting up monuments has been continuous since the war, the end of the Cold War, the collapse of the Soviet Union and the fiftieth anniversary celebrations of the wars that Finland had to fight in 1939–45 have all boosted the demand for new marked sites indicative of this narrative of the nation. The contemporary monuments and heritage sites have been built not only for the veterans and their relatives but to serve the needs of post-war generations, too. Since the outbreak of the Winter War, the most important theme related to the memory of war has been the unity of the nation. This narrative of a political healing process, steadfastness and stubborn defence in a just cause, together with the sacrifices that were needed, has been the dominant theme of the memorials and memory of the wars. During the last decade the narrative of unity and sacrifice has been supplemented with the theme of victory. Nowadays the Finns are celebrating not only the survival of their country but a successful defence and a symbolic victory over their eastern neighbour.

Sixty years after the outbreak of the Winter War, the memory and commemoration of the Second World War is still a vital part of Finnish national identity. The national will and need to memorialize these difficult times has not come to an end, but rather the process is developing and taking on new forms. In addition to traditional war cemeteries, monuments and battlefield relics, the new heritage museums, replica landscapes and simulacrum milieux connected with the Second World War have become visible tourist attractions during the last decade. The personal sites of family mourning and of tributes to the memory of companions-in-arms, created to serve the generation that participated in the war, have been transformed into collective historical sites telling of heroic deeds and wartime sacrifices, for their successor generations.

Notes

1 J. Young, *The Texture of Memory: Holocaust Memorials and Meaning* (New Haven, 1993).
2 L. Haataja, 'Rauha ja sota', in L. Haataja (ed.) *Ja kuitenkin me voitimme: sodan muisto ja perintö* (And Still We Win: A Memory and Legacy of War) (Helsinki, 1994), 7–10.
3 A. Paasi and P.J. Raivo, 'Boundaries as barriers and promoters: constructing the tourist landscapes of Finnish Karelia', *Visions in Leisure and Business* 17.3 (1998), 30–45.
4 S. Zetterberg, 'Itsenäistyvä Suomi', in S. Zetterberg (ed.) *Vuosisatamme Suomi* (Our Century's Finland) (Porvoo, 1992), 49–52.
5 B. Pitt (ed.) *The Military History of World War II* (London, 1994), 16–17.
6 A. Tuomisto, *Suomalaiset sotamuistomerkit* (Finnish War Memorials) (Jyväskylä, 1998).

7 P. Visuri, 'Suomen sotahistorian kiinnekohtia', in H. Tiilikainen (ed.) *Muistoja Pohjolasta: Opas suomalaisten sotamuistomerkeille ja taistelupaikoille* (Memories from the North: A Guide to the Finnish War Memorials and Battle Sites) (Jyväskylä, 1999), 10–23.

8 M. Ignatieff, 'Soviet war memorials', *History Workshop* 17 (1984), 157–63.

9 T. Vihavainen, 'Hyvinvointi Suomi', in S. Zetterberg (ed.) *Suomen Historian Pikku Jättiläinen* (A Little Giant of Finnish History) (Porvoo, 1988), 867.

10 Tuomisto, *Suomalaiset sotamuistomerkit.*

11 A. Reitala, 'Sodan jäljet Suomen taiteessa', in Haataja (ed.) *Ja kuitenkin me voitimme*, 103–20.

12 M. Räsänen, 'Suomalaisten arki 1920- ja 1930-luvulla', in Zetterberg (ed.) *Vuosisatamme Suomi*, 177.

13 For example, *The Unknown Soldier* uses this metaphor. The film version of the book (made in 1955) actually starts with a scene where Finns are retreating and leaving a body of their fellow-soldier behind the lines – a powerful metaphor that is still widely recognized among Finns.

14 H. Hakala and R. Lipponen, *Luovutettujen alueiden sotilashautausmaát muistomerkkeineen* (Military Cemeteries and Memorials of the Ceded Areas) (Unpublished memorandum, 1999).

15 Tuomisto, *Suomalaiset sotamuistomerkit*, 276.

16 Reitala, 'Sodan jäljet Suomen taiteessa', 113–14.

17 J. Mayo, 'War memorials as political memory', *Geographical Review* 78.1 (1988), 62–75.

18 Young, *The Texture of Memory*, 2.

19 Tuomisto, *Suomalaiset sotamuistomerkit*, 277.

20 A. Baker, 'Introduction: on ideology and landscape', in A. Baker and G. Biger (eds) *Ideology and Landscape in Historical Perspective: Essays on the Meanings of Some Places in the Past* (Cambridge, 1992), 1–15.

21 See, for example, the argument that the battles of summer 1944 were a successful defence operation by the Finnish army (and thus a victory for Finland), rather than a triumph for the Red Army (and thus a disastrous defeat for Finland), in U.E Moisala and P. Alanen, *Kun hyökkääjän tie suljettiin* (When the Attacker's Way Was Closed) (Helsinki, 1989); and the discussion of the war crimes of Soviet partisans in eastern Finland, in V. Erkkilä *Vaiettu sota: Neuvostoliiton partisaanien iskut suomalaisiin kyliin* (Concealed War: Soviet Partisan Assaults on Finnish Villages) (Helsinki, 1998).

22 Hakala and Lipponen, *Luovutettujen alueiden sotilashautausmaat muistomerkkeineen.*

23 D. Lowenthal 'Age and artefact', in D.W. Meinig (ed.) *The Interpretation of Ordinary Landscapes: Geographical Essays* (Oxford, 1979), 104–28.

24 E. Diller and R. Scofidio, 'Suitcase studies: the production of a national past', in E. Diller and R. Scofidio (eds) *Visite aux armées: tourismes de guerre / Back to the Front: Tourism of War* (Basse-Normandie, 1994), 32–107.

25 My argument here is founded on empirical material concerned with Finnish war memorials which I have collected and analysed during the years 1997–99. My database includes information about war cemeteries and 546 other sites of commemoration related to the memory of the Second World War in Finland. This chapter is the first published account of the findings of this research, which pioneers a new approach to the topic.

26 E.T. Linenthal, *Sacred Ground: Americans and Their Battlefields* (Chicago, 1993).

27 *Sisu* is an ancient Finnish word with no exact translation into English, that describes the spirit of courage, fortitude and endurance against all the odds.

28 *'Längelmäen Perinnekorsu'* (The 'Heritage Dugout' of Längelmäki), a leaflet.

6 Remembered/replayed

The nation and male subjectivity in the Second World War films *Ni Liv* (Norway) and *The Cruel Sea* (Britain)

Peter Sjølyst-Jackson

Cinema, as a mass-medium, responds to, reflects upon and explores central issues of popular experience and consciousness. As Nicholas Pronay says in his survey of films relating to the Second World War made in Britain during the post-war era, cinema may have offered 'ways to tackle deeper problems without direct pain ... in the aftermath of a major war and especially of a total war which involved most people directly in some form or another'.[1] This chapter explores two war films that were produced in the post-war period in Norway and Britain, Arne Scouen's *Ni Liv* (Nine Lives) of 1957 and Charles Frend's *The Cruel Sea* (1953). Taking this dual focus, I seek to develop ways of thinking about male subjectivity, memory and war trauma in the context of different national frames of story-telling.

While *Ni Liv* belongs to a tradition in Norwegian fiction known as the 'Occupation Drama' which focuses on the collective struggle on the home front, *The Cruel Sea* paints on a different historical and geographical canvas. The British war narrative, as Graham Dawson explains, constructs civil life and military life as two distinct spheres: 'This rhythm – separation, convergence, and then a separating-out once again of the civilian and military wars as the Blitz comes to an end – provides the distinctive shape to Britain's experience of the Second World War.'[2] Echoing this rhythm, *The Cruel Sea* depicts both the wider collective struggle and that of individuals caught up in a continual process of separation and convergence. According to Gunnar Iversen, *Ni Liv* signals a shift in the development of the Norwegian Occupation Drama, from an exclusive focus on the collective struggle to individuals within that struggle: heroes and their 'inner life'. Alongside this emphasis on singular characters however, *Ni Liv* also presents its individuals explicitly in terms of collective, national belonging. Individuals, Iversen says, become 'something greater', 'icons' or 'symbols' of the collective. In *The Cruel Sea* the question of self-definition – individual, collective and national – is never made as explicit, a fact which should be understood in relation to the differing geographical proximity of the invading forces. While focus on such differences may highlight cultural specificity, there are also important similarities between the two films. Both display a concern with masculinity, the psychological strains placed on male combatants. As such the films address – via their singular male characters – legacies of war relating both to the traumas of individuals and trauma at a broader societal level. This chapter elaborates upon the complex ways in which war-narratives both approach and disavow such fraught problems.

* * *

Both *Ni Liv* and *The Cruel Sea* depict events during the war but were made in the post-war period and are thus part of much wider discursive formations which take place *after the event*. Significant, also, is how such forms of retroaction are publicly encouraged and supported. As Iversen points out, *Ni Liv*'s making received funding, and hence ideological endorsement, from the Church and the Education Department of the Norwegian state.[3] Moreover, when it was released in 1957, the press received the film with rapturous reviews. As one enthusiastic critic in the national paper *Dagbladet* put it:

> Let us absolutely send this film to Cannes, we can be responsible for it with pride. But it can't be avoided that a lot of Norwegian feeling lies in evaluating a work like this, – the pride of the man, the men, of our outrageous mountains …[4]

Its Oscar nomination as best foreign film of 1959 is a rare achievement by Norwegian standards, indeed, as recently as 1991 it was declared 'the best film of all time' during peak viewing hours on national television. In its canonical status, *Ni Liv*'s popular legacy towers over other Occupation Dramas of the period such as *Shetlands Gjengen* (The Shetland Gang, 1954) or *Kontakt!* (Contact!, 1956). *The Cruel Sea*, though immensely popular in its time, has not achieved similar posthumous recognition. Compared to films like *Bridge on the River Kwai* (1957) or *The Dam Busters* (1954), which seem perhaps the closest to achieving similar status to *Ni Liv* in Norway, *The Cruel Sea* appears relatively marginal. Yet it has struck me as significant how both the film, and Nicholas Monsarrat's novel upon which it was based, are always – in accidental conversations with people who recall it or when mentioned in critical literature – *vividly* remembered.

Some of *The Cruel Sea*'s emotive significance is surely down to investments in national identity. It was, of course, such issues Harold Wilson was appealing to when, as President of the Board of Trade (1947–51), he called for a British cinema which could 'genuinely show *our* way of life' as opposed to the 'gangster, sadistic and psychological films of which we seem to have so many'.[5] In terms of policy, this involved various strategies to curb the import of American films, although, ultimately, this did little to underpin growth in British film-production. It is interesting, therefore, to note that *The Cruel Sea* was the top grossing film of 1953 in Britain, constituting a significant critical and commercial success,[6] a success the novelist Len Deighton, in a recent article, does not hesitate to put down to its 'quality' and 'Britishness'.

> *The Cruel Sea* was a British production using a British book about a British subject with a script by a British writer … Perhaps if Britain didn't share

> with Hollywood the English language and its demand for world-wide 'product', we would still have a film industry.[7]

Might Deighton's fervent patriotic endorsement of *The Cruel Sea* testify to a certain unifying appeal of the film? Although I can't underpin such a claim with sociological evidence, it could perhaps be suggested that *The Cruel Sea*, like *Ni Liv*, is the sort of film whose popular endorsement, both with critics and audiences, cuts across traditional divisions between 'mass product' and 'quality drama'.

As Benedict Anderson has argued, in the collective imagining of nationality, 'regardless of the actual inequality and exploitation that may prevail in each, the nation is always conceived as a deep, horizontal comradeship'.[8] The war genre, it follows, with its appeals to unity in order to combat the antagonistic 'outside', is particularly powerful in its ability to awaken such national imaginings. However, as Pronay's formulation – that films may offer 'ways to tackle deeper problems without direct pain' – suggests, fiction and cinematic imagery involve more complex processes than straightforward nationalistic celebration.

In order to develop a more careful understanding of this, I now turn to Kaja Silverman's *Male Subjectivity at the Margins* (1992) which develops a model for the historical analysis of masculinity and war trauma. Feminist in orientation, Silverman draws upon Althusserian Marxism and Lacanian psychoanalysis and is, in this sense, also situated within the wider theoretical concerns of ideology critique, national identity and sexual difference as evidenced in the work of Klaus Theweleit, Ernesto Laclau, Chantal Mouffe and Slavoj Zizek.[9]

My focus is upon Silverman's analyses of male subjectivity in Hollywood films of post-war America. Central here, for Silverman, are the ways in which films such as *It's a Wonderful Life* (dir. Frank Capra, 1946) and *The Best Years of Our Lives* (dir. William Wyler, 1946) disrupt the received images of male subjectivity. The issue at stake here is how 'the dominant fiction', that is, the 'images and stories through which a society figures consensus', is being challenged.[10] In her readings, she argues that these post-war Hollywood films dramatize a profound 'historical trauma' which is evident in their 'preoccupation with male lack' and 'loss of faith in the familiar and the self evident'.

> The hero no longer feels 'at home' in the house or town where he grew up, and resists cultural (re)assimilation; he has been dislodged from the narratives and subject-positions which make up the dominant fiction, and he returns to them only under duress.[11]

For Silverman, the representation in *The Best Years of Our Lives* of war veterans returning to everyday life in post-war America involves no less than a 'forced confrontation of the male subject with castration', dramatizing 'his resulting dislocation from the paternal position and hence from the dominant fiction'.[12] According to Silverman, ideological belief is re-articulated, not by disavowal, but rather through the explicit representation of masculine suffering.

Silverman argues that the 'historical trauma' threatens the male subject with 'a *psychic* disintegration – the disintegration, that is, of a bound and armoured ego, predicated upon the illusion of coherence and control'.[13] Invoking Freud's elaborations on war neuroses, Silverman suggests that 'the historical trauma' emerges as 'an *internal* response to an *external* danger'. The historical trauma is thus related to what Freud, in *Beyond the Pleasure Principle*, called 'the death drive', a relentless and aggressive movement towards unbinding and death. This, for Silverman, relates to the very real sense in which the death drive operates 'in a dramatically exteriorized form', 'in the violence of the battlefield', 'the social orchestration of mass death'.[14]

Throughout her readings of post-war Hollywood films, she relates this specific sense of the death drive to images and stories of displaced men. In one such example, she reads a key scene from *The Best Years of Our Lives* as pushing the negativity of male lack to its limits, rather than covering it over. The scene in question is when one of the main characters, Fred, a former bombardier in the armed forces of the USA, 'imagines himself aloft once again on a bombing mission' as a result of coming across 'a field filled with hundreds of World War Two bombers'. During this brief moment, whilst Fred is in the grip of a vivid and overpowering reminiscence of his war experience, Silverman says, 'he escapes social rationality, opting for the non-ego over the ego, the threatening outside over the coherent inside, death over life'.[15] Moving at the edges of what is explicable, Silverman's psychoanalytic reading is at its best highly sensitive to the affective and sensory intensities of images and stories evoked in film. Her readings alert our attention to the fictionality of reality, the reality of fiction, in such a way as to illuminate aspects of the social and emotive significance of trauma. In the following pages, I develop a critical response to some of Silverman's articulations with special emphasis upon questions around memory and historical change. Before dealing with Silverman more directly, however, I seek to widen my current scope in order to include some of the valuable insights – historical and theoretical – offered in some studies of popular memory in Norway and Britain.

In an article about history writing on the Second World War, in the collection *National Fictions: World War Two in British Film and Television*, Graham Dawson notes the apparent fixity and persistence of war mythology in Britain. There is, he says, a popular investment in the mythic 'which is not amenable to "objective" or "factual" correction' since 'Dunkirk and the Battle of Britain have survived two generations of scholarly scrutiny!' In a follow-up article from the same collection, Graham Dawson and Bob West describe British conservative patriotism as 'the celebration of a glorious national past and its apotheosis in the present'. This, however, does not mean that these 'dominant meanings which have sedimented into naturalized, taken-for-granted conceptions' are fixed and unchanging. According to Dawson:

> written histories of all kinds, in taking their orientation from the mythic, are in fact addressing such taken-for-granted meanings; recycling, reworking, challenging, transforming them. This continues to be an active process

today. And it is this engagement with an active, fluid semantic field that guarantees the 'popularity' of war histories.[16]

Although drawing upon Dawson and West's work in the Popular Memory Group, Anne Eriksen's study of popular memory of the Second World War in Norway, *Det Var Noe Annet Under Krigen* (It was Completely Different During the War), does not place special emphasis on the transformative structure of war myths. In a telling articulation, she describes the structural openness of myth – in Barthes' terms 'made of yielding, shapeless associations'[17] – as indeed being open for 'different understandings' but only insofar as these are 'so imprecise that any conflicts and disagreements are never brought to light'.[18]

Eriksen's emphasis, I would argue, is not simply to be taken as a different reading of Barthes' myth-theory, but also as indicative of the differing dynamics of war mythology in Norway and Britain. Eriksen's study emphasizes the very acute sense in which stories of the occupation in Norway are unquestioningly received as 'important' and universally 'relevant'. The perceived gravity of the war story in Norway is conveyed in the popular expression: 'the five dark/murky years' (*de fem mørke år*); and the even more sharply delineating phrase, 'the dark brackets of our history' (*den mørke parantes i vår historie*).[19] Eriksen holds that Norwegian war mythology, by designating the occupation not simply as a 'past event' but as 'a completely different time' (*en helt annen tid*), works to close off and prevent any significant reconsideration or re-evaluation of it in the present.

Dawson and West call for closer consideration of memories excluded from the public realm, particularly the 'private' memories of people who lived through the war such as 'those of death and loss and the horrifying violence and suffering of war', which could 'provide the disorganized fragments of a potentially oppositional popular memory'. They suggest that 'the reunion of family or friends, or the more formal gatherings of servicemen ... offer opportunities to *reassess* the past in the light of subsequent experience and new information'.[20] If a film like *The Best Years of Our Lives* is capable of bringing into the public realm those forms of memory which have typically been confined to the 'private' sphere, Silverman does not explore such an avenue. This is perhaps surprising given her overall concern with the question of opposition and re-negotiation of the dominant fiction 'as a way of saying "no" to power'.[21] In line with this objective, she consistently focuses upon the *disruption* of the dominant fiction brought about by the historical trauma. According to Silverman, *The Best Years of Our Lives* 'never resolves the crisis it sets in motion' and, even though closing with marriage, it is more a 'declaration of ideological disbelief'[22] than a reaffirmation of the dominant fiction. At this point, however, Silverman does not discuss the film as potentially oppositional but rather emphasizes how critical response to the film (e.g. that of Robert Warshow) 'disavows the conspicuous display of male castration' and thus 'shores up the dominant fiction'.[23] In this way, she effectively suggests that putting the historical trauma to oppositional ends is impossible. Stating that *The Best Years of Our Lives* 'even four decades after its release ... remains profoundly disturbing, and at times almost unwatchable',[24]

Silverman on the one hand affirms its disruptive qualities yet, on the other, places the historical trauma *beyond* any political or ideological negotiation.

There is, therefore, a certain determinism in Silverman's approach, a tendency to answer questions before they have been put to the films.[25] My suggestion here is not to reject Silverman's model, but rather to consider the historical trauma with more careful attention to the dynamics of memory and historical change. We need, in short, to consider the complex ways in which films operate under the sign of retroaction. In Silverman's text, memory is referred to as performing certain 'binding' functions which bring trauma under control. She here draws upon the work of Laplanche and Pontalis but only in a limited way, overlooking, and I will return to this, the highly significant points they make about Freud's conceptions of memory as retroactive. Of traumatic memories, Silverman asserts that 'memories characterized by high degrees of affective and sensory intensity' are bound, 'anchored to signifiers' and brought within 'linguistic control' through the displacement of mnemic traces. Building on this analysis of how traumatic memories are bound within the individual psyche, she further suggests that:

> ... when the process of binding takes place on a larger, cultural level, it involves not only the transformation from one mnemic system to another, but also the gradual reaffirmation and reconstitution of the dominant fiction.[26]

Silverman then rejects the notion that the war, 'an event of monumentally tragic consequences, provided a privileged agency of social transformation' since, 'as we all know, the opposite has been the case'.[27] Though placed immediately following her theorization of memory, the past tense of this latter sentence completely omits any sense of the Second World War *as memory*, that is, as a secondary elaboration taking place *after the event* through discursive means such as cinema. In this respect, Dawson and West's emphasis upon popular memory as a transformative, potentially oppositional practice taking place in the present, is instructive.

However, while Dawson and West theorize memory essentially as an *active* process, I would argue that a psychoanalytic perspective draws attention to how the work of memory can surmount active appropriation. The tension here, between the conception of the past as something *which is returned to* via the ever-changing grids of the present, and the past as an unchanging, fixed 'event' or 'scene' *which returns* to traumatize and paralyse the present, can be elaborated through Freud's term *Nachträglichkeit*.

The French psychoanalyst and translator of Freud, Jean Laplanche, explains that the term *Nachträglichkeit* – 'deferred action' in Strachey's standard translation – although significant, never actually developed into a fully fledged concept. In its most basic sense, Freud used the adjective *nachträglich* in the purely temporal sense of 'later' or 'subsequently'.[28] In *The Language of Psychoanalysis*, Laplanche and Pontalis stress how *Nachträglichkeit* explicitly addresses itself to the problematic of retroactivity, which is to say, traumas belonging to a person's history are always in some sense being worked upon, revisited and revised in the present. In respect of this, they argue that Freud does not reduce 'the subject's history to a linear determinism envisaging nothing but the action of the past upon the present'.[29] In his more recent work however, Laplanche has argued that the sense of 'retrospective' in Freud's work is relatively rare, overshadowed by a 'conviction that what comes before determines what happens after'.[30]

What are the consequences of such a view? Is it too deterministic and reductive? Or might a rejection of 'determinism' deny the gravity of subjective experience? Consider the testimony of a British veteran of the Queens Royal Regiment, who, during the retreat from Dunkirk, saw civilians under his protection repeatedly strafed by German machine gun fire from the air: 'I will never forget it. Even sixty years on I can wake up and still see it vividly in my mind, I can relive every single moment and always will.'[31] From a theoretical standpoint, we cannot ignore this sense of *Nachträglichkeit* on the basis that it is 'determinist'; rather, we need to recognize these affective, sensory intensities as highly significant in the dynamics of war mythology and popular memory.

A striking example of this is evidenced in the controversial psychiatric work of William Sargant who, in Britain during the Second World War, endeavoured to establish the most efficient programme possible for curing servicemen suffering from war neurosis. Arguing that the most important lesson from Dunkirk and the Battle of Britain was 'never to let a neurotic pattern of thought or behaviour remain fixed in the patient's brain',[32] Sargant rejected Freudian psychoanalysis as too time-consuming and set about curing patients with a liberal use of drugs and electro-convulsive therapy. However, in spite of his anti-Freudianism, Sargant did establish a 'talking cure' of sorts. Through the use of drug therapy and suggestion, he would encourage patients to recall and relive traumatic events of war.

> When it came to the crucial moment in the event, such as the death of a close friend, Sargant would ask the patient to focus all his anger and grief on the German machine-gunner who had done it, or the man who fired the mortar, thus channelling all the suppressed fear and anger at the enemy.[33]

As an instance of binding the affective sensory intensities of trauma, we witness here a powerful reaffirmation of war mythology. Nevertheless, as the work of *Nachträglichkeit* and, in a wider cultural arena, popular memory implies, however inevitable the return of the trauma, there is nothing inevitable about the specific signifiers, cultural and ideological, that bind it. In response to Dawson's suggestion that the popular investment in war myths lies in the active, fluid process of the present, we should also stress the paradoxical durability of war myths as predicated on the inexplicable traumas they bind and rebind. Without resorting to Silverman's determinism, a psychoanalytic perspective certainly sees Dawson and West's notion of constructing an oppositional practice out of personal fragments as a far more precarious endeavour. Indeed, if we are to take Freud's notion of 'the death drive' seriously, any discourse aiming to recuperate such publicly repressed war memories is fraught with trauma, compulsion and aggressivity.

Below I engage in a critical reading of *Ni Liv* and *The Cruel Sea* bearing in mind the tensions of *Nachträglichkeit*. The preceding theoretical discussion should not be viewed as prescriptive for film analysis, rather, it serves to articulate a set of queries which structures my investigation. The question here concerns the operation of trauma in film, the consequences of its 'binding' and the politics of its articulation.

* * *

Andy Medhurst suggests that *The Cruel Sea* 'is the closest that British 50s war films came to making an anti-war statement'. Crucially, this follows from the key scene of the film when Captain Ericson (Jack Hawkins) of *HMS Compass Rose* chooses to deploy depth-charges – even though this will kill a number of British sailors struggling in the water, cast off from a previously sunken ship – due to the suspected presence of a U-boat in the depths beneath.

> The film here stresses the confusions, the anxieties, the sheer messiness of the war, all elements which had no place in the construction of heroic myths. Narratively Hawkins is excused by later sinking two U-boats, but there is no feeling of triumph.[34]

While this moment forms one of the principal articulations of the historical trauma in *The Cruel Sea*, the anti-war sentiment is complicated by a set of complex personal and political relations. As noted by many, it is the relationship between Ericson and his trusted 'number one' Lockhart (Donald Sinden) which takes centre stage. Closing with the end of the Second World War, the film shows Ericson and Lockhart, among the few original crew who survive *Compass Rose*, reflecting upon past events and looking hopefully to the future. The war genre, Medhurst says, allows 'a licensed space for the otherwise inexpressible', an exploration of 'the male–male relationship'.[35] The closure, however, leaves that relationship more or less confined to 'the past' while investing future hope in the traditional heterosexual couple. On account of his romance with Julie (Virginia McKenna), Lockhart – in contrast to Ericson, one supposes – has 'something to live for'. During the narrative itself, however, the Lockhart–Julie couple remains superficial and unconvincing in comparison to the Ericson–Lockhart relationship. Viewed against the backdrop of the 1950s – when the nuclear family as the dominant mode of social organization tightened its stranglehold, pushing homosexuality further away from what was legal and ideologically acceptable – it seems that Julie's character functions to shore things up. Her references to 'David and Jonathan' (Ericson and Lockhart) are riddled with repression and anxiety, bearing witness to the film's faltering attempts at containing the homoerotic economy: 'Women don't often have that relationship, and if they do it's not usually about something important … like running a ship or fighting a war.'

Though tenuous, the concluding hopefulness of *The Cruel Sea* relies upon a certain exorcism of various tensions and traumas relating to events before the sinking of *Compass Rose.* Here, a number of sub-plots put in place apparently irreconcilable conflicts of class and gender. These sometimes occur simultaneously, as when the menacing First Lieutenant Bennett's (Stanley Baker) background as a second-hand car salesman is set up against Officer Morrell's (Denholm Elliott) peace-time occupation as a barrister. Later, when Bennett falls ill with an apparent ulcer, he is confined to the position of the 'weaker sex' as Morrell, alluding to one of Bennett's previous insensitive gibes, suggests: 'Perhaps you have a bun in the oven?' Once Bennett is relieved of his duties and consequently removed from the story, the sub-plots revolve around the servicemen's relations with women on land. In contrast to Julie's self-effacing loyalty, Morrell's wife is a self-serving and unfaithful seductress preoccupied by parties, clothes and make-up, offering no understanding or affection. Morrell's death with the sinking of *Compass Rose* removes his anxieties and depressions from the centre stage of the story. The same fate is assigned to the two crew members Jim and Bob, whose future dreams are shattered when Gladys, the maternal, homely character, Bob's sister and Jim's wife-to-be, is killed in her house during an air raid. By 'allowing' Morrell, Jim and Bob to go down with *Compass Rose*, the film's hopeful ending seems to 'forget' rather than deal with tensions and anxieties relating to class and gender.

From the very outset *The Cruel Sea* focuses attention on the officers' ranking and competence. Although, in most respects, it rejects inflated forms of heroism, Ericson is ultimately shown to be more than adequate to the task of leadership, an obligation which weighs heavily on his shoulders. Anchored in 'the present', Ericson's voice-over narrative establishes that most of the crew 'had never been to sea before. As for the officers, I was the only professional senior among them. The rest were amateurs.' The sense of class-bound responsibility is only heightened when a crew member on the lower decks cries 'Bloody murderer!' at Ericson following the attempted attack on the U-boat involving the deaths of helpless 'British boys'. The depiction of harrowing war episodes introduces a complexity of emotion which cuts across, yet strongly reinforces, the film's world of class division which is underpinned by the Queen's English of the officers contrasted with the regional accents of the 'working-class' crew members below decks. While the profound sense of anguish problematizes the position of Ericson and the officers, the 'necessity' of that position is never in question.

In a brief but significant moment in the film, these burdens are connected to British nationality. Three foreign captains, French, Polish and Norwegian (!) previously rescued by *Compass Rose*, visit Ericson to offer their gratitude. Alluding to the unbearable incident earlier that day, the foreign captains offer their sympathy, proposing a release for the pain: 'naturally there will be thoughts … and for thoughts there is gin?' Ericson's uneasiness with the situation is underscored by a later scene in which Lockhart finds Ericson on the upper deck, sorrowful and depressed.

Lockhart:	Are you alright?
Ericson:	No … I don't mind telling you … I'm not.
Lockhart:	You've got to forget all about it. It's no good worrying about it now, you can't change anything.
Ericson:	There was a submarine there, I'm sure of it. It's where a submarine making an attack would have been. And if it was a submarine how many more ships and men would it have killed? I had to do it. Anyway, it's all in the report.
Lockhart:	It was my fault, I identified it as a submarine. If anyone murdered those men I did.
Ericson:	No one murdered them, it's the war, the whole bloody war. We've just got to do these things and say our prayers at the end.

Shortly after, as if to 'excuse' the emotional impact, the film has Ericson slumping into a chair, falling asleep in a drunken haze. Yet, this does little to change the content and the profound anguish of what has been said. Remarkable, certainly in British cinema of the period, this scene conveys Lockhart's aspiring identification with Ericson as paternal authority and, by consequence, his willingness to take on the 'burden' carried by that authority. This gesture, indicated by Lockhart's offer to take the blame on himself, puts in place a repetition-through-identification of Ericson's guilt-feelings coupled with unquestioning affirmation of the British, class-specific, paternal position.

While the Second World War certainly put Britain in a position of responsibility, *The Cruel Sea* firmly connects this to British upper-middle-class servicemen and hence disconnects it from the working classes and foreigners. Positioned as moral authorities, as internationally responsible, this image of British servicemen is intimately bound up with mythologies of Empire.

The retrospective assumption of such responsibilities placed upon British servicemen, however, was by no means inevitable. In 1946, the documentary, *A Defeated People – A Film About the Government of the British Occupied Zone of Germany* (dir. Humphrey Jennings), went out of its way to address a moral climate more apt to disavow than make commitments to the Germans who 'lost the war of their own making'. The documentary went on to argue that 'We [Britain] have an interest in Germany that is purely selfish: we cannot live next to a disease-ridden neighbour.' According to Pronay, the documentary displays the attitudes 'that war is hell; that it took German insanity to want a war; that it led to them behaving like murderous automatons'.[36] Seven years on, *The Cruel Sea* retains the sense – indicated by its documentary footage of large ships, convoys and the machinery of war – that to prevent the bloody war, one must succumb to the insanity of that war. Increasingly hardened in his attitude, Ericson is unapologetic about his duty: 'Now it's just a matter of killing the enemy', he tells Lockhart. Resolution in such circumstances – 'a cruel sea, made more cruel by man' – seems to hinge upon the battered grace of the British officer class, that is to say, an affirmation of Britain's 'colonial responsibility'.

Though dwelling upon, in Ericson's words, the 'dehumanized' aspects of war, the film also emphasizes the potential 'binding' which is allowed following an historical trauma. Near the end of the film, when 'the enemy' is finally seen, in the form of helpless Germans being rescued by the very ship that has sunk their U-boat, Lockhart says to Ericson: 'They don't look very different from us do they.' It is no less significant that this moment echoes a scene earlier in the film, when one of the crew members from the lower ranks is saving a British serviceman from the water: 'It's all right mate, we've got yer.' The repetition of this compassionate moment, this time when rescuing a German towards the end, not only forges a link across the narrative, it also establishes a moment of identification across the ranks and – albeit cautiously – an identification across the dividing lines of nationality and between allies and enemies.

* * *

In Norway, the historical trauma takes different forms, assuming other political functions. Rather than focusing on the machinery and dehumanization of war, *Ni Liv* focuses upon the ambiguities of occupation, enemies within and the 'collective struggle'. As Anne Eriksen argues, collective memory constructs the occupation as a time – 'the dark brackets' – in which Norwegians 'glimpsed their existential foundations and essence'.[37] *Ni Liv* tells the story of how a heroic Resistance soldier, Jan Baalsrud (Jack Fjellstad) must rely upon the collective effort of a Norwegian farming community for his salvation. The opening credits state:

> Over a hundred countrymen, known and unknown, helped him to survive. This film can't include every individual and therefore names no one by their real names. Those who participate stand as symbols for everyone.

The film opens from the point of view of Baalsrud who, in a Swedish hospital, must provide a 'full report' to a delegate from London of his escape from the Nazis, across Tromsø in the north of Norway into the neutral territory of Sweden, in order to avoid arrest for his Resistance activities. During his escape, Baalsrud becomes snow-blind in the stormy mountains and eventually falls into the helping hands of a young couple from the farming community, Agnes (Henny Moan) and Martin (Alf Malland). At this point, the film develops a significant sub-plot in which themes of Norwegian nationality and cross-generational trust are played out. Whilst concealing Baalsrud from the Germans, Agnes and Martin don't let 'Grandfather' (Joachim Holst-Jensen) in on their secret. Increasingly suspicious, Grandfather is disappointed with their mistrust of him, but nevertheless proves his trustworthiness: 'Don't forget the important details my boy, I've had to tidy up after you, stopping people's mouths.' Emphasizing this theme of cross-generational trust and collective dependency, Grandfather is constantly seen in connection with Agnes and Martin's newborn child. These themes and images owe much to Norwegian National Romanticism, a literary movement including writers like Bjørnstjerne Bjørnson whose canonized *Synøve Solbakken* (1857) remains set reading on the national curriculum.

Anne Marit Myrstad, who has observed this aesthetic in Norwegian silent cinema from the 1920s onwards, identifies Synøve from Bjørnson's story as epitomizing National Romantic imagery, 'fair-haired, beautiful and mild but with physical and moral strength, a farmer's daughter'. Myrstad notes how this is connected to 'true' Norwegian culture which 'was to be found in farming communities in remote valleys'.[38] Whereas such stories typically concern the courtship of the young, Agnes and Martin of *Ni Liv* represent the next logical step, that is, the patriarchal family unit.

From the very outset, *Ni Liv* delineates the position of authentic Norwegians in terms of a distinction, not from the German occupants but, rather, from the 'enemy within'. Indeed, Baalsrud's flight is caused by someone's treachery. Upon contacting a supposed ally, shoemaker Hansen, Baalsrud and his fellow Resistance soldier are not recognized. Struck by fear, the shoemaker asks them to leave immediately; however, Baalsrud has already found a piece of a paper on the floor which reads: 'CONTACT WITH THE ENEMY ... IS PUNISHABLE BY DEATH.'

Baalsrud:	Who are you?
Shoemaker Hansen:	Hansen.... the shoemaker!
Baalsrud:	And who is the enemy?
Resistance soldier:	And who are we?

The *old* Shoemaker Hansen, due to his Resistance activities, has been captured by the Germans. The new shoemaker, who subsequently reports the Resistance to the Germans has, not insignificantly, the *same name.* In her account of the Norwegian war myth, Eriksen argues that disloyal Norwegians provoke far more fear and resentment than the occupying forces themselves. Germans were in fact often perceived as 'only human' and 'not to blame', while 'the real Nazis' were the defecting Norwegians. Part of this legacy is the abusive term for cowardice and treachery common in both Norwegian and English, *quisling.* The specificity of the term lies in its derivation and close association with the Norwegian Nazi leader, Vidkun Quisling, who was executed for treason (*landsforæderi*) following the post-war judicial hearings.[39]

The difficulty of distinguishing Shoemaker Hansen from 'true' Norwegians testifies to the profound ambivalence and uncanniness of occupation. The hazardous delineation of Norwegian from Nazi, moreover, is manifest in a broader cultural arena. As Myrstad points out, 'Nazis drew heavily on the national symbols and the notion of the protection of the true Norwegian soul, inherited ideologically from national romanticism.' The filmic adaptation of Knut Hamsun's *Growth of the Soil* (*Markens Grøde*) of 1921, Myrstad notes, 'was an inspiration for the Nazi "Blood and Soil" ideology'.[40]

The uncanny associations and connotations of National Romanticism with Nazi ideology are thus ripe in the post-war period, which helps us to understand the historical weight of these ambiguities and how *Ni Liv* is shaped as a response. Part of this response, as Eriksen argues, is an over-emphasis on the Norwegian

Resistance which, in relative magnitude, formed a minimal part of people's lives during the occupation. As a retrospective way of coping with the ambiguities of occupation, I would argue that the maintenance of an unproblematical Norwegian national identity in the post-war period relied – and still relies – on the repression, if not foreclosure, of the historical associations between Norwegian National Romanticism and Nazi ideology. It is with the rigid distinction of the traitor, the 'real Nazi', from the 'true' Norwegian that *Ni Liv* is able to construct its heroic narrative of the 'authentic' collective struggle. In this respect, an historical trauma is bound into a narrative of 'authentic' Norwegian culture. Indeed the very format of Occupation Dramas is always, in some sense, about claiming back something that has been wrongfully taken away; the 'true' Norwegian nation.

Yet *Ni Liv* contains an element which, although marginalized in its narrative, can be seen as agitating the very framework upon which the film is constructed. After a number of failed attempts to smuggle Baalsrud across the border to Sweden, he is left physically damaged and increasingly delirious in a snow-hole in the mountains for twenty-seven days. At this point, an unexpected twist occurs; a group of Sames (Laplanders) come to his salvation and transport him safely across the border with reindeer. The Sames – an 'ethnic minority' whose home, Lapland, extends across northern Scandinavia (Sweden, Finland and the Kola Peninsula of Russia) – are a historically suppressed and colonized people in Norway. Within the ideological universe of *Ni Liv*, the appearance of Sames is kept to the margins and they seem altogether excluded from the 'collective struggle'. At one point they seem little other than scapegoats when the heroic Norwegian Resistance soldiers (represented by uncannily Aryan, fair-haired men) fail to locate Baalsrud's snow-hole in order to get him across the border. Following one such incident, one of the Resistance soldiers complains about the Sames who were guiding them through the mountain terrain: 'They swore and cursed in Norwegian, Swedish and Finnish, there's no time for aping around (*apestreker*).'

Although it is the Sames who save Baalsrud in the end, the film's emphasis lies firmly with the 'collective struggle' which eventually, and I will return to this, becomes the hero's 'inner battle'. In the narrative resolution when Baalsrud, in the Swedish hospital, takes his first tentative steps unaided by the nurses and walks away down the corridor, the effort of the Sames seems more or less erased. Due to their lack of a unified national belonging, a lack made explicit in derogatory terms in the film ('They swore and cursed in Norwegian, Swedish and Finnish'), the Sames are not easily assimilated into those 'over a hundred *countrymen* (*landsmen*)', those 'symbols for everyone' who took part in the collective effort to save Baalsrud.

Although their appearance seems marginalized, perhaps encouraging a certain degree of forgetting, the Sames are nevertheless present. In fact, their basic narrative function is crucial for Baalsrud's survival, a function which seems entirely out of place within the ideological thrust of the film. In this respect, the presence of the Sames agitates and makes problematical Baalsrud's 'triumph'. In its unresolved state, this representation testifies perhaps to the place of Sames – or, more precisely, to the way in which they have been placed – in Norwegian post-war history. The state's policy of 'Norwegianizing' them through the obligatory learning of the Norwegian language (and prohibition of their traditional singing, '*joik*', in schools) was common throughout the 1950s and is indicative of prevailing tensions. On a more fundamental level, what these policies display is the profound unease, if not antagonism, that Norway – with its deeply held ideologies of national belonging – has towards elements which complicate this notion. Despite its residual antagonism and suppression of Sames, *Ni Liv* in fact leaves their place as an open question, an opening which, in an optimistic reading, might be seen as indicating something more heterogeneous than prevailing notions of nationhood. Interestingly, Per Høst's documentary film about the life of Sames, *Same Jakki*, released the same year as *Ni Liv*, achieved comparable financial success.[41] This interest, however, remained on the level of curiosity which, in the context of *Ni Liv*, simply renders them as an anomaly, falling outside of the central logic and ideological tensions of National Romantic imagery.

*   *   *

In his survey of films relating to the Second World War made in Britain during the post-war era, Pronay argues that, collectively, these films 'allowed people in the audience to re-live vicariously their experiences, the fears, guilt and dilemmas of their own particular war; and to catharsise psychological sores still festering'.[42] Yet for all its particularity, as Silverman suggests, it is somehow the imagery of masculine suffering that takes precedence in popular representations of historical trauma. I will end this chapter by discussing the two films' explorations of male subjectivity and trauma.

Whilst *Ni Liv* in one sense disavows historical traumas which are disruptive of national belonging, in a different sense, it approaches those traumas metaphorically through the imagery of masculine suffering. On account of Baalsrud's total dependence on the collective, Iversen suggests that *Ni Liv* problematizes myths of the autonomous hero.[43] Yet, Baalsrud's 'lack' is treated with a peculiar sense of 'seriousness'. In the opening of the film, when the female delegate, to whom Baalsrud must provide a 'full report', notices his bandaged feet, she falls silent, as if embarrassed, provoking an abrupt response from Baalsrud, 'Nothing to worry about!' The sense of prohibition of direct utterance regarding his lack is also played out in the scene of the actual amputation which the film conveys by cross-cutting between Baalsrud's tortured face, clenched fist and the stormy landscape outside. This logic of alluding to, but not showing, finally dissolves towards the end of the film when, after a long wait in a snow-hole in the cold mountains, Baalsrud's physical deterioration is paralleled by his psychic deterioration. Hearing sounds of footsteps in the snow, Baalsrud sees an image of himself, on crutches, amputated at the knees. This hallucination is followed by images of his own funeral procession, which he then greets with manic laughter. This suggestion and dramatization of 'madness', however, does little to problematize male subjectivity. Although we witness his deterioration and confrontation with castration, the structure of battlefield narrative is simply moved to a psychological level, putting masculinity to the test, rather than questioning it as such.

I have suggested that the narrative rebinding in *The Cruel Sea* stems from the trope of 'British (colonial) responsibility' figured in the officer class, a structure which is inseparable from ideologies of the Empire which were increasingly being questioned through the 1950s. Although 'forgetting' an array of social tensions developed in the first half of the film, the second half increasingly portrays the war as machine-like, dehumanized and beyond the mastery of the crew. Also Ericson, the paternal authority and moral centre of the film, becomes obsessive in his relentless search for German submarines. This shift is signalled when *Compass Rose* goes down, but also by a scene of irresolvable tension and *Nachträglichkeit* when Ericson is inspecting his new ship. Here, the sight of communication pipes triggers a traumatic memory of an event during the sinking of *Compass Rose* when Ericson, via the communication pipes, heard the frenzied screaming of helpless crew members in the lower decks.

Cross-cutting between close-ups of Ericson's face and the pipes, the sequence enacts a repetition of the original event, transposing the sounds of screaming and yelling on to the present scene. The association between the new ship's communication pipes and the dying crew members of *Compass Rose* takes us deeply into Ericson's position, helpless in the grip of an overpowering memory. Having been compelled to close the pipes in the original event, Ericson now repeats his action *nachträglich* in the present setting. Immediately after cutting off the traumatic memory by quite literally putting a lid on it, Lockhart appears and Ericson regains his composure. Thus, while the characters are involved in a disavowal of the historical trauma, the film nevertheless *shows* the return of the repressed, the incessant repetition of trauma in the present.

In *Beyond the Pleasure Principle*, Freud considers repetitive acts, such as child's play, recurring dreams/nightmares, regular topics of conversation and the recounting of stories, as manifestations of 'the compulsion to repeat'. As technologies of reproduction, and hence repetition, films facilitate the repetition compulsion insofar as they can be replayed up to and throughout our present moments. The point at which active pleasurable repetition, which functions to bind and rebind, turns into its 'daemonic' opposite, the unbinding death drive, in which repetition becomes increasingly compulsory and involuntary, is never simply transparent. Yet it seems that both *Ni Liv* and *The Cruel Sea* contain elements of both, and it is in this precise Freudian sense, I argue, that war films do not necessarily recuperate historical traumas within the dominant fiction. The reason why *Ni Liv* remains a more repeatable, pleasurable and hence canonized film, lies perhaps with the issue of mastery and unbinding. *Ni Liv*'s climactic ending is predicated upon Baalsrud regaining mobility. Having been through hell, he triumphs over 'madness', both literally and metaphorically. *The Cruel Sea*, by contrast, lacks a closed narrative in which trauma acts as a prelude to catharsis. Its representation of the return of the repressed, the compulsion to repeat past traumas in the present, offers no simple reaffirmation of dominant fictions.

Notes

Many thanks to Bob Bennett for encouraging me to consider my dual nationality as a resource in this project. Thanks also to Katrine Skretting for guiding my initial Norway-based research. I would also like to record my debt to Val Hill, Clare Lewis and the editors of this volume for their useful criticism and advice.

1 N. Pronay, 'The British post-bellum cinema: a survey of the films relating to World War II made in Britain between 1945 and 1960', *Historical Journal of Film, Radio and Television* 8.1 (1988), 39.
2 G. Dawson, 'History-writing on World War II', in G. Hurd (ed.) *National Fictions: World War Two in British Film and Television* (London: BFI, 1984), 4.
3 G. Iversen, 'En indre kamp: Om Arne Skouens *Ni Liv*' (Inner Battle: on Arne Skouen's *Nine Lives*), *Z filmtidskrift* 1 (Oslo, 1993), 46–8; my translation. For a condensed and more generalized account of *Ni Liv* published in English, see Iversen's chapter on Norwegian film history in T. Soila, A. Söderberg-Widding and G. Iversen, *Nordic National Cinemas* (London: Routledge, 1998), 121–30.
4 Quoted from L.T. Braaten et al., *Filmen i Norge: Norske Kinofilmer Gjennom Hundre År* (Film in Norway: A Hundred Years of Norwegian Cinema) (Oslo: Gyldendal, 1995), 212; my translation.
5 Harold Wilson quoted from C. Barr, 'Amnesia and schizophrenia', in C. Barr (ed.) *All Our Yesterdays: 90 Years of British Cinema* (London: BFI, 1986), 14; my emphasis.
6 Charles Barr, *Ealing Studios* (London: Cameron & Tayleur / Newton Abbot: David & Charles, 1977). See also S. Harper, *Picturing the Past: The Rise and Fall of the British Costume Film* (London: BFI, 1994), 147–53.
7 L. Deighton, 'Sand and sea', *Sight and Sound* n.s. 5.1 (Jan. 1995), 33.
8 B. Anderson, *Imagined Communities: Reflections on the Origins and Spread of Nationalism* (revised edn, London: Routledge, 1991), 7.
9 See K. Theweleit, *Male Fantasies, vol. I: Women, Floods, Bodies, History*, trans. S. Conway (Minneapolis: University of Minnesota Press, 1987); E. Laclau, *Politics and Ideology in Marxist Theory* (London: Verso, 1977); E. Laclau and C. Mouffe, *Hegemony and Socialist Strategy* (London: Verso, 1985); S. Zizek, *The Sublime Object of Ideology* (London: Verso, 1989).
10 K. Silverman, *Male Subjectivity at the Margins* (London: Routledge, 1992), 30.
11 Ibid., 53.
12 Ibid., 106.
13 Ibid., 62.
14 Ibid., 56.
15 Ibid., 80–6.
16 Dawson, 'History-writing on World War II', 2.
17 R. Barthes, 'Myth today' in Susan Sontag (ed.) *Barthes: Selected Writings* (London: Fontana, 1983), 105.
18 A. Eriksen, *Det Var Noe Annet Under Krigen* (Oslo: PAX, 1995), 19; my translation.
19 Ibid., 23–5; my translations.
20 G. Dawson and B. West, 'Our finest hour? The popular memory of World War II and the struggle over national identity', in Hurd (ed.) *National Fictions*, 9–11.
21 Silverman, *Male Subjectivity at the Margins*, 389.
22 Ibid., 67, 87, 89.
23 Ibid., 89–90.
24 Ibid., 67.
25 This manifests itself elsewhere in Silverman's discussion, such as when she links the masculine 'illusion of coherence and control' with 'the stable core around which a nation's and a period's reality coheres' (ibid., 41–4). Though suggestive, this remains undeveloped and hence too reductive in terms of its conception of 'nations' and the operations of fictions therein.

26 Ibid., 64.
27 Ibid.
28 J. Laplanche, *Seduction, Translation, Drives*, trans. M. Stanton (London: ICA, 1992), 218.
29 J. Laplanche and J.-B. Pontalis, *The Language of Psychoanalysis*, trans. D. Nicholson-Smith (London: Hogarth Press, 1973), 112.
30 Laplanche, *Seduction, Translation, Drives*, 220.
31 Quoted from W. Holden, *Shell Shock, The Psychological Impact of War* (London: Channel 4 Books/Macmillan, 1998), 77.
32 W. Sargant, as quoted in ibid., 80.
33 Ibid., 82.
34 A. Medhurst, '1950s war films', in Hurd (ed.) *National Fictions*, 36–7.
35 Ibid.
36 Pronay, 'The British post-bellum cinema', 43.
37 Eriksen, *Det Var Noe Annet Under Krigen*, 51; my translation.
38 A.M. Myrstad, 'National Romanticism and Norwegian silent cinema', in R. Dyer and G. Vincendeau (eds) *Popular European Cinema* (London: Routledge, 1992), 181–7.
39 Eriksen, *Det Var Noe Annet Under Krigen*, 85–9.
40 Myrstad, 'National Romanticism and Norwegian silent cinema', 189–92.
41 Iversen, 'En indre kamp', 48.
42 Pronay, 'The British post-bellum cinema', 51.
43 Iversen, 'En indre kamp', 50–1.

7 Postmemory cinema

Second-generation Israelis screen the Holocaust in *Don't Touch My Holocaust*

Yosefa Loshitzky

The memory of the Holocaust and its victims has always been a locus of controversy in the Israeli public sphere. The Zionist view of the Holocaust is predicated on the perception of the state of Israel as the most suitable monument to the memory of the Jews of Europe, and a form of secular redemption for their suffering. Indeed, the United Nations' decision in 1947 to divide Palestine into two states, one Arab and one Jewish, was directly connected to Western guilt regarding the extermination of Europe's Jewish population, as well as an attempt to solve the problem of Jewish refugees. Furthermore, the Israeli state views itself as the representative of the destroyed European Jewry, and therefore as the custodian of the 'official' memory of the Holocaust. Numerous speeches delivered by Israeli politicians regarding the relationship between the establishment of the Zionist state and the Holocaust bear testimony to the success of the state in nationalizing and monopolizing its memory, and constructing this as part of the Zionist narrative of redemption.[1]

This narrative has a mythical function which privileges Zionist pioneering over life in the Diaspora. Zionism is based on the belief that life in the Diaspora is unworthy and that Jews should live only in Israel. Zionist ideology therefore negates the value of the life of those Jews who want to live outside of Israel: within its framework their condition is seen as unnatural, and they are expected to return to their homeland. As Charles S. Liebman observes, 'there can be no conception of Zionism without a conception of diaspora'.[2] The state of Israel, the ultimate goal and fulfilment of the Zionist project, is regarded as the antithesis and negation of 2,000 years of Jewish life in the Diaspora. From a Zionist perspective, the logical culmination to the Jewish Diaspora in Europe was the Holocaust. James Young notes that early leaders of the state, like David Ben-Gurion, Israel's first Prime Minister, 'regarded the Holocaust as the ultimate fruit of Jewish life in exile; as such, it represented a diaspora that deserved not only to be destroyed, but also forgotten'.[3]

This notion of the 'negation of exile' is crucial to Zionism and its memory of the Holocaust. As Amnon Raz-Krakotzkin explains, 'negation of exile' refers to a mental attitude and ideological position that sees the present Jewish settlement in, and sovereignty over, Palestine/Eretz Israel as the 'return' of the Jews to a land considered to be their homeland, and described as empty.[4] Negation of exile is seen to be the 'normalization' of Jewish existence, the fulfilment and 'solution' of Jewish history. According to this point of view, the cultural framework that Zionists wished to actualize was an 'authentic' Jewish culture, one that could not develop under exilic circumstances, and which was also considered to be in direct continuity with the ancient, 'pre-exilic' past. As such, the negation of exile demanded cultural uniformity and the abandonment of the different cultural traditions through which Jewish identity had previously been determined. A phrase attributed to Ben-Gurion posits the objectives of Zionism as 'getting the Jews out of the Diaspora and getting the Diaspora out of the Jews'.[5]

These conceptions also shaped the Zionist image of the 'new Jew', defined in antithesis to the rejected 'old Jew', or 'exilic Jew' of the ghetto. According to Zionist ideology, the old Jew – rootless, cosmopolitan, unproductive, weak and passive – is to be negated and replaced by the new Jew, who is productive, rooted in his own land, strong and active. The new Jew is not necessarily a *Sabra*, someone who was born in Israel; he or she could also be an immigrant from the Diaspora. Nevertheless, the figure of the mythic *Sabra* epitomizes the ideal of the new Jew.[6] Immigrants to Israel, both before and after the establishment of the state in 1948, were required to suppress their language, culture and identity (and in many cases to change their names); and to integrate into the Hebrew, *Sabra*-oriented culture which despised the values and attitudes of the old diasporic Jew. This official 'melting-pot' policy of the new state (especially during Ben-Gurion's time) attempted to assimilate all the different identity groups, particularly the masses of immigrants who came to Israel from Europe and North Africa in the 1950s, into the imaginary ideal of the new Jew. Both Yiddish (the language of most of the *Ashkenazi* immigrants from Eastern Europe) and Arabic (the language of the *Mizrahim* or Oriental Jews,[7] who usually migrated from the Arab countries) were rejected and suppressed (sometimes even violently) by the Zionist establishment. Many of the Zionist leaders, including Ben-Gurion, were *Ashkenazi* Jews whose first language was Yiddish. Yet they fought Yiddish as the language of the hated Diaspora and insisted on promoting Hebrew as the only recognized language in Israel.[8]

During the 1950s, the days of mass immigration from Europe and North Africa, alienation and suppression characterized the foundational experiences of both the *Ashkenazi* Holocaust survivors and the *Mizrahim* in Israel. Both groups competed with each other for resources and also came into conflict with the new Jew, raised on the ethos of the negation of exile and, along with this, of diasporic Jewish culture. The shocking encounter between the ideal of the *Sabra* and the *Mizrahim* was not significantly different from that between the *Sabra* and the Holocaust survivors. This latter encounter was traumatic because the survivors epitomized for the *Sabra* the ideological construct of the old Jew. They symbolized, in particular, Jewish weakness, and there were real fears that the new, heroic Jews would be 'infected' with this by the survivors.

Although the state laid claim to the memory of the Holocaust, then, this did not refer to the real survivors and their suffering: Israel remembered the

Holocaust but forgot the survivors. In the early years of Israeli statehood, the dominant tendency was to unify, standardize and nationalize the memory of the Holocaust. The dominant position, best expressed by the Minister of Education and Culture during the Second Knesset (Israeli Parliament) (1951–55), Ben-Zion Dinur, 'gave preference to national and united Holocaust commemoration over and above the particularistic commemoration of individual movements'.[9] Consequently, during the 1950s the Holocaust occupied a marginal place in Israeli public discourse, and survivors, who epitomized the old Jews, were subject to alienation, latent repression and self-suppression. While most of the murdered European Jews were not Zionists, Israel presented itself as their sole inheritor. In the reparation agreement with Germany, for example, compensation was paid to the government of Israel and not individually to the survivors, most of whom received very little money (if any). Currently, many old and sick Holocaust survivors still live in poverty in Israel with hardly any state support. In a fragile post-war Israel where survival, economic growth and military heroism were the priorities, Israelis did not want to hear about the Holocaust. The survivors' stories could not be heard in case they turned the country into 'a big insane asylum'; and therefore the survivors became 'immigrants in their own land'.[10]

A turning point in Israel's attitude towards the Holocaust occurred as a result of the Adolf Eichmann trial in 1961. 'Holocaust survivors were finally given the chance to speak', and, overnight, 'they became the focus of national attention'.[11] As Haim Gouri observes, the Eichmann trial, ' "historic" in the fullest sense of the term, compelled an entire nation to undergo a process of self-reckoning and overwhelmed it with a painful search for its identity'.[12] A further new phase in the perception of the Holocaust was opened by the June 1967 War, followed by an ideological polarization between the left and the right which reached its peak with the rise to power of the right-wing Likud party in 1977. Since the early 1980s the memory of the Holocaust has been institutionalized within the curriculum of the state's educational apparatus, and ritualized in Israeli official discourse. Organized visits of Israeli and Diaspora Jewish youth to the death camps in Poland (known as 'the March of the Living'), and the appropriation of the symbols of the Holocaust , such as the yellow Star of David, by racist, ultra-right-wing movements (*Kach* and *Kahana Hay*), have both become a part of Israeli civil religion.[13]

In the 1990s, Israeli public discourse on the Holocaust has shown greater openness towards the more contradictory and disturbing aspects of the Holocaust. The emergence of the Israeli 'new historians',[14] with their revisionist reading of the history of the Israeli–Palestinian conflict; the corrosive effect of the *Intifada* (the Palestinian popular uprising) on the Israeli public; and the start of the peace process with the Palestinians and Israel's Arab neighbours, have all worked to synthesize the concerns of Israel's artistic, intellectual and academic communities, encouraging the Israeli public at large to re-examine the space occupied by the Holocaust in the grand narrative of Zionism. Israeli attitudes towards the Palestinian 'Other', the disturbing charges of collaboration between Zionism and Nazism,[15] and the traditionally arrogant and patronizing attitude of

Israelis towards the Jewish Diaspora, have all been publicly discussed and debated. Furthermore, the abandonment of the melting-pot ideology and growing awareness (and acceptance) since the 1980s of the multicultural, multi-ethnic composition of Israeli society, is generating an increasing fragmentation and segmentation of the memory of the Holocaust along lines of ethnicity and/or religion.

As in Germany – where most of the discussions on Nazi crimes against the Jews first took place in the cultural and aesthetic spheres – so in Israel, art preceded purely political debate. Films such as Orna Ben-Dor's *Because of That War* (1988), as well as the music of Yehuda Poliker, an Israeli popular rock-singer whose Greek parents are Auschwitz survivors (Greek Jews are considered to be *Mizrahim*), opened Israeli public discourse to the suppressed plight of the Diaspora survivors, and to the effect of their lived experience on their sons and daughters, who have come to be known as 'the second generation'. In a play such as *Arbeit Macht Frei vom Toitland Europa* (Work Frees from the Death-land of Europe) (Mayan/Acre Theatre Group, 1993–96), the Israelis' and the survivors' selves are disturbingly mirrored in their Palestinian Other.

The Israeli historian, Dina Porat, delineates three stages in the evolution of perception within Israel's culture of Holocaust memorialization.[16] In the 1950s, Israeli society perceived the survivors in terms of an amorphous mass composed of an incomprehensible number of '6 million' dead who were commemorated by state ceremonies. Later, during the 1970s, those '6 million' were conceived in terms of communities, which are more defined publics. They were memorialized by ceremonies in schools, by the 'March of the Living' tours, and by the *Yizkor* books of the destroyed communities. In contrast, the current, third stage of memorialization, that began in the 1980s, is characterized by a personal attitude, informed by the understanding that (in the words of an Israeli popular song) 'each person has a name', and that each of the 6 million has her or his own story. The emerging memoir culture of the survivors is part of the process which places the individual and her or his story at the centre of memorialization. The gradual change in the attitude of Israeli society to the Holocaust and the survivors, as Porat and others have observed, includes also a more complex view of notions of heroism, survival and resistance. This view is more receptive to the survivors' voices and the complexities and multiplicities of their experiences. The survivors' voices found, perhaps, their most sensitive response among the second generation.

In a moving article, Marianne Hirsch, Professor of French and Comparative Literature at Dartmouth College and a child of Holocaust survivors from Romania who migrated after the war, first to Vienna and then to the United States, says that, along with many European Jews of her generation, she shares a sense of exile from a world they have 'never seen and never will see, because it was irreparably changed or destroyed by the sudden violence of the Holocaust'.[17] Yet, she adds, this world – those countries from which their parents were exiled – will always be, for the second generation, 'the place of identity, however ambivalent'.[18] Children of survivors, according to Hirsch, live at a greater temporal and spatial remove from that decimated world.

> The distance separating them from the locus of origin is the radical break of unknowable and incomprehensible persecution; for those born after, it is a break impossible to bridge. Still, the power of mourning and memory, and the depth of the rift dividing their parents' lives, impart to them something that is akin to memory.[19]

Searching for a term that would convey its temporal and qualitative difference from survivor memory, Hirsch has chosen to call this secondary or second-generation memory, 'postmemory'.

> [Postmemory] is a powerful form of memory precisely because its connection to its object or source is mediated not through recollection but through an imaginative investment and creation. That is not to say that memory itself is unmediated, but that it is more directly connected to the past. Postmemory characterises the experience of those who grow up dominated by narratives that preceded their birth, whose own belated stories are displaced by the stories of the previous generation, shaped by traumatic events that can be neither fully understood nor re-created.[20]

Holocaust postmemory, according to Hirsch, attempts to bridge more than just a temporal divide.

> The children of exiled survivors, although they have not themselves lived through the trauma of banishment and the destruction of home, remain always marginal or exiled, always in the Diaspora. 'Home' is always elsewhere ... This condition of exile from the space of identity, this diasporic experience, is characteristic of postmemory.[21]

For many second-generation Israelis, film has become the main vehicle through which they articulate their Holocaust postmemory. These postmemory films work against the attempt to standardize and routinize an official, monolithic memory of the Holocaust. They privatize memory rather than nationalize it. Hence they reflect the current schism of Israeli society and its conflicts of identities. Israeli postmemory films thus constitute a mechanism of identity-formation.

Although, as Roni Perchak observes,[22] many Israeli film-makers come from survivors' families, for many years the Holocaust occupied a marginal space in Israeli cinema. Most documentaries on the Holocaust were produced for Israeli television, were educational and were designed for use as illustrative material in schools. Except for indirect allusions to Jewish refugees, and a few feature films that directly deal with the Holocaust, Israeli cinema has tended to ignore this traumatic event in Jewish history. The relative silence of Israeli cinema can be explained by its role as both a mirror of, and complicit participant in, the construction and perpetuation of dominant Zionist ideology, including the negation of exile; and by the association of the Holocaust with diasporic consciousness. Another possible explanation is that, because cinema has been categorized traditionally as entertainment, it was not regarded as a suitable medium to deal with this topic.

As Judd Ne'eman points out,[23] Holocaust representations in Israeli cinema appear in two historically separate stages: an early post-war stage and a later one initiated by the production of Ilan Moshenzon's *Wooden Gun* in 1978. What Ne'eman calls the 'new wave' Holocaust cinema consists of a small number of films, such as *Tel Aviv-Berlin* (Tzipi Trope, 1986), *Because of That War* (Orna Ben-Dor, 1988), *The Summer of Aviya* (Eli Cohen, 1988), *Hugo* (Yair Lev, 1989) and *New Land* (Orna Ben-Dor, 1994).[24] In the films of the late 1980s, a change can be traced in the attitude of Israeli cinema to the Holocaust. The documentaries, *Hugo* and *Because of That War*, and the fiction film, *The Summer of Aviya*, began to deal with the trauma of the survivors and their children. This tendency has intensified in recent years with four new documentaries. In three of them – *Choice and Destiny* (Tzipi Reibenbach, 1994), *Will My Mother Go Back to Berlin?* (Micha Peled, 1993) and *Dady Come to the Luna Park* (Nitza Gonen, 1994) – the second-generation film-makers focus both the movie camera and their own gaze towards their survivor-parents' experience and the memory of the Holocaust.

The explorations and investigations undertaken by these films are similar to those in the contemporary drama and fiction of the second generation in Israel. As Yael Feldman observes, the writers of this work share in common 'a rejection of the collective model of representation that they inherited from their parents and cultural mentors', and the search for 'a close subjective encounter with the experiences that the ideology of that model of necessity suppressed'.[25] For example, David Grossman's postmodern novel on the Holocaust, *'Ayen 'Erekh: Ahavah* (See Under: Love) (1986), uses carnivalesque humour and fantasy in its effort to imagine this traumatic event, and to resist the official Zionist discourse of redemption.

> By emplotting [the] long journey into the heart of the death camps as an intensely personal odyssey, triggered by a writer's block and traced to the particular pathology of (single) children of survivors, [Grossman] has attempted to release the Shoah from the shackles of the collective and reclaim it as a subjective experience.[26]

In a similar spirit, second-generation documentary films express rebellion against Israel's nationalization of the Holocaust and its appropriation as collective property. These films attempt to denationalize the Holocaust by retrieving the private and expressing it as unique and personal. Eventually, however, this attempt to privatize memory results in a reconciliatory compromise between the private and the historical, the intimate and the collective. In a style reminiscent of the *cinéma vérité* tradition, the movie camera is used in these films as a valuable catalytic agent, a revealer of inner truth which begins with the return to the private only to find out that the intimate is ultimately the collective. The camera in these films, as Perchak perceptively observes, is the instrument which shapes the family, its speech and the intergenerational encounter, and which finally expands the definition of the second generation by including the spectators in this shared destiny, as well as in their role as witnesses.[27]

In the rest of this chapter I focus on a fourth recent documentary by an Israeli second-generation film-maker: Asher Tlalim's *Don't Touch My Holocaust* (1994). This assigns another task to the medium of film. It documents the production of the play, *Arbeit Macht Frei vom Toitland Europa*, and its performance during 1993–96 by the Acre Theatre Group in Acre, Israel, and on a trip to Germany. The film is a mélange that includes direct presentation of some scenes from the theatrical performance, interviews conducted with the actors, a re-enactment of the actors' encounters with people from their past, and a documentation of the group's visit to Germany. But the film also traces the roots and origins of the creators of these performances, and conducts a second trip, this time to Morocco, to explore the birthplace and Jewish-Moroccan roots of both Dudi Mayan, the playwright, and Asher Tlalim, the film director. The actual act of making this film, as well as the content of the theatrical performance itself, link the discussion of the Holocaust to both the *Mizrahim* and the Palestinians. Making these groups part of the Holocaust experience goes against a Zionist dichotomizing mechanism which distinguishes between those who experienced directly, or through their families, the horrors of the Holocaust, and those who remain outside that experience. The highly charged locus of the film is a cluster of seemingly unrelated and insignificant scenes, in which Tlalim's camera, while freely floating in the streets of Casablanca, stops on a film board of the movie, *Casablanca*, and later on a small movie theatre in Tangier, where Tlalim was born. In this way Tlalim's camera testifies that cinema is not only the global producer of images, fantasy and fiction, but also the matchmaker of dislocated fragments of identity that on the surface seem different and disconnected.

The theatrical performance of *Arbeit Macht Frei* was itself a participatory journey through seven installations and tableaux. Part museum tour, part coach trip, part school play, part confession, part concert recital, part reconstructed ghetto and part testimonial, it used an intimacy only conceivable in live performance to lead its audience through an overwhelming five-hour experience of total theatre. The performance exploited all the senses and made its audience involuntarily complicit in the disturbing theatrical experience by making it an uncomfortable participant-observer. The people who made this performance are 'total' people who mingle life with art, pushing both to the extremes. Their encounters with the audience affected the performance and were assimilated into each individual show. Political and current events were continuously integrated into the play as well, thus making it more relevant and responsive to Israel's stormy political life.

The personal journey into her own past of Smadar (Madi) Yalon-Mayan, the principal actress (and the playwright, Dudi Mayan's then-wife), as part of the work on the play, epitomizes the totality of the performance and its blurring of the boundaries between life and art, reality and performance. Her self-training for this role involved working through her private mourning for her father, a Holocaust survivor who, after his liberation from the camps, migrated to Israel (where, according to Madi, 'he wasted his life').[28] In Tlalim's film, Madi speaks about her experiments with self-starvation in preparation for her role in the play as a Mussulman.[29] She lost so much weight that she had to be hospitalized. She also tattooed a number on her arm, a significant number which indicated the date of her father's death. In the first scene in the theatre space she touches this tattoo, either caressing it or trying to erase it.

The group members worked on the performance for six years beginning in 1990. They collected materials for three years and performed the play for another three. *Arbeit Macht Frei* ran for three consecutive years, twice a week in the theatre in Acre, originally an Arab town which today has become an Arab ghetto. Only thirty people were allowed to attend each performance and tickets had to be reserved a year ahead. At three o'clock in the afternoon the thirty lucky spectators were driven in a luxury bus to the Holocaust museum of *Kibbutz Lohamei Ha'Getaot* (Fighters of the Ghettoes Kibbutz). Dudi Mayan, the leader of the Acre group, said that it was essential for the audience to go through the museum first because this functions as a transitional stop to prepare them for their imminent entrance into the unknown labyrinthic space of the theatre, which recalls the sewage system of the Warsaw ghetto.[30]

In live peformance, Madi Mayan, in the role of Zelma, Holocaust survivor, guides the spectators among the museum's exhibits.[31] In a fake and wearily didactic tone, and in hybrid language composed of broken Hebrew, German, Yiddish and English, she discusses the exhibits and makes comments about the process of extermination, and the rise of fascism. 'It is not a new invention', she says. 'We, too, know a chosen people, yes? Who can give me a contemporary example of a ghetto?' she asks the audience. 'There is no need to be shy, nobody can see us here', she adds. 'Soweto', somebody says. 'Gaza', says someone else.[32] 'Very interesting', Zelma says, and comments: 'Once in one of the groups I had a child, who said that "Here, in my heart, I have a ghetto." '

Later, Zelma takes the spectators to another hall where they can see wood sculptures of Mussulmans. 'This Mussulmanchik', she says (in funny, broken Hebrew), 'whose stomach is glued to his back; I would have given a fortune to know where he hides a piece of bread. I'll give millions to know where his scream comes from.' Hours later, towards the end of a typically Israeli dinner shared by the actors and the audience as part of the performance, the audience see Zelma as a Mussulman lying naked on the dinner-table and pulling out of her vagina a piece of bread. Then Haled, a Palestinian (played by the actor, Haled Abu-Ali, a Palestinian citizen of Israel) [33] takes over the role of the tour guide from Zelma and explains what happened in Treblinka using a wooden model of the camp. 'I', he says 'did not know about the Holocaust until four years ago. Only after visiting Yad Vashem,[34] I began to believe that it really happened. That people can indeed kill like that. I wish you could see me once guiding [Arab] youth from the surrounding villages, who are not familiar with the Shoah.' The text that Haled is performing is not a fictional one. He did indeed become aware of the Holocaust only a few years ago; and since then he has guided around the museum groups of young people from Arab villages in the local area of Galilee. One of these guided tours is documented in Tlalim's film. And the Arab youth say to the camera that the Jews suffered less than the Palestinians: that they [the Jews] at least were exterminated immediately, while the Palestinians are being slowly tortured.

According to Amnon Raz-Krakotzkin, the uniqueness of the performance of *Arbeit Macht Frei* – described as 'one of the most overwhelming experiences provided by Israeli culture in recent years'[35] – is that it subverts and challenges the established memory of the Holocaust, while acknowledging the fact that the Holocaust is the primary, formative factor in the shaping of Jewish-Zionist consciousness. Furthermore, its point of departure is the creation of a sense of responsibility towards this memory. *Arbeit Macht Frei* enables its Israeli spectators to have a different kind of memory, one which is critically shaped against the existing one, and yet which grows out of it. The unique use of the theatrical medium enables the creation of a dialogic relationship between the repressing present and the repressed past, which turns the memory of the Holocaust into a basis for a completely different ethics.[36]

Memory in *Arbeit Macht Frei* is performed by the characters of Zelma and Haled, who lead the narrative, while constantly changing and transforming themselves and their roles in the course of the performance. Zelma, with a middle-European accent and broken language, represents the victim, the one who came from 'there', and whose memory has been absent from the traditional Israeli ideological framework which refers to the murder but not to the victims. Her broken language signifies a cultural positioning which goes beyond the fact that she does not speak standard Hebrew: it is linked to her interpretation of the exhibits in the museum, which offers new and provoking insights regarding the Israeli/Palestinian conflict. When Zelma is later replaced as guide by Haled, the identification of the Palestinian with the suffering of the Jews challenges the monopoly established by Zionism over the memory of the Holocaust, and the particularist Jewish lessons which it draws from this. The inclusion of Haled in the ritual of remembrance assigns to the ritual a broader meaning, and gives the spectator a fresh perspective from which to understand some key ideological assumptions of Israeli society.

The play's challenge is most powerfully evident in its ending: Zelma and Haled are united in an image of memory which develops out of a torture scene, described by the Israeli journalist, Tzvi Gilat, as 'the closest thing to a nightmare which I have ever experienced in relation to any work of art'.[37] The scene begins after the audience, along with the actors, have finished their dinner. They have ascended one by one through a very narrow door, that has been opened in the ceiling of the room where most of the performance takes place, into another space. This is a hybrid space evoking associations of both discotheque and gas chamber. A festival of 'Israeli folk dancing' is projected on huge video screens along with images of the singer, Shoshana Damari, who in Israeli public consciousness is known as 'the singer of the wars'. Surrounding the video screens are various photographs and exhibits related to the Holocaust. In one corner of this space, a naked actress, playing a Holocaust victim, is obsessively eating the remains of the dinner to the point of vomiting; Madi Mayan, as Zelma, is crouching naked in another corner in a Mussulman posture; and Haled, also completely naked, is dancing on a wooden table in another corner of the hall. It was on a similar table, Haled had explained to the audience only a few hours before in the *Lohamei Ha'Getaot* museum, that Jews were beaten and tortured every day by the Ukrainian guards in Treblinka before being rushed (naked) to the gas chambers. Now Haled is beating himself with a club similar to the one that was used by the Ukrainians to beat the Jews in Treblinka, and he invites members of the audience to come to the table and join in the beating. Horrifying as it is, there are people (both in Israel and in Germany) who accept his 'invitation' and beat him. Others pull the bottle-opener which hangs from Haled's neck in order to open bottles of beer.

The horrific party culminates with a slowly-rising, high-pitched sound emanating from the loudspeakers, that changes from wild drumbeats into Israeli popular songs of the 1950s celebrating the Zionist project of settling the land; all this while the eyes of Holocaust victims and of Herzl, the founder of Zionism, stare at the audience from images hanging on the walls. A tough attendant leads the audience to the exit (or entrance?) gate which bears the words, *Arbeit Macht Frei* (the infamous slogan which welcomed the victims to Auschwitz). Haled is still frantically dancing and beating himself, his naked body flooded with strong beams of blinding light thrown by powerful projectors. On the way to the exit the members of the audience see Haled, exhausted from the physical and mental effort, collapse and cry cathartically, while Zelma, still naked, hugs and calms him down in a *Pietà*-like posture.

This *Pietà* image invites the spectators to acknowledge the victimhood of the Palestinian as well as that of the survivor, whose repression in Israeli culture derives from his or her epitomizing the ultimate icon of the diasporic Jew, rejected by the Zionist ideology of the negation of exile. Thus, as Raz-Krakotzkin observes, the memory of the victim becomes a focus for Jewish identity, which paradoxically enables the opening up of the memory of the Palestinian past. Yet no memory, neither Jewish nor Palestinian, can fully retrieve the past. The common memory of victimhood, which this last scene so powerfully establishes, can only create the conditions for a change of consciousness about the boundaries of discourse on, and of, memory.[38] This opens the possibility for acknowledgement and discussion of previously taboo issues: for example, that not only Jews were victims of the Nazis, that the Jews victimize the Palestinians, and that being a victim does not entitle the Jewish collective to victimize others.

The last scene of *Arbeit Macht Frei* establishes a mirroring effect (which is echoed by Tlalim's 'The Mirror Way' section in *Don't Touch My Holocaust*). To look at oneself in the mirror is also to look at the Other; and this act of looking is precisely the point of junction where the Jewish Holocaust survivor and the Palestinian victim of Zionism merge into a hybrid image of victimhood.

However, it is important to remember (as Raz-Krakotzkin reminds us) that the role of Haled, the Palestinian, does not represent an indigenous 'Palestinian consciousness', despite the dialogism and openness towards the Other that characterizes this work of theatre.[39] Rather, Haled's role constitutes part of a play that was written by Jews, mostly for Jews, and that deals with Jewish consciousness. The actor may present 'his' (Palestinian) perspective, but what is actually being represented by this is the consciousness of the Jew from the point of view of the Other. The distinction here is not between a 'particularistic' versus 'universal' memory, but between two patterns of memory which the language of Israeli culture creates. The Palestinian actor makes two essential points: that it is the fate of the Palestinians that the Holocaust occupies a more significant role in their collective consciousness than is the case for any other nation bar only the Jews, the Germans and the Poles; and that this is so because the memory of the Holocaust is 'forced' upon them in order to justify not only Jewish settlement in Palestine, but also the Palestinians' expulsion and the negation of their right to freedom and self-determination.

This dialectic process, based upon a diasporic consciousness, opens up the possibility of seeing the extermination of the Jews as the point of departure in forging a new attitude towards the continuous oppression of the Palestinians, without reverting to the simplistic and senseless comparison between two kinds of national catastrophes ('holocausts') which is made by the Arab youth in Tlalim's film. This rupture in the existing Jewish consciousness, as Raz-Krakotzkin observes, opens up a space for a new definition of Jewish existence in Israel/Palestine. It is a definition which assumes full responsibility by acknowledging the wrongs done to the Other, and identifying with the repressed.[40]

As the Israeli critic, Michael Handelzaltz claims, the final scene of the play, in which the naked Haled is beaten by 'volunteers' from the audience, creates a reality of humiliation. (Tlalim's film shows the beating at a performance in Germany.) The spectator who does not interfere to stop the suffering is as indifferent as were the bystanders in the Holocaust. But had the spectator interfered, he would have collaborated in the activity of the theatre. The spectator, thus, is both an innocent and a complicit observer. You (the audience), Handelzaltz concludes, 'are both observer and collaborator, there is no way out'.[41] Haled tells how, from time to time, people who have beaten him on the wooden table call and ask him for forgiveness. Only now, they explain, have they realized what they have done. 'What kind of forgiveness are you asking for?' Haled asks them; 'You have become Nazis yourselves.' 'But you gave us a club and asked to be beaten', they respond. 'And if I had given you a knife?' he replies.

The significance of Tlalim's film, *Don't Touch My Holocaust*, is that it is not a simple documentation of the theatre performance, but incorporates this into a filmic investigation of memory and identity. As the director himself acknowledges: 'All the films that I make are about memory.'[42] Tlalim perceives *Don't Touch My Holocaust* as a continuation of his series of six films on the 1973 War in which he tried to examine the trauma of war. In this film he confronts the question of how Israelis deal with memory: with how they remember and how they forget. These questions, as Tlalim explains, were triggered by the personal trauma of the loss of his brother during the 1973 War. His film, *The Missing Picture* (1989), explores the notion of war trauma and its influence on its victims' daily lives. After having confronted the trauma of 1973, Tlalim felt that the time was ripe for him to touch 'the ultimate trauma which overshadows all the static memories of our daily life: the Holocaust'.

At that time Tlalim had already made the section on the Holocaust in *Amud Ha'esh* (Pillar of Fire), an ambitious series produced during the early 1980s by Israeli television on the history of Zionism and the state of Israel. Tlalim explains that, during the period when this series was conceived and produced, the Holocaust was not a popular topic in Israel and consequently film-makers were reluctant to touch it. Yiggal Lossin, the producer of *Amud Ha'esh*, begged Tlalim (who had withdrawn from normal life and was living in a tent and working as a shepherd) to direct the section on the Holocaust. Tlalim could not resist the pressure because his wife at that time was the only daughter of survivors who had lost all their family in Auschwitz, and, while getting to know her, he had realized what it meant to be a second-generation child. Tlalim himself has no direct relationship with the Holocaust. He was born in Tangier and raised in Spain. Yet he has always felt that the division between the children of survivors and non-survivors, as well as between *Ashkenazim* and *Mizrahim*, is unjustified, and in relation to the third generation is no longer significant. 'For the generation which has grown up here, the memory of the Holocaust is a part of them; and it does not matter if they were there or not', he claims.

Don't Touch My Holocaust begins with a 20-minute section, captioned 'The Mirror Way', in which Tlalim tries to explore the meaning of the memory of the Holocaust for Israelis. He creates in this introduction what he calls 'an inventory of the memory of the Holocaust'. It is a retrospective tour of the lexicon of the Holocaust as it has been shaped by Israeli collective memory, and is composed of the exploration of some key words and phrases, such as '*Arbeit Macht Frei*', that have come to symbolize the Holocaust. Yet as Tlalim insists, this inventory of institutionalized memory is represented 'with love, from within', not in order to criticize and satirize Israeli society but in order to place it before a critical mirror. The film, as Tlalim explains, is structured like a psychoanalytic journey. 'It is a journey deep into the soul.'

Dudi Mayan and Tlalim became soulmates right from their first meeting. Mayan was born in Casablanca and Tlalim in Tangier, and their shared *Mizrahi* diasporic roots turned them immediately into close friends. In the film, the seeming paradox of Mayan and Tlalim, two Oriental Jews commemorating the extermination of *Ashkenazi* Jews, is formulated as a question: 'What does a Moroccan have to do with the Holocaust?' This introduces Tlalim's documentation of the visit to their respective childhood hometowns. The attempt 'to penetrate deeper and deeper into memory in order to extract the beginning of it all' led, as Tlalim explains, 'in unexpected directions'. The whole film, according to Tlalim:

> is about looking at ourselves in the mirror. We look in the mirror and see ourselves as we are without putting the blame on the Germans and without adopting a victim's perspective. It is a film without Germans. We look at ourselves and try to see what we really look like. What is behind this film and all the other films that I have made is the belief that we are all ill. That we all suffer from dark dreams. We all suffer from some childhood trauma and the only way to face it is to examine it; to look at ourselves, and to make fun of ourselves. This is the way in which Madi [Mayan] is looking at us, cruelly and yet lovingly. This is the way we look at our perversions and dreams, painfully, laughingly. I tried in this film to observe Israeli society mercilessly yet affectionately in an effort to understand what we are made of.

The dilemma behind the film, Tlalim asserts, is how to remember, both personally and collectively. Perhaps, he suggests, it would be better to live in amnesia, but this is impossible because the past always creeps into our lives.

> I have a problem with institutionalized ways of memorialization, but the question is: what is the alternative? I see my children, the third generation, and the ways they are touched by the Holocaust. We need to relate to memory here and now. To see ourselves as we live it now and not as it really happened. The film is structured like an internal journey and each spectator is invited to look at his own life. I call it 'The Mirror Way'.

One of the problems of Israeli society, according to Tlalim, is that it deals with the Holocaust only on the ceremonial level. This has brought about the trivialization and banalization of the memory of the Holocaust. In addition, the Palestinians of Israel have been excluded from the memory of the Holocaust. This exclusion is also related to the division between *Ashkenazim* and *Mizrahim*. This attempt at division, according to Tlalim, has resulted in a monopolization of the memory of the Holocaust by the children of survivors. This exclusionist attitude has damaged the Arabs in particular. The message of his film is that 'the memory of the Holocaust belongs to us all'. It was therefore very important for Tlalim to document in his film one of Haled Abu-Ali's guided tours of the *Lohamei Ha'Getaot* museum, conducted in Arabic for the Arab youth of Galilee. 'I was shocked by their reactions', Tlalim says. 'But I do not blame them, but the educational system that does not understand that every Israeli citizen has to confront this issue. Only then people will not make these silly comparisons between Israeli and Nazi soldiers, and the *Intifada* and the Holocaust.'

Tlalim's attempt to emphasize that the Holocaust belongs to us all lies behind the filming of his and Dudi Mayan's quest for their own *Mizrahi* roots and memories in Morocco. Additionally, in 'The Mirror Way' section which opens the film, he introduces the story of Muni Yosef, one of the actors, who was born in Iraq and was not familiar at the beginning of the production with the phrase, '*Arbeit Macht Frei*'. Yet during the research period some of Muni's repressions disappeared, and he recalled that his childhood in Moshav Mazor, an agricultural settlement in north Israel, was spent amongst the survivors who founded it. Tlalim's film also documents his visit to Mazor and his meeting with one of the neighbours who tells him that she can still see Mengele performing a selection.

Participating in *Arbeit Macht Frei* was an unsettling, disturbing experience, especially for an Israeli audience. The use of pornographic iconography, the evocation of Palestinian victimhood and Israeli racism in conjunction with the Holocaust, and the complicity of the audience in the theatrical performance created a moral dilemma for the ideal, imaginary (and imagined) Israeli identity. The same is also true of Tlalim's film. Indeed, of all the second-generation Holocaust films, *Don't Touch My Holocaust* is the most 'remote' from the Holocaust. Both Tlalim and Mayan were initially attracted to the subject through their first wives, who both happened to be daughters of survivors. Both used the Holocaust as a 'tool' in their search for their own repressed *Mizrahi* identity, and both focused on a phantasmagoric Holocaust based on imagination rather than on testimonies or fact-oriented evidence.

It should be emphasized time and again that both *Arbeit Macht Frei* and *Don't Touch My Holocaust* are cultural products generated by Israeli second-generation sensibility, and are primarily aimed at second-generation audiences. The second-generation sensibility, being spatially and temporally distant from the Holocaust, is based on the desire to fill a void, to imagine what David Grossman in *'Ayen 'Erekh: Ahavah* (See Under: Love) calls '*Eretz Sham*' (The Country Over There).[43] Thus, the 'Holocaust' of the play and film (like 'Grossman's Holocaust', and consciously so) exists in the realm of the imaginary and strives to explore, not what really happened 'there', but how what happened 'there' affects us here and now; how our imagined and fantasized Holocaust has been assimilated into our consciousness.[44]

Don't Touch My Holocaust is dominated by second-generation voices and completely ignores the real survivors' voices, which are transmitted in this film through the sometimes perverse prism of their ventriloquist children, whose Holocaust remains in the realm of morbid fantasy. This narrative intertwining, of the second generation's voices with the imagined voices of the survivors, is related to the orchestration of a gender/parenting economy. It is fascinating to examine who speaks in this film and who keeps silent, who is absent and who is present, who is visible and who is invisible. Madi Mayan's father is the ghost behind it all, occupying the space of a structuring absence since there are no other fathers or mothers in the film. This stands in sharp contrast to the realistic style adopted by, for example, *Because of That War*, the first Israeli film to deal with the effects of the transmission of the Holocaust trauma to the second generation, which gives an equal voice to the survivors and their children; at once respecting the survivors' memories and showing compassion towards their children too. By using the children of survivors as protagonists in a Holocaust film for the first time, *Because of That War* helped to establish the second generation as a distinct identity group. *Don't Touch My Holocaust* has contributed to the further development and acceptance of the distinct voices of the second generation, who now speak as secondary victims (both direct 'victims' of their survivor parents, as well as indirect victims of the Holocaust).[45]

Although second-generation Holocaust films go against the attempt to nationalize the Holocaust, the collective spirit of the 'Israeli tribe' is stronger than any attempt at individuation. To paraphrase the famous slogan of the feminist revolution, in Israel the private is always public. Consequently, the phenomenon of second-generation Israelis going 'back' to Poland (in most cases with their survivor parents) to visit the death camps in order to perform a 'secular ritual' of 'Holocaust tourism'[46] has become part of Israeli civil religion. Similar journeys in search of lost diasporic identity are undertaken by second-generation *Mizrahi* youth who travel to Morocco to find their roots. The conflation of the two trends is expressed in Tlalim's *Don't Touch My Holocaust*, which explores the complex issues of identity, memory and ethnicity through the conjunction of the Holocaust with both the *Mizrahim* and the Palestinians.

Identities in contemporary Israel (as in most of the contemporary world) derive from a multiplicity of sources, including nationality, ethnicity, social class, gender, sexuality and religion. These may conflict and lead to a society marked by fragmentation and polarization, which is the current state of affairs in Israel. However, within this complexity, three major sources of identity can be distinguished which, in my view, are the foundational sites of the struggle over Israeli identity: the Holocaust; the question of 'the Orient' (concerning the inclusion within the national collective of Jews of non-European origin); and – in an ironic echo of the historical 'Jewish Question' – the so-called 'Palestinian Question'. *Don't Touch My Holocaust* raises fundamental questions about the identities that have developed from all three of these sources: those of the Jewish Holocaust survivors and their children, the 'second generation'; of the *Mizrahim* (particularly the second generation of Israeli *Mizrahim*); and of the Palestinian citizens of Israel. Yet the film does not treat each identity group as a separate and coherent entity; rather, it attempts to see the conflation, interplay and conflict between them.

What connects all the identity groups to which *Don't Touch My Holocaust* gives voice is a strong sense of victimhood, in particular of being victimized by the Zionist state.[47] All the identities represented in the film are marked by victimhood. Indeed, to a large extent Israeli society has become a society of victims or perceived victims; and contemporary Israeli identity politics is now based on demands for recognition and acknowledgement of victimization. The recent appeal by Ehud Barak (the leader of the Israeli Labour Party) for forgiveness from the *Mizrahim* for the way they were treated by the former Mapai (a predecessor of today's Labour Party) is a case in point. Lately the hegemonic Israeli-Jewish narrative itself has been oscillating between the former myths of voluntary sacrifice and heroism, and the emerging appeals for recognition of involuntary past suffering. This was clearly evident in the case of the Likud-dominated government of the late 1990s, which viewed itself as a coalition of past victims; or, in the language of contemporary Israeli journalistic rhetoric, as a coalition of minorities composed of previously oppressed and stigmatized groups (the ultra-Orthodox, the *Mizrahim*, Russian immigrants and the Israeli right). This coalition points to the centrality of appeals to victimhood in contemporary Israeli identity politics.

Paradoxically and tragically, however, identification with the Jewish victim increases as acknowledgement of Israel's primary victims, the Palestinians, wanes and is denied. The tragic irony of this is increased by the fact that the basis of the identities of the *Mizrahim* and the Holocaust survivors, like that of the Palestinians, is the experience of geographical, cultural and spiritual displacement. The contemporary Israeli dialectic of victimhood, perceptively captured by *Arbeit Macht Frei*, recalls the American elevation of the Holocaust into a master moral paradigm. In place of a proper acknowledgement of the real victims in the United States – namely, Native Americans and African-Americans – this substitutes an 'easier' identification with the Jewish Holocaust.[48]

Identity, particularly in the case of extreme forms of national and/or ethnic conflict, is marked by processes of inclusion and exclusion, which constitute an 'us' and a 'them'. However, the three main identities represented in *Don't Touch My Holocaust* are viewed as constantly playing against each other, as well as against an imaginary and imagined Israeli collective identity, the *Ashkenazi*-dominated *Sabra*. Hence, Israel is represented in this film not only as a site of contention and struggle but also as a site of meeting and negotiation, voluntary or otherwise. Historical realities and necessities have turned Israel into a meeting place, or even a melting pot, of conflating as well as conflicting identities. Yet as Stuart Hall has argued in a different context, these identities, 'far from being eternally fixed in some essentialized past … are subject to the continuous "play" of history, culture, and power'.[49]

Notes

This chapter is adapted from a longer analysis in my forthcoming book, *Conflicting Projections: Identity Politics on the Israeli Screen* (University of Texas Press, 2001).

1 J. Young, 'Israel: Holocaust, heroism, and national redemption', in his *The Texture of Memory: Holocaust Memorials and Meaning* (New Haven, CT and London, 1993), 209–18. See also S. Friedlander, 'Memory of the Shoah in Israel', in J. Young (ed.) *The Art of Memory: Holocaust Memorials in History* (New York, 1994), 149–58; T. Segev, *The Seventh Million: The Israelis and the Holocaust* (New York, 1993); E. Sicher (ed.) *Breaking Crystal: Writing and Memory after Auschwitz* (Urbana and Chicago, 1998), particularly Part 2: 'The View from Israel', 91–181.

2 C.S. Liebman, 'Restructuring Israeli–Diaspora relations', *Israel Studies* 1 (1996), 318–19.

3 Young, *The Texture of Memory*, 211.

4 A. Raz-Krakotzkin, 'Galut betoch ribonot: Lebikoret "shlilat ha'galut" batarbut ha'yisraelit' (Exile within sovereignty: towards a critique of the 'negation of exile' in Israeli culture), *Theory and Criticism* 4 (1993), 23–56; continued in *Theory and Criticism* 5 (1994), 113–32 (in Hebrew).

5 See Young, *The Texture of Memory*, 211.

6 On images of the Zionist 'new Jew' versus the diasporic 'old Jew', see, for example, S. Gilman, *The Jew's Body* (London, 1991).

7 *Mizrahim* is the self-chosen name, adopted into American English, for Jews from Arab countries, who originated in Spain (*Sepharad* in Hebrew). The term covers a wide

spectrum of ethnic groups. The majority of *Mizrahim* in Israel are from Morocco and arrived during the 1950s. On the stigmatization (as being 'foreign' or inferior) of groups seen as 'ethnic', see W. Zenner, 'Ethnic factors in Israeli life', in I. Lustick (ed.) *Books on Israel*, vol. 1 (Albany, New York, 1988), 48.

8 On the controversy surrounding Ben-Gurion's attitude towards the Holocaust, the refugees, the reception of the survivors in Israel and the Yiddish language, see Ben-Gurion's official biographer, S. Tevet, *Ben Gurion and the Holocaust* (New York, 1998).

9 Y. Weitz, 'Political dimensions of Holocaust memory in Israel', in R. Wistrich and D. Ohana (eds) *The Shaping of Israeli Identity: Myth, Memory and Trauma* (London, 1995), 133.

10 Z. Solomon, 'From denial to recognition: attitudes to Holocaust survivors', *Journal of Traumatic Stress* 8 (1995), 215–28. In recent years the question has repeatedly arisen whether the survivors could have told their stories immediately after the Holocaust. The prevailing view today, inspired by growing criticism of the reception and absorption of the immigrants (both from Europe and the Arab countries) by the young Israeli state, holds that the survivors faced indifference and contempt that kept them silent. The attitude towards the survivors has recently become a focus of heated debates in Israel. See D. Porat, *The Blue and the Yellow Stars of David: The Zionist Leadership in Palestine and the Holocaust 1939–1945* (Cambridge, MA, 1990); H. Yablonka, *Ahim Vezarim: Nitzolei Hashoah Bemedinat Yisrael 1948–1952* (Brothers and Strangers: Holocaust Survivors in Israel 1948–1952) (Jerusalem, 1994) (in Hebrew).

11 H. Gouri, 'Facing the glass booth', in G.H. Hartman (ed.) *Holocaust Remembrance: The Shapes of Memory* (Oxford, UK, Cambridge, MA, 1994), 154.

12 Ibid., 155.

13 On the appropriation of the Holocaust by the right wing, see L.J. Silberstein, 'Others within and others without: rethinking Jewish identity and culture', in L.J. Silberstein and R.L. Cohn (eds) *The Other in Jewish Thought and History: Constructions of Jewish Culture and Identity* (New York and London, 1994), 1–34.

14 For a further discussion of this issue, see 'Politics and collective memory: the debate over the "new historians" in Israel', *History and Memory* 7 (1995), special issue on 'Israeli Historiography Revisited' (ed. G. Ne'eman Arad), 17–23.

15 For a recent publication on this issue which has generated heated debate in Israel, see D. Frenkel, *Al Pi Tehom: Hamediniyut Hatziyonit Veshelat Yehuday Germanya, 1933–1938* (On the Abyss: Zionist Policy and the Question of the German Jews, 1933–1938) (Jerusalem, 1994) (in Hebrew).

16 D. Porat, 'Leolam lo darchi ha'acharona' (It is never my last road), *Ha'aretz* (15 January 1997), 5. This is a book review of F. Zandman, *Darchi me Vichy le Vichy: Meneourim Bashoah Lebagrut shel Mada Vetasiya* (My Way from Vichy to Vichy: From Adolescence in the Holocaust to an Adulthood of Science and Industry) trans. A. Karmel (Tel Aviv, 1997) (both in Hebrew).

17 M. Hirsch, 'Past lives: postmemories in exile', *Poetics Today* 17 (1996), 662.

18 Ibid.

19 Ibid.

20 Ibid.

21 Ibid.

22 R. Perchak, 'Haintimi hu hakolektivi: al kolnoa Yisraeli venose Hashoah' (The intimate and the collective: on Israeli cinema and the Holocaust), *Dimuy* 13A (1997), 38–41 (in Hebrew).

23 J. Ne'eman, 'The empty tomb in the postmodern pyramid: Israeli cinema in the 1980s and 1990s', in C. Berlin (ed.) *Documenting Israel* (Cambridge, MA, 1995), 117–52.

24 Ibid., 130.

25 Y.S. Feldman, 'Whose story is it, anyway? Ideology and psychology in the representation of the Shoah in Israeli literature', in S. Friedlander (ed.) *Probing the Limits of Representation: Nazism and the 'Final Solution'* (Cambridge, MA and London, 1992), 238.

26 Ibid. David Grossman (b. 1954), an Israeli-born novelist and documentary commentator, is not himself a child of Holocaust survivors, yet he belongs to the age-group of the Israeli second generation.

27 Perchak, 'The intimate and the collective', 38–41.

28 Interview with Madi Mayan in *Don't Touch My Holocaust.*

29 'Mussulman' was a term widely used by concentration-camp prisoners to refer to inmates who were on the verge of death from starvation, exhaustion and despair. A person who had reached the Mussulman stage had little, if any, chance for survival and usually died within weeks. The origin of the term is unclear.

30 T. Gilat, 'Mesibat h'azvaa' (The horror party), *7 Yamim* (*Yediot Ha'aronot* weekly supplement), (4 November 1994), 47 (in Hebrew).

31 All the quotations in the account that follows are taken from a live performance of the play.

32 In Tlalim's film Madi Mayan discusses visual similarities between Arab Acre and the ghetto, in particular the Warsaw ghetto.

33 The term, 'Palestinian citizen of Israel', is used here in preference both to the Zionist term, 'Israeli Arab', and the oxymoron, 'Israeli Palestinian'. On the ideological implications of these terms, see D. Rabinowitch, 'Nostalgia Mizrahit: ech haphchu HaPalastinim learviye Israel' (Oriental nostalgia: how the Palestinians have become Israeli Arabs), *Theory and Criticism* 4 (1993), 141–51 (in Hebrew). For the highly complex relationship between the Arabs (including the Palestinians) and the Holocaust, see A. Bishara, 'Ha'aravim vehashoah: nituha beayatiyuta shel ot hibur' (The Arabs and the Holocaust: an analysis of the problematics of a letter of conjunction), *Zemanim* 53 (1995), 54–71 (in Hebrew). Recently there has been a growth of interest in the Holocaust in Palestinian circles. Some major Palestinian intellectuals, like Edward Said, Mahmud Darwish and others, have appealed for the withdrawal of analogies made by some Arabs and Palestinians between the Holocaust and the Palestinian *Nakba* (Catastrophe), the name adopted by the Palestinians to describe the tragedy of 1948, and for recognition of the Jewish Holocaust. In return they have asked that Israelis recognize the Palestinian *Nakba.*

34 Yad Vashem is Israel's national Memorial Authority of the Holocaust. It is an amalgamation of outdoor monuments, exhibition halls, massive archives, a research centre, etc. It orchestrates the state's official activities and performances related to the Holocaust.

35 Raz-Krakotzkin, 'Exile within sovereignty', 121–2.

36 Ibid.

37 Gilat, 'The horror party', 44.

38 Raz-Krakotzkin, 'Exile within sovereignty', 121–2.

39 Ibid.

40 Ibid.

41 M. Handelzaltz, 'Tzofim vemeshatfim peula' (Observing and collaborating), *Ha'aretz*, Literary Supplement (19 April 1991), B2 (in Hebrew).

42 All the citations from Tlalim here and below are taken from a lecture that he gave to my class, 'Identity politics in Israeli cinema', at the Hebrew University of Jerusalem, 8 January 1997.

43 This term, which appears throughout the novel, makes a play upon *Eretz Israel* (the Land of Israel) – the Biblical religious name for the land of Palestine. By using the notion of '*sham*' (literally, 'from there, not from here'), Grossman emphasizes the alienation between the diasporic survivors and the *Sabras* (such as Mumik, the name of the protagonist of his novel) who were born 'here' (in *Eretz Israel*).

44 Similar sets of concerns and sensibilities were raised with regard to 'Within the Line', an exhibition by Ram Katzir (a grandson of survivors), which opened at the Israel Museum on 21 January 1997 and became the focus of controversy and survivors' protests; as well as to the even more scandalous exhibition, 'Live and Die Like Eva Braun', by Roi Rosen (a son of survivors), which opened at the same museum in December 1997 and provoked a stormy public debate.

45 It is understood by psychologists that the trauma of the Holocaust is transmitted to the second generation by their survivor parents. Many survivor parents who tried to reconstruct their feelings of identity through their children transformed the second generation into 'memorial candles' for their lost relatives. See among others: D. Wardi, *Memorial Candles: Children of the Holocaust*, trans. N. Goldblum (London and New York, 1992); H. Epstein, *Children of the Holocaust: Conversations with Sons and Daughters of Survivors* (London, 1988, c. 1979); M.S. Bergmann and M.E. Jucovy (eds) *Generations of the Holocaust* (New York, 1990); H. Cooper, 'The second generation "syndrome" ', *Journal of Holocaust Education* 4 (1995), 131–46; G. Glassman, 'The Holocaust and its aftermath: how have the survivors fared during the last fifty years?' *European Judaism* 29 (1996), 52–63.

46 J. Kugelmass, 'Why we go to Poland: Holocaust tourism as secular ritual', in Young, *The Art of Memory*, 175–84.

47 According to Ella Shohat the *Mizrahim* (like the Palestinians) are certainly the victims of Zionism. See E. Shohat, 'Sepharadim in Israel: Zionism from the standpoint of its Jewish victims', *Social Text* 19–20 (1988), 1–35. In the case of *Mizrahim* and Holocaust survivors, the anger and the frustration is usually not directly channelled against 'Zionism' or the Jewish state but is displaced and disguised in different forms. However, among intellectuals and artists of the second generation of *Mizrahim* and Holocaust survivors, the anger is sometimes openly directed against Zionism; or at least, against the 'historical mistakes made by the state' with regard to the absorption of immigrants from the Islamic countries, and Holocaust survivors from Europe.

48 For discussion of the place of the Holocaust in American culture, see P. Novick, 'Holocaust memory in America', in Young (ed.) *The Art of Memory*, 159–65.

49 S. Hall, 'Cultural identity and diaspora', in S. Hall and P. du Gay (eds) *Questions of Cultural Identity* (London, Thousand Oaks, New Delhi, 1996), 52.

8 Hauntings

Memory, fiction and the Portuguese colonial wars

Paulo de Medeiros

Memory is both more and less than history, and vice-versa.

(Dominick LaCapra)[1]

Portugal has recently celebrated the twenty-fifth anniversary of the revolution of 25 April 1974, which put an end to almost fifty years of dictatorial rule and soon afterwards to thirteen years of gruesome colonial wars in Africa. This has naturally given rise to a series of commemorations and some reflections on the role of the revolution and its promise of a new future for Portugal as a progressive society. Instead of the socialist utopia dreamed of by so many and feared by as many others, the nation seems to have found its way through European integration, based on a solid embrace of the cardinal virtues of marketing and consumerism. Nonetheless, the fundamental precepts of a liberal democracy can be said to be solidly anchored in Portuguese society. After the initial chaos brought about by the revolution and the ensuing attempts by the extreme left and right to seize control, the political process has been marked by decades of stability aided by significant economic support from the European Union. Wide-scale land reform and government takeover of the banking system and key industries have yielded to market pressures and were undone by re-privatization. The economy is seen as solid, unemployment rates are among the lowest in the European Union, and an entire generation has been able to grow unencumbered by censorship. It would seem as if by abandoning its empire Portugal has also freed itself of the questionable position of singularity which Salazar's *Estado Novo* – the dictatorial regime installed in 1928 – had forced it to assume. No longer the sick man of Europe, Portugal seems indeed to have resolved its contradictions and exchanged its pariah role for a healthy unexceptionality. And yet, it is as if in the rush to put the past behind it, the nation had decided simply to forget the wound at the base of both its past and present conditions: the decision in 1961 to pursue war rather than entertain alternatives to accommodate the desire for autonomy by its African colonies.

The year of 1999 saw a series of public commemorations of the revolution, from travelling exhibitions to series of lectures in Portugal and abroad. Images of soldiers in combat position on the streets of Lisbon, surrounded by children, or with carnations planted on their silent guns – symbolizing the fact that, even though the revolution occasioned a radical rupture in the nation's history rather than inaugurating a transition, it did so with a minimum of violence and only a few accidental deaths – have been continuously displayed in public. Yet images of the colonial war, or any other forms of discourse on the war, have been almost entirely absent from such displays, even though both the revolution and the colonial war were carried out by the same military. The special 'dossier' published by the influential weekly *O Expresso* to commemorate 'twenty-five years of April' can be seen as symptomatic of such disparity.[2] It includes a series of interviews with witnesses of the revolution as well as with prominent figures such as the last Minister of Foreign Affairs of the old regime – under the headline, 'If it was good for Portugal it was awful for Africa' – and with Angola's first Prime Minister after independence, who is asked to reflect on the significance of the revolution for his country. *O Expresso* also ran a prominent feature on a new film being produced to 'narrate the fantastic history of the twenty-four hours of that unique April day' and a piece of humorous fiction with the title 'The 25th of April never happened', in which life in Portugal is described as if the revolution had failed. Although there are some mentions of the war – in that fictitious story it still goes on – there is no dedicated feature on it, even though there is one covering 'the fashion of 74'. From this issue of *O Expresso* it seems clear that the effects brought about in Portuguese society by the revolution have taken root and one can afford to play games with its memory. Indeed, one could say that the memory of the revolution is so firmly established, so institutionalized, and its consequences for national identity so present everywhere, that it has assumed a sort of invisibility and it is only by pretending to negate it that one can hope to bring it further into the public light. The fictitious story is a well-realized, gentle satire of notable developments in the period immediately following the revolution – which was marked by severe instability – as well as of prominent figures of the political and cultural elite. Its conclusions, however, are stunning for it would seem that the worst consequences of the revolution's failure, in the writer's imagination, would simply be that access to the coveted consumer goods of other Western European countries would be delayed, above all portable phones, whereas the war in Africa would still be going on even after the most repressive signs of the regime – political prisoners, a brutal secret police, complete censorship of all media – had all but completely dissolved through sheer inertia. Not only is such a commemoration of the revolution significant for its inability to address the effects of the colonial war on the nation, it is also revealing in terms of what collective identity it projects. It insists on the positive, if somewhat unexpected, role of the armed forces in putting an end to dictatorial rule within Portugal, while suppressing another, more conventional one: the army's role in prosecuting the colonial war. Put in other terms, the revolution, inasmuch as it is seen as a factor which has enhanced national identity, has been assimilated and historicized, whereas the colonial war, with its inherently complex and largely negative connotations, has been largely avoided.

In this chapter I want to explore the question of how processes of memory have been deployed to attempt to surmount this collective amnesia and the fundamental role played by fictional narrative as the privileged, if not exclusive, instrument for anamnesis. My focus is limited to post-revolutionary Portuguese texts. Not only would the conditions and effects with regard to the African nations who fought to gain independence from Portugal be quite different, but in most cases, even if the war of liberation was successfully won, it was not replaced by peace. Indeed, I think it would be rather faulty to speak of war in terms of memory in the case of either Angola and Mozambique, or even, if differently, in that of Guinea-Bissau or East Timor.[3]

That the revolution and the colonial wars are intrinsically linked is hardly disputed, many observers going so far as to consider the disastrous conditions faced by the Portuguese armed forces in Africa as a direct cause of the actions of the officers who removed Salazar's successor, Marcello Caetano, from power. In the words of Mário Soares, the former Portuguese President and leader of the Socialist Party:

> The 25th of April resulted, essentially, from the fact that the military did not want to continue, indefinitely, the so-called missions of war overseas … The colonial war caused the regime to rot which would necessarily lead to the 25th of April and to decolonization.

That is also the view of Manuel dos Santos, former commander of the anti-aircraft units of the Movement for the Independence of Guinea-Bissau and Cape Verde (PAIGC) who occupied several ministerial posts after independence and for whom the use of Russian SAM 7 ground-to-air missiles was 'decisive for the end of the dictatorship in Portugal. … the 25th of April, in a certain way, was conceived here in Guinea-Bissau'.[4]

If 25 April marks a decisive caesura in the history of twentieth-century Portugal, the colonial wars were no less significant for the nation. Although numbers and statistics by themselves are limited tools they can serve to sketch an idea of what the war effort meant for the Portuguese, a nation of 9–10 million. João Paulo Guerra notes that the armed forces mobilized a total of 820,000 young men for combat and that war expenses averaged 43 per cent of all public expenditures during those thirteen years, at the end of which there had been 8,831 dead, around 30,000 wounded, close to 4,500 mutilated and over 100,000 soldiers affected with post-traumatic stress disorders.[5] As João de Melo notes, 'the days and long years of the colonial war belonged to all the Portuguese'.[6] And yet, discussion of the colonial war has been almost completely absent from the public domain. Even if by now one can no longer say that there have been no attempts to reflect on the experience of the war, this is nevertheless the claim made in the few studies that have been published. The editors of the second published collection of iconographic documents relating to the war experience, note that they had to do without material from official archives (much of which remains secret), relying instead on other centres of documentation and photographs taken by individual soldiers who had participated in the war.[7] Moreover, not only do the editors explicitly argue for the validity of their efforts as a countermeasure to the lack of knowledge about the war, 'banished from memory by many and unknown by others', they also frame their effort within a process of memory, citing Pierre Nora to express the hope that their efforts at documentation will provide a stimulus and base for future historical processing of the war. It is clear from their introductory remarks that they see the necessity of preserving what they refer to as 'live memory' in order to prepare the ground for historical analysis, that is, that in the absence of historical reflection, the recourse to memory is an ethical imperative.[8]

In addition to their charge that minimal public attention has been paid to the colonial war and that official bodies perpetuate this by remaining silent about the events and conditions that so marked the nation, all of those who have written on the subject, interestingly, make a very explicit link between their efforts and questions of memory. João de Melo, in the preface to the extensive anthology of fiction and essays that he published in 1988, traces a first sketch of what would become the literature of the war. And in the editor's note, the conjunction between the pioneering role of such work and the issue of memory is already succinctly exposed:

> *Os Anos da Guerra – 1961–75*, is, fourteen years after the start of decolonization, the first collective attempt to get a picture, however imprecise, of what was the presence of the Portuguese in Africa in the crucial years during which the colonial war was undertaken on three fronts ...
>
> *Os Anos da Guerra – 1961–75* is a first step. But it was impossible to postpone it. Because it would be absurd to pretend that that which memory has not yet ceased to remind us of has already been forgotten.[9]

That he ultimately argues for restricting the anthology to the recent writings centred on the war campaigns starting in 1961 is both understandable and critically logical. This anthology brings together forty-six different authors, some represented by several entries; these pieces can be quite short, but some constitute entire narratives or essays. Prose fiction has been almost exclusively the genre chosen to reflect on the war and one can think of several reasons for this, from the fact that the novel is widely granted more recognition, to the relative neglect of any autobiographical forms, to the lack of direct engagement on the part of historians and other critics who might pursue a different discourse. One notable film, Manoel de Oliveira's *No, or the Vain Glory of Command* (1990), has attempted to engage the issues of the war by inserting it into a larger historical framework and seeing it in terms of Portugal's historical myths of overseas expansion, but, although it received critical acclaim, its abstract approach failed to engage the audience. Indeed, at a general level, one can say that the film has been dismissed as a viable critique of the war. Moreover, narrative fiction can also be said to allow more freedom for selective memory, even if it remains strongly autobiographical. The writers represented were, for the most part, directly involved in the war. Alongside some of the greatest names of contemporary Portuguese literature one also finds others whose work is of a markedly different quality. And besides Portuguese authors, including some who used an 'African' pseudonym, there are also important African writers such as Luandino Vieira, Pepetela, and José Craveirinha. The argument behind this of course is that the war affected everyone – albeit in different ways – and that all those who opposed the war, or were forced to participate in it, were actually on the same side, that of freedom, and against oppression in all of its forms, be it colonialism or fascism. Without wanting to deny some validity to this view, and keeping in mind that the colonial wars were never simply an issue of race, as social class for instance also played an influential role in attitudes and experiences, I think nonetheless that one ought to keep some critical distance. To maintain such a distance is naturally difficult for those who have suffered directly from the effects of the war, and who thus often feel not only a pressing need to keep its memory alive, but also a certain form of authority derived from personal experience.[10]

Although the question of authenticity would seem to be a minor one, it has been invested with key significance by all of the writers and even by a number of critics. That this is not a simple point can be seen from the fact that such a sense of positioning sometimes takes extreme forms and raises considerable doubts. For instance, the need to delimit the literature of or on the colonial wars, leads quite a few to imagine that only direct participation in the war authorizes one to write about it. This is perhaps most evident in the position taken by Rui de Azevedo Teixeira, whose 1998 book on the colonial war and Portuguese novels is the first sustained critical study of the most important narratives about the war. Teixeira, himself marked by active duty in Angola from 1973 to February 1975, when considering the 'cardinal question of combat experience' comes up with as many as six classifications for writers, from those who saw actual combat, to those who were neither in the armed forces nor in Africa.

What is relevant about such a typology is its very necessity and the fact that one of the six categories, that of women, is considered 'special'.[11] Although I will come to this question in more detail later, it is important to note that this is given relevance by Teixeira. In reference to one of the most complex novels of the war, Lídia Jorge's *The Murmuring Coast*, he must remark that in it 'there is no direct description of combat, no *empirical* elements of the war whatsoever', something he views as a 'capital lacuna' and which would, in his opinion, cause a self-gratifying 'obsessive reflection over the war phenomenon'.[12] Clearly much of this can be explained by keeping in mind how those writers and critics felt themselves to be trapped by the war. Yet the desire to see themselves as much victims of colonial oppression as the liberation fighters must be regarded as highly problematic, especially when it takes the form of assuming that the one could represent the others. Without wanting at all to fall into a simple dichotomy between good and evil, oppressed and oppressors – and in many cases this is a strategy which the writers themselves employ, in reverse, so that the white Portuguese assume the role of villains in contrast to the black heroes of the guerrilla movements[13] – accepting the possibility that a Portuguese writer could speak for the African would be at best critically naïve, and at worst another form of oppression in itself. For, if experience to those writers appears so fundamental as grounds for authority, authenticity and memory, speaking for the Other instead of promoting understanding, is merely a form of ventriloquism whose consequence would be the elaboration of a false memory.

The articulation of memory, the (somewhat frustrated) search for a diversity of memories, and the attempt to evoke a collective memory from the integration of individual memory was the explicit purpose of what may have been the most significant (and to my knowledge the only) public attempt to discuss the effects of the colonial war: the colloquium held on 5 November 1993 in Coimbra. Maria Manuela Cruzeiro, in her opening words, makes a succinct but clear exposition of the conditions under which such an effort was organized ('the total lack of support from circa seventy institutions approached') and of the apparent shame involved in talking about the colonial war, relegating it to the level of a discursive and existential taboo. It is against such a silencing, and against the drive for commemoration on the part of the state and its apparatuses (media and schools), that Cruzeiro imagines the discussion that took place before 300 observers.[14]

Silencing is not just a tool of the state. It is also a consequence of the fact that, for many, memory is still primarily – and perhaps always will be – a source of shame, an unbearable pain whose power prevents many from speaking in public. Repression and denial are the terms under which Cruzeiro sees such an evasion of the responsibility to bear witness. And so, in Cruzeiro's apt phrasing, those who have come forth to write and speak about the war, either through fiction or through analysis, would be those who have 'the courage of memory'.[15] Indeed, the situation in Portugal, in spite of democratization and a rapid adaptation to contemporary European standards which the previous regime had tried to obstruct, has meant that the public sphere has been less than welcoming to the necessary process of exorcizing the national phantoms. As I have been arguing, this put the burden of memory on narrative fiction, the one medium which perhaps more readily lent itself to the dual attempt of exorcizing private ghosts and forging a national consciousness that would not succumb to amnesia.[16]

Whereas João de Melo in his initial exploratory essay of a decade ago necessarily surveyed a wide range of materials to establish what he considered the war generation, which in his eyes also constituted a 'literary generation', more recent works, either Ribeiro's 1998 essay or Teixeira's 1997 book, have narrowed analysis to a smaller number of the most representative novels. Teixeira's extensive analysis turns on his contention that the novels of the colonial war perform a cathartic function, allowing their authors to confront their anguished memories of the war and of their own forced participation in it. Ribeiro, achieving a greater distance from the texts studied, distinguishes precisely between those which stay focused on personal experience and those which are able to transcend it and thus contribute towards a rethinking of national identity. She ends her essay with precisely the question of what kind of future one might imagine for Portugal, based on the novels of the war. In what follows, because what concerns me at present is the deployment of processes of memory, I will limit myself to the consideration of two novels: António Lobo Antunes's *Os Cus de Judas*, first published in 1979 and translated in 1983 as *South of Nowhere*, and Lídia Jorge's 1988 *A Costa dos Murmúrios*, whose translation as *The Murmuring Coast* only appeared in 1995.[17] These two novels are not only, arguably, the most important of all fictional thematizations of the war, they also problematize conditions of memory and forgetting, showing in the process, ways of addressing the destruction of individual identity and imagining new forms of national identity. And, to a certain degree, one could say that they are correlated, Jorge's novel both serving to expand on the limitations exposed by Antunes, not necessarily answering them, but provoking new questions, and suggesting different models both for existence as well as for remembrance.

Until José Saramago received the Nobel Prize in 1998, he and Lobo Antunes were considered the likeliest Portuguese candidates for it. A psychiatrist who served in the war in Angola, Lobo Antunes's novels have caused both admiration and shock among the Portuguese public and his international reception has been considerable, especially in Germany. His work revolves around the issues of the colonial war, even if they are not the only theme or preoccupation of the author. Indeed, one could just as well say that the Portuguese national identity, the repressive conditions of bourgeois life, the rottenness of the fascist past and the impossibility of happiness in human relations all constitute the core of his work. But the war in one form or another always seems to make an appearance and, especially in the earlier works, is undoubtedly the factor under which everything else is subsumed. Or, to be more precise, it is the remembrance of the war, the impossibility of forgetting the horror and the trail of human destruction that is its legacy, which is so striking in Lobo Antunes's novels. Critical attention to Lobo Antunes's novels has been sparse and tends to focus precisely on the question of the war although without paying attention to the issues of memory. In a necessarily brief and sweeping overview of his works, however, Henry Thorau aptly comments that 'the concept *memory*, which appeared in the title of his first novel, *Memória de Elefante*, published in 1978, can be said today to have become the programme for all his further novels'.[18]

In some important ways it could be said that *Os Cus de Judas* is the same novel as *Memória de Elefante*: in both the reader is confronted with a doctor as main protagonist who saw combat service in Angola, whose marriage collapsed, and who is haunted by the memories of the war, desperately trying to escape his solitude and the appalling mediocrity of his society. The similarities between the two novels are many, even down to some of the incidents related. But *Os Cus de Judas* is a much more refined narrative, written from beginning to end as an obsessive monologue spoken to a voiceless woman, in which the functions still attributed to other characters in the previous novel – for instance, the dismantling of the main protagonist's self-pity through the sarcasm of his listeners – are combined in the same voice. And *Os Cus de Judas* is written in altogether another key, more tragic and more lucid, for the real weight of the war experience and the haunting power of memory are given primacy here. What in the first novel was suggested as the cause of individual and collective collapse, is brought out in the open, and the depiction of Portuguese society as an inescapable realm of ghosts is explicit. There are the ghosts of the previous regime, 'The spectre of Salazar, our glorious leader, hovered over the white washbasins, protecting us from the gloomy and suspect idea of socialism', and the 'desperate, afflicted ghosts inhabiting [his] body' that constitute the recollection of bourgeois childhood. There are the ghosts of the 'dead companions persecuting [him] in [his] sleep, begging [him] not to let them rot in their lead coffins', and even those who return alive are spectral: 'The plane that flies us to Lisbon transports a cargo of slowly materializing ghosts, officers and soldiers yellow with malaria, fastened to their seats, looking out the window at blank space.'[19]

The images evoked bring to mind Vietnam. For one thing, both war efforts were extensive and prolonged, roughly thirteen years. Both were waged from a similar ideological perspective, as heroic efforts to contain the spread of international communism and uphold Western civilization, both had their share of infamous massacres, and both were doomed to failure. Also, the position of the soldiers returning home was at least ambiguous and at worst a hellish complex of guilt and torn feelings. However, whereas Vietnam forced a repositioning of American consciousness and its memories have been explored from a variety of points of view and in a variety of media, the Portuguese colonial wars have been displaced by the events of the revolution in Portugal. Consequently, Portuguese soldiers never really faced public condemnation for their actions in war, and were celebrated instead as the heroes who had rescued the nation from authoritarian rule. But that move has also meant that there is no public space to work out the extensive trauma occasioned by the colonial war.

The silencing of memory suggested in the conclusion of *Memória de Elefante* as a possible means of survival is denounced unequivocally in *Os Cus de Judas*: 'Why the hell won't they talk about it? I'm beginning to think that the million and a half Portuguese who passed through Africa never existed and I am narrating for you a cheap, implausible novel, a contrived story composed of one third bullshit, one third alcohol and one third tenderness, you know what I mean?'[20] In both novels the shattering of identity caused by the war is inescapable. An important difference, however, is that in the second novel the narrator maintains one constant element and that is the certainty of failure when facing family expectations. This is relevant inasmuch as the war here is explained as a form of last hope on the part of his family for the transformation into manhood of the narrator. And this apparently familiar drama can be seen as a national allegory, in which a generation formed by fascist ideals, repressive and hypocritical morals, and imperial propaganda mired in a supposedly grandiose past of conquest, justifies the war with all its horror and sacrifice as an instrument of national regeneration. At the very end of the novel Lobo Antunes depicts a scene in which the narrator quietly explodes such a myth. The confrontation between the narrator and his family not only allegorizes a generational clash, a clash accentuated by the experience of the war and the unavoidable gulf between propaganda and reality, but suggests an answer to the official silencing of memory. For in the careful depiction of the bourgeois surroundings Lobo

Antunes not only puts forth class markers, he embodies a whole national/colonial past. The passage should be cited at length:

> I visited my aunts a few weeks later, wearing a shabby suit I had bought before the war. I waited standing next to the piano with the candelabra between an imperial cabinet full of framed pictures of dead generals and an enormous clock ticking softly. The curtains in the windows waved evasively, the silver sparkled from the cabinets in the dark. My aunts turned on the lamp to observe me better and the light suddenly revealed faded Arraiolos carpets, Chinese carpets with dragons, the curiosity of the maids who peeked through the door, wiping their fat hands on the kitchen aprons. Instinctively I assumed the stiff serious pose reserved for park photographers or for standing at attention when I was a cadet in Mafra, in front of the moody captain who always frowned arrogantly. The room smelled like camphor, naphthalene and Siamese cat piss, and I wanted very much to walk out of there to the Rua Alexandre Herculano and look at the sky. A bamboo cane traced a disdainful arabesque before it was stuck in my chest, while a weak, raspy voice said 'You're thinner than when we last saw you. I always hoped the army would make a man out of you, but I guess there's nothing to be done.'
>
> And the pictures of the generals in the cabinet seemed to confirm the evidence of the disgrace.[21]

Yet, as Lobo Antunes also makes clear, silencing memory is at best a temporary expedient, for although the narrator goes about emptying ashtrays and pretending that all is normal, he will always maintain the memory of Africa inside him, ready to intrude upon the appearances of a stable present: 'You never know, Tia Theresa [the black woman in whose hut he had looked for solace from the war] might come by and pay me a visit.'[22] And ultimately silencing memory is not only a betrayal of one's own dead comrades, and of those who were killed, tortured or executed by the Portuguese, it is also a monstrous denial of reality:

> Everything is real, especially my hangover, the pain at the back of my neck, my arms as difficult to control as prosthetic devices. Everything is real except the war that never was; there were never colonies, or fascism, or Salazar, or the Tarrafal prison camp, or the PIDE [secret police], or the revolution, none of that ever existed, you understand, none of it, time in this country stopped so long ago ...[23]

Lest one think that such a denial is just imaginary, one only has to note that in 1993 it is still possible for one of the principal men implicated in the colonial wars, General Kaúlza de Arriaga, military commander in Mozambique from 1969 to 1973, to deny the reality of much of the tragedy of the war. Not only does he deny the significance of the massacres of civilians by the troops under his command, he also denies the disastrous conditions of the war, preferring to see the events of 1974 and the subsequent independence of the colonies as acts of political treason:

> There was no military problem … Where was the war lost? Not the war, where were the overseas provinces lost? They were lost here in Lisbon … The coup in Lisbon was made because we did not lose the war overseas. And this was prepared by the communists … [24]

In the continuing absence of historical reflection on the war, memory and the voicing of memory through narrative are the only available remedy against not only the fading of time but the attempt to cover up the 'disgrace' in imaginary shortcomings of manhood or equally imaginary conspiracies.

Memory, of course, as Dominick LaCapra aptly notes, 'is both more and less than history' and as such it is also highly individual and can play all kinds of tricks. One of the problems inherent in the memory of the colonial wars and their narratives is that they tend to reflect primarily, if not exclusively, the point of view of those who, even if against their will, took an active part in the war, who both suffered the effects of the war and were part of the colonial army. Written by a woman who – although she was in Africa for a period as the wife of an officer – had no further direct experience of the war, Lídia Jorge's *A Costa dos Murmúrios* necessarily exhibits a very different perspective. More important, however, is the fact that in this novel not only are memories of war the motivating force of the narrative, but the narrative itself constitutes a prolonged reflection on the war and reality and on the processes of memory through which such experience is filtered and framed.

The very structure of the novel indicates the extent to which it is a metanarrative problematization of the processes of memory and of the authority of the survivor, as well as of the relationship between reality and representation. The novel opens with a short narrative, entitled 'The Locusts', written from an omniscient point of view, which relates the marriage in Mozambique of a young enlisted officer to a woman called Evita, and how he later dies by suicide. This is followed by a much longer text in which Eva – that is, Evita twenty years later – addresses the voiceless and nameless 'author' of the first text in a monologue. The second narrative does not so much question or expand on the first, as deny any transparency to language, thus questioning the effects of representation, and 'annulling' the first text. As the concluding words of Eva explicitly state: ' "Little by little the words detach themselves from the objects they designate, then the sounds separate from the words, and of the sounds only murmurs remain, the final stage before erasure", said Eva Lopo laughing. Handing back, annulling "The Locusts." '[25]

It is not surprising that practically all the critics who have approached this novel have commented on its questioning of history and of the historical process. Ana Paula Ferreira, for instance, regards Lídia Jorge's narrative as a postmodern text, critical of Enlightenment values and of those narratives about the war, such as 'Lobo Antunes' *Os cus de Judas*, which are written as though the real is knowable' and whose 'ideological and philosophical assumptions on which their claims to "truth" are grounded ... reproduce the same masked, secretive, irrational violence that they wish to denounce'.[26] Even if Ferreira is right to read the two novels against each other from a gendered perspective, Jorge's novel is not the only one to be critical of historiography or of representation. Lobo Antunes's novels do that as well, even if in the end the choice of the cynical male narrator whose bitter perspective is a direct result of war experience might still create a form of heroic anti-hero, something which Lídia Jorge is able to avoid completely. However, the implication of the involvement of the Portuguese army in the atrocities, which in Antunes's work cannot but be attenuated by focusing on the figure of the resisting doctor, is in Jorge's novel emphasized by the voice of Eva, especially as she becomes conscious of her young husband's participation in the massacres. That in itself would already constitute a large gap between the two texts and might indeed lead to an observation of how Jorge's novels seem in many ways to be in dialogue with those of Lobo Antunes, carrying on where his narrative had been forced to stop by the limitation imposed by his own participation in the war and by his gender. But one way in which the novels really differ is in their treatment of memory.

To start with, one should note that whereas for Lobo Antunes's novel the war is the compelling trauma bringing forth memory which erupts in narrative, in the case of Jorge's novel it is narrative itself – 'The Locusts' – which brings forth memory and in turn produces the subsequent narrative, which both reflects on the previous narrative and its effects. Of course the war is still the motivating force behind the two narratives and the memory recalled in the second narrative is a memory of the war. But this strategy also allows the narrator more space for reflecting on the process of memory itself. And, even if in some crucial points the two novels necessarily coincide in their view of memory and the place of forgetting, in between there are universes of difference.

In the two novels one can see some form of a therapeutic function of narrative, as if by voicing their memories of the war the characters were looking forward to a kind of validation of their experiences; a recognition that indeed they had gone through those experiences, and that in spite of them, or because of them, they were who they were. This is something which both Dori Laub and Susan Brison, among others, have insisted on as being crucial to the rebuilding of identity after trauma.[27] But a major difference has to do with the fact that in Lobo Antunes's novel the audience is a silent woman whereas in Lídia Jorge's novel the audience is an unidentified male who had 'authored' the initial short narrative, 'The Locusts', with which the novel opens and against which Eva's monologue develops. In other words, whereas in Lobo Antunes's text memory is deployed to perform before a silent audience whose only role is to listen passively, thus enabling the narrator to constitute himself into a subject, in Lídia Jorge's novel the narrator's memory is also replacing another type of memory, a more coherent but fundamentally limited and perversely blind memory, which had already tried to establish itself as the 'truth' about the war. Jorge's text is quite complex. The novel opens with a short narrative written from the point of view of an unidentified male which narrates the wedding of Evita to a young officer stationed in Mozambique and his death shortly afterwards. Then follows the novel proper, narrated by Eva – who frequently reminds the audience that she is Evita twenty years later – which reacts to the previous narrative and expands on it. This narrative strategy has several advantages and one of them is that it allows the author to present a critique of the role of narrative memory whilst at the same time avoiding a direct representation of the relationship between narrator and audience. As such, it forces the reader to question issues of positioning when reflecting on memory. A major difference between the two novels resides not only in the fact that they are separated by ten years – after all both authors belong to the same generation and became known after the revolution – but also in the way that gender issues are approached.

Lobo Antunes does not hide the relationship between gender and war – after all, the aunts' hopes that the narrator might become a man through the war are a sort of frame for the novel. Moreover, love, but especially sex, occupies a prominent role in his novels, as both an escape from, and as an act of, violence.[28] And yet, he perpetuates a view of women as objects in a male symbolic order.[29] For the judging aunts simply enforce that order; and the black lover – with whom the narrator temporarily found solace from the brutality of the war and who is ultimately gang-raped and executed by the secret police who accuse her of collaboration with the liberation forces – becomes a symbol both of the narrator's impotence in relation to the regime in whose army he serves, and of an idealized and victimized Other.

The narrator in the opening part of Lídia Jorge's novel appears to have subscribed to a similar logic when young – taking on a black lover instead of accepting the offer of seduction by Helen, the wife of the captain under whose command Evita's husband served – but otherwise the reader is given a perspective that resists not only war, but the male glorification of it which underlies so much traditional national historical narrative. In this perspective, as Ferreira and others have noted, male 'heroics' appear juxtaposed with female courage, the shining scar that is the Captain's pride is made obsolete and contrasted to the disgraceful scar which one of the soldier's wives incurs when, having been refused admittance to the clinic, she delivers a dead child and ruptures her sphincter muscle. By making visible the presence of women in the war scenario, even if still kept away from the combat zone, Lídia Jorge certainly seriously problematizes traditional conceptions of what constitutes a war narrative. That she does so without falling into yet another dichotomy with women opposed to men in an eternal good and evil struggle – Eva and Helen obviously are charged names – is another way in which the writer demonstrates the theoretical coherence of her narrative. Indeed, as Barbara Freeman asserts, referring specifically to the stereotypical opposition between men and women on the issue of war, 'What is crucial for an effective resistance to war is not a reversal of the same themes, but a displacement of the very structure that produces it.'[30] Jorge's novel also has the effect of displacing the traditional structure of the role of memory and of its relation with history: memory in *A Costa dos Murmúrios* is not simply opposed to silencing, nor does it provide the cathartic deliverance that it is imbued with by Lobo Antunes.

Before going further with an exploration of the differences that separate the two novels on the issue of memory, it should be noted that they also exhibit three crucial similarities. First, there is the fact that memory is in both cases directly linked with the authors' own experiences of war and their memories of it. Second, both novels at times put forward a need to forget as an essential element for survival. And, lastly, both authors view memory as an essential antidote to collective historical erasure. Nonetheless, even such similarities already imply radical differences as well. One of the most marked, of course, is that the experience of war of the authors is clearly different, so that even though for both the autobiographical content is necessarily elevated, the nature of the memories cannot but diverge. Thus it is not unexpected to find a concentration of details in Jorge's narrative that might have appeared insignificant to Lobo Antunes, just as Lobo Antunes's direct involvement with the carnage of the battlefield will have resulted in more traditional memories. However significant such a difference, it does not put into question in any way, the 'authority' of either narrative. Teixeira's attempt to distinguish between the two on the basis of actual combat experience cannot be endorsed. When dealing with issues of memory anyway, only a naïve reading would presume that actual experience of any kind is a guarantee for the fidelity of memory. As Lynne Hanley aptly remarks in the introduction to her seminal study, 'our fictions have power, they shape our memories of the past and they create memories of pasts we have never had, of experiences not even remotely like anything that has ever happened to us'.[31] In a sense, by providing a neat and coherent, but as ultimately revealed false, narrative of the past in 'The Locusts', Lídia Jorge presents the opportunity to recognize such a difficulty, not so as to provoke complete relativism in the reader, but rather to avoid an all-too-idealized view of memory's reliability.

The need to forget as an element for survival, which both authors at times put forward, is also handled differently in each narrative. In the two books by Lobo Antunes the desire to forget, that is, to stop the irrepressible flow of memory from disrupting daily life, is only ironically advanced as a possibility for ensuring survival. The reader knows well that the promises which the main protagonist in *Memória de Elefante* makes in the conclusion of the novel, to forget the war experience and accommodate himself to a life of mediocrity, are false, even a form of self-delusion. Likewise, the gesture of emptying the ashtrays in the conclusion of *Os Cus de Judas*, which can be read as signifying a break with memory, is denied through the unrealistic hope voiced by the narrator that he might be visited by the black woman whose house he frequented in Africa.

In Lídia Jorge's narrative, however, forgetting the past is never just an ironic but ultimately empty gesture. For one thing, in the figure of Alex, Evita's young husband, forgetting his past as an idealistic student of mathematics is indeed key to his adaptation to the new role as soldier. Even though I would not go as far as Maria Cabral in assuming that the themes of memory and forgetting 'are, above all, fictionally demonstrated in the character of the bridegroom', I certainly agree with her view that 'his trajectory can be entirely read as a process of amnesia'.[32] The narrator Evita remembers having been shocked by the chaos seemingly opened up by such radical forgetting:

> at some point we always lose the memory of what we once desired, and the groom might as well have lost it then, but in fact it complicated matters quite a bit for him to have forgotten it like that. Because if we started to forget what we wanted to discover, and after that our names, and after that what country we came from, how could we agree on what time to go out, or when to go shopping?[33]

But Eva also, in her distrust of the efficacy of memory to establish an accurate account of the past, or even simply to survive the erasure of time, can be deeply sceptical of remembering: 'What historic memory, what testimony? Forget it again, forget it (said Eva Lopo)'.[34] This distrust is not just an expression of bitterness in the light of the lack of collective consciousness but indicative of a necessary split in identity signalled constantly by the naming of the narrator as both Evita and Eva in recognition of the passage of time. Here one could think, even though safeguarding the obvious differences, of the need expressed by survivors of limit-situations to effect such a split between their past and present selves, to the point of assuming what cannot but be seen as a double identity. For instance, Susan J. Brison, writing on Holocaust survivors, quotes Charlotte Delbo:

> No doubt, I am very fortunate in not recognizing myself in the self that was in Auschwitz ... I live within a two-fold being. The Auschwitz double doesn't bother me, doesn't interfere with my life. As though it weren't I at all. Without this split I would not have been able to revive.[35]

In a way, the narrator of Lídia Jorge's novel is also literally such a double character, whose identity split enables her to go on with her life. This would be another radical difference between Jorge's narrative and those by Lobo Antunes. For in the latter the characters still keep going back to infancy in search of answers to their present situations. The war experience and its memories constitute a caesura but not a break in the identity of the characters who maintain unchanged their basic characteristics. However, Eva also obviously does not follow her own injunction to forget and almost immediately also urges her audience to go and look up the military archives, 'No, I'm not making it up. Look in the military archive. Go to the museum entrance, turn in your card to the soldier ... Ask for it – they're always nice people, the caretakers of History ... '[36]

As much as she might feel the futility of memory and laugh at traditional history – of the repeated jabs at history, the description of her Contemporary History class at university is perhaps the most explicit, as the lecturer, a priest, when confronted by her view of relativity, explodes with a tirade about God being above all[37] – Eva nonetheless is not like Helen, the Captain's wife, one who is 'innocent, without memory'.[38] Helen knows much more about the atrocity of the war than Evita and it is through Helen's revelations that Evita comes to know the extent to which her own husband had become a sadistic killer. It is Helen who shows to Evita the photographs documenting the massacres. But Helen had already lost her former self. As one of her servants dies from poisoning Helen seeks solace with Evita who, however, can be lucid in the face of Helen's grief: 'Helen is getting ready, she holds on to my hand … From her face flows a torrent of tears … I have no doubt that the person she mourns is herself lost in her reflection in someone else.'[39] And when Evita finally refuses to be seduced by Helen, who imagined such a relationship as an act of vengeance on their husbands and, by extension, on the army, Eva once again has recourse to a similar metaphor, saying, 'Between you and me, identity is a mirror that reflects us and implacably isolates us.'[40]

The young woman who is slowly confronted with the horrors of the war and with the chaos that threatens the foundations of her self is represented in the linear, coherent and false, narrative, 'The Locusts', as believing 'that memory had no end'.[41] But as Eva discloses in her narrative, she grew more and more ambivalent towards memory. Eva certainly does not hold a simple concept of truth or of reality and is well aware that history is made of wars and massacres that no one remembers anymore. Eva oscillates between the conviction that memory is but 'a hoax meant to deceive dust-colored oblivion', and the necessity to escape 'the deep grave of forgetfulness'.[42] She also knows that the past is made of ruins and ghosts but insists that 'It does no good to wish there to be ghosts among the rubble. Why wish that ghosts remain amid the rubble?' As she continues, it is obvious that she lucidly perceives that although those who left Africa and returned to Portugal might go on living as if time had stopped at the moment of their passage through Africa, time indeed has not stopped. And that lucidity is perhaps the one distinctive quality that still unites the narrator's two selves, Evita and Eva.

The split between Eva and Evita, and the split between the two narratives in Lídia Jorge's novel, has further implications in terms of the construction of Portuguese national identity and the role given to memories of the colonial war in it. If the split between Eva and Evita is a necessary strategy to ensure survival at an individual level, perhaps a similar split is also necessary in order to ensure, not so much collective survival, as the survival of a certain kind of national identity. My interpretation of the split in the personality of Lídia Jorge's narrator was facilitated by a comparison to the statements made by Susan Brison when analysing trauma survivors and, in particular, Holocaust survivors. Obviously, this is a comparison which should not be pushed too hard as there are significant differences; nonetheless, I think it applies, and a further brief comparison between the construction of national identity in Portugal and Germany might make its relevance clearer. For in the aftermath of the Holocaust and the splitting of Germany into East and West, German national identity has been clearly not only split but also re-made in two different ways, according to either a

Western model which necessitated public and official mourning to atone for the hideousness of the Nazi crimes, or an Eastern model which preferred rather to exalt a memory of heroic resistance to fascism in the past and consequently suppressed much of its collusion with that past. This is precisely what Gerd Gemünden explores in a recent article. As he states, Germans now feel that they are someone again but wonder who it is that they are.[43]

In the case of Portugal none of this would seem to apply. Portugal, in contrast to Germany, is a mature nation-state, indeed its national unity (linguistic and territorial) has been in place and hardly changed since the thirteenth century. Furthermore, Portugal remained neutral during the Second World War, always a difficult balancing act between its oldest ally, England, and its ideological preference for Germany. And in the case that matters now, the colonial war, Portugal seems to have decided almost from one day to the next to stop it unilaterally. As a result there was not a profound split in the country's identity, except, perhaps, for older generations, to whom the process, no matter how necessary it was, seemed like treason, a betrayal of Portugal's cherished imperialist myths. And perhaps it is here that one can begin searching for the reasons why there is a disparity between the commemorations of the revolution, and the difficulty in preserving even the memory of the colonial war.

Eduardo Lourenço, one of Portugal's most respected intellectuals, has advanced the thesis that, unlike other nations, Portugal does not have any doubts as to its national identity. In a 1984 lecture he asserted that Portugal was characterized by a hyper-identity, and decolonization had only served to intensify such a feeling. In his view, the colonies were always judged as extraneous to Portugal. The notion that the colonies were indeed a key component of Portugal existed only in the ideology of the old regime, so that once that was gone, the disappearance of the colonies, rather than being felt as a loss, would confirm Portugal's true identity, an identity rooted in its history spanning eight centuries.[44] Although there is much that is fascinating in Lourenço's thought, this thesis presents a series of difficulties, including a return to a concept of Portuguese exceptionality which Lourenço would have been the first to criticize in another context. If I mention this thesis, however, it is because in wanting to link national identity with memory (the words of the lecture's title) Lourenço effectively collapses memory into history and all too easily dismisses the issue of the colonies or of the colonial war and its memory. In doing so, and by appealing to Portugal's long history, Lourenço ends up attempting to suppress those memories. This might provide us with the necessary insight to understand the discrepancy between commemorations of the revolution and the absence of discourse on the colonial war. What is evident is that in order to claim a unified national identity, Lourenço has to insist on a certain humanistic tradition involving Portugal's overseas experiences while completely silencing the colonizing aspects of such an enterprise. In much the same way, the Portuguese armed forces will insist on being celebrated as having freed the nation from despotism and conveniently forget to mention their role in the colonial war.

If the resurgence of the novel in Portugal is due in part to the need to refashion and re-imagine Portuguese identity after the revolution, as Ellen Sapega has argued,[45] then it is to be expected that precisely those novels focusing on the memory of the colonial war would assume special relevance. That they have not is indicative of the fact that, for many, like Lourenço, the desire to erect a uniform national identity modelled on glorious achievements, be it the voyages of discovery in the sixteenth century or the end of fascism in the recent past, necessitates forgetting the ghosts of imperialism and of the colonial wars. History thus becomes the celebrated realm of Portuguese achievement and the grounds for a uniquely positive and secure national identity, whereas memory, of the colonizing enterprise, or more to the point, of the colonial wars, is negated. Just as in the case of trauma victims, this might be a necessary strategy for survival. Portugal in this way can celebrate the revolution and ensure its participation in the material comforts of the European Union without having to question itself. The memory of the colonial war is still experienced too much like a wound in itself and not just the recollection of that wound. Portugal has shown that it does not lack the ability to mourn. When its most famous Fado singer died in 1999, the government decreed an official three-day period of national mourning. But then again, such mourning readily fits in with the desire to ensure the unalterable nature of Portuguese national identity, whereas the memory of the colonial war would threaten to destabilize it. By contrast, although a monument to remember the dead of the colonial wars has now been erected, its official opening on 5 February 2000 met with protests from families of the deceased, who found the date to have no 'patriotic' significance and the site itself to be peripheral. They would have preferred the place where the bodies of the dead soldiers had lain in wake upon returning to Portugal, a church aptly named the 'Church of Memory'. Thus, memories of the wars do not lend themselves easily to internalization; and the novels which insist on preserving them will still – but for how long? – remain a necessary antidote to institutionalized history. Especially when, as in the case of Lídia Jorge's novel, they do not limit themselves to a cathartic function but go further, denouncing not only the colonial war and the structures that enabled it, but at the same time probing the insufficiency of memory and narrative either to counteract war or to hold up the mirror of the past to future generations as if in possession of ultimate truth. Her narrative might well represent 'the final stage before erasure' of the public memory of the colonial wars. At the same time, in its criticism of language and representation, and its evocation of the fragmentation of individual memory, it compels us constantly to revise history so that it both remembers and remains haunting.

Notes

1 D. LaCapra, *History and Memory after Auschwitz* (Ithaca and London, 1998), 20.

2 The issue in question is number 1,381 of 17 April 1999. It is available on-line at <http://www.expresso.pt/ed1381/abr.asp>, through which access is also granted to all of the other special issues published in previous years to commemorate the revolution.

3 As Patrick Chabal notes in his 'Introduction' to *The Postcolonial Literature of Lusophone Africa* (London, 1996), the first and most useful overview of the African literatures in Portuguese since independence, 'The fourth factor which distinguishes the modern history of Portuguese-speaking Africa from that of most of the non-Portuguese colonies is the character of its nationalism and, in Angola and Mozambique, the impact on society of civil war' (p. 21). He continues:

> the impact which continued civil war since independence has had on Angolan and Mozambican society cannot be minimized. Although contemporary Angolan and Mozambican literature is rarely explicit about the conflict which has ravaged these two countries, it is the ever-present backdrop against which all writers and indeed all citizens have had to live. It is too early to say how literature will eventually assimilate the experience of civil war.
>
> (p. 24)

Although Guinea-Bissau has been less ravaged, war broke out again in 1999. And Timor, which was annexed by Indonesia after the Portuguese left, not only suffered massacres but, on the eve of autonomy under UN supervision, violence again increased significantly.

4 Both comments come from interviews with João Paulo Guerra. The interview with Mário Soares took place on 29 Dec. 1995, the one with Manuel dos Santos on 27 Jan. 1991. See João Paulo Guerra, *Descolonização Portuguesa* (Lisbon, 1996), 40 and 130. Here and subsequently, except where noted, all translations from the Portuguese are mine.

5 J.P. Guerra, *Descolonização*, 32–4.

6 João de Melo, 'Uma imagem vale por milhões de palavras … ', in Renato Monteiro and Luís Farinha (eds) *Guerra Colonial: Fotobiografia* (Lisbon, 1990), 8.

7 Interestingly, even though this volume is prefaced by João de Melo, who had edited a two-volume compilation of texts with a good number of photographs – *Os Anos da Guerra 1961–1975: Os Portugueses em Africa: Crónica, Ficção e História* (Lisbon, 1988) – the accent here is still on the need to bring out unknown material, as the following statement makes clear: 'With this work, we want before anything else, to contribute to the dissemination of an unpublished, unexplored, fragmentary, and dispersed world of documents', (Monteiro and Farinha, 'Nota introdutória', 13).

8 Monteiro and Farinha, 'Nota introdutória', 11–13.

9 Melo, *Os Anos da Guerra 1961–1975*.

10 Margarida Ribeiro in a recent survey article, 'Percursos Africanos: a guerra colonial na literatura pós-25 de Abril', *Portuguese Literary & Cultural Studies* 1 (1998), 136, calls attention to the fact that:

> Until the beginning of the 90s the literature of the colonial war was the object of a self-reflexive critique, that is a critique made by those who had also written poems or novels about the war and who had been profoundly involved in the war, so that the distance required by criticism was unattainable.

11 See Rui de Azevedo Teixeira, *A Guerra Colonial e o Romance Português* (Lisbon, 1998), 105, note 130.

12 Teixeira, *A Guerra Colonial e o Romance Português*, 334–5. Lídia Jorge's *A Costa dos Murmúrios* (Lisbon, 1988), was translated as *The Murmuring Coast* (Minneapolis and London, 1995) by Natália Costa and Ronald W. Sousa.

13 Ribeiro, 'A guerra colonial na literatura pós-25 de Abril', 139.

14 Maria Manuela Cruzeiro, 'Guerra colonial: entre o recalcamento e a denegação', *Vértice* 58 (1994), 5–7. This special number of *Vértice*, the most influential cultural journal on the left includes a series of the essays presented at the Colloquium and

others which were added to them. The essays are short and reflect their oral presentation as some are actual transcriptions of oral comments. Two directly address the literary representations of the war. One, by Pires Laranjeira, 'Perspectivas da literatura africana de guerrilha', 8–10, is confined to a very brief overview; the other, by Carlos Vale Ferraz, 'Guerra colonial e expressão literária: falta de memória? Falta de talento? Ou nós somos mesmo assim?', 13–17, assumes the position that not only have there been few works to focus on the colonial war but that their quality is also far from optimal. From this he generalizes that the Portuguese do not rise to the needs of great historical events, a view which might still echo the general complaint about the lack of attention given to the colonial war, but stands in clear opposition to that of most other observers.

15 Cruzeiro, 'Guerra colonial:. entre o recalcamento e a denegação', 5.

16 Teixeira's subtitle, *Agony and Catharsis* is quite indicative of this view of the novel as a medium for working out the haunting traumas of the war. At one point Teixeira also explicitly notes how narrative fiction became the privileged medium in the absence of either public (official) debate or extensive interest on the part of film and television producers (the rare exceptions are mentioned). As Teixeira argues:

> it is then to literature, in its diverse disciplines and different degrees of quality, that falls the cathartic function in relation to the Colonial War. And it will be in the novel that that therapeutic purgation will reach its highest artistic level.
>
> (*A Guerra Colonial e o Romance Português*, 98–9)

17 António Lobo Antunes, *Os Cus de Judas* (Lisbon, 1979); translated as *South of Nowhere* by Elizabeth Lowe (New York, 1983). The Portuguese title, literally, 'The Asses of Judas' means the end of the world; the English title, although catching this meaning, necessarily leaves aside the overdetermining associations both with treason and with eschatology. Although Teixeira has very little to say on processes of memory, he also singles out these two novels, together with another one by Wanda Ramos, *Percursos (Do Luachimo ao Luena)* (Lisbon, 1981), as 'novels of memory' in a subdivision of his book, *A Guerra Colonial e o Romance Português*, 123–35.

18 Henry Thorau, ' "Die Toten wecken und das Zerschlagene zusammenfügen". Die Romane des António Lobo Antunes', in Henry Thorau (ed.) *Portugiesische Literatur* (Frankfurt, 1997), 521–40. This article is part of an anthology on twentieth-century literature. The passage cited here, in my translation, 525.

19 For ease of reference citations are taken from the published translation, even though it has significant differences – especially the distribution of chapters as each corresponds to a letter of the alphabet and the Portuguese does not use k, w, and y – *South of Nowhere*, 9, 14, 121, 152–3.

20 *South of Nowhere*, 51.

21 *South of Nowhere*, 154.

22 *South of Nowhere*, 154. This is the very last sentence of the novel.

23 *South of Nowhere*, 152.

24 See J.P. Guerra, *Descolonização*, 55–6. The interview was conducted on 10 February 1993. 'Ultramar' (that is, 'overseas') is the term used by the general to refer to the colonies, keeping in line with the desire of the Salazar regime to view the colonies as 'provinces' of a unified, pluri-continental Portugal. When asked about Wiriyamu, the most infamous massacre in Mozambique (which was denounced in *The Times*, and other international press), and whether this had been the cause of his having been recalled by Lisbon, Kaúlza de Arriaga not only denied it but also tried to minimize the extent of the massacre, claiming that there had only been sixty dead on both sides. He also at the same time claims that as military commander he would only know of a handful of operations, since practically all of the 150 daily operations

would have been planned by the field officers and not by the Commander-in-Chief (J.P. Guerra, *Descolonização*, 47).

25 L. Jorge, *The Murmuring Coast*, 274.

26 Ana Paula Ferreira, 'Lídia Jorge's *A Costa dos Murmúrios*: history and the postmodern she-wolf', *Revista Hispanica Moderna* 45.2 (1992), 269. Besides this article, still one of the most theoretically cogent and developed analyses of Jorge's novel, others have also foregrounded the question of history: Helena Kaufman, 'Reclaiming the margins of history in Lídia Jorge's *A Costa dos Murmúrios*', *Luso-Brazilian Review* 29 (1992), 41–9; Ronald W. Sousa, 'The critique of history in Lídia Jorge's *A Costa dos Murmúrios*, or Helen of Beira meets Luís of Troy', *Cincinnati Romance Review* 16 (1997), 135–43. The issue of memory, although alluded to has been relatively neglected. Renate Hess in 'Schreiben heisst: sich erinnern. Zum Werk von Lídia Jorge' in Thorau, *Portugiesische Literatur*, 501–20, for instance, in spite of the suggestive title, assumes simply that the first narrative provides the stimulus for the memory that leads to the elaboration of the second narrative. Although true such a statement does not begin to analyse the importance of memory processes in the novel. Maria Manuela A. Lacerda Cabral has published an essay, '*a Costa dos Murmúrios de lídia jorge* – inquietação pós-moderna', in *Revista da Faculdade de Letras do Porto* 14 (1997), 265–87, which is based on her MA thesis, the title of which promises a direct linkage between history and memory ('A história como memória em *A Costa dos Murmúrios* de Lídia Jorge', Porto, 1996). In this essay, however, only a few pages focus on these issues (280–7). Nonetheless, and in spite of some reservations as to the view of memory put forward by Cabral, this study is key and Cabral's focus on 'forgetting' as the counterpart to memory is important for my own views in the present chapter. In my article, 'Memória infinita', in *Portuguese Literary and Cultural Studies* 2 (1999), 61–77, I try to address some of those questions and some of the views expressed there will be mentioned in my present argument. This issue of *Portuguese Literary and Cultural Studies* is dedicated to the works of Lídia Jorge, including several articles in English and a bibliography of published research on the author.

27 See for instance Susan J. Brison, 'Trauma narratives and the remaking of the self', in Mieke Bal, Jonathan Crewe and Leo Spitzer (eds) *Acts of Memory: Cultural Recall in the Present* (Hanover and London, 1999), 39–54. Brison builds on the work of Dori Laub as when she cites the following passage: 'Bearing witness to a trauma is, in fact, a process that includes the listener' (p. 46). See also Shoshana Felman and Dori Laub, *Testimony: Crises of Witnessing in Literature, Psychoanalysis and History* (New York, 1992), 70.

28 Luís Madureira in 'The discreet seductiveness of the crumbling empire – sex, violence and colonialism in the fiction of António Lobo Antunes's, *Luso-Brazilian Review* 32 (1995), 17–29, offers a mordant critique of the deployment of sex in the novels of Lobo Antunes. However, even if in agreement with many of his points, I would nonetheless point out that Lobo Antunes is also conscious of the chauvinism he imparts to his narrator and that scenes of Portuguese soldiers raping men are used in an attempt to directly counter standard images of male heroics. In other words, the conjunction of sex and violence in Lobo Antunes's novels is not simple and unreflective.

29 Although in diverse ways this is also the view put forward by Ferreira, 'Lídia Jorge's *A Costa dos Murmúrios*' and Sousa, 'The critique of history in Lídia Jorge's *A Costa dos Murmúrios*'.

30 Barbara Freeman, ' "Epitaphs and epigraphs": the end(s) of man', in Helen M. Cooper, Adrienne Auslander Munich and Susan Merrill Squier (eds) *Arms and the Woman: War, Gender, and Literary Representation* (Chapel Hill, NC, 1989), 308.

31 See Lynne Hanley, *Writing War: Fiction, Gender, and Memory* (Amherst, 1991), 3–4.

32 Cabral, '*A Costa dos Murmúrios* de Lídia Jorge', 280. Of course one could argue that in order to adapt to the combat situation, Alex had first to develop the lust for senseless violence and bestiality. The 'sport' of mowing down flocks of birds with automatic fire leads to cutting off heads of civilians and impaling them at the top of their huts.

33 Jorge, *The Murmuring Coast*, 42.
34 Ibid., 200.
35 Charlotte Delbo, *Days and Memory*, translated by Rosette Lamont (Marlboro, Vermont, 1985), 3, as quoted by Susan J. Brison, 'Outliving oneself: trauma, memory, and personal identity', in Diana Tietjens Meyers (ed.) *Feminists Rethink the Self* (Boulder, CO and London, 1997), 20.
36 Jorge, *The Murmuring Coast*, 225.
37 Ibid., 202–3.
38 Jorge, *A Costa dos Murmúrios*, 93. In the English translation, curiously, this was left out.
39 Jorge, *The Murmuring Coast*, 213.
40 Ibid., 236. On the relationship between Helen and Evita and on the question of identity see R. Sousa, 'The critique of history in Lídia Jorge's *A Costa dos Murmúrios*', and Paula Jordão, '*A Costa dos Murmúrios*: uma ambiguidade inesperada', *Portuguese Literary* & *Cultural Studies* 2 (1999), 49–60.
41 Jorge, *The Murmuring Coast*, 20.
42 Ibid., 70, 235.
43 Gerd Gemünden, 'Nostalgia for the nation: intellectuals and national identity in unified Germany', in Mieke Bal et al., *Acts of Memory*, 120:

> A headline in *Der Wochenspiegel* from July 1993 read: 'Wir sind wieder wer! Aber wer?' ('We are somebody again! But who?') Indeed, the fall of the Wall in 1989 and German unification the following year have had a paradoxical effect on Germans, instilling in them a euphoric sense of national pride, but also triggering a deep crisis about what precisely it is that one ought to be proud of.

44 Eduardo Lourenço, 'Identidade e memória: o caso português', in his *Nós e a Europa ou as duas razões* (Lisbon, 1990), 9–15. For a radically different view of Portuguese society in the period after the revolution see Boaventura de Sousa Santos, 'State and society in Portugal', in Helena Kaufman and Anna Klobucka (eds) *After the Revolution: Twenty Years of Portuguese Literature, 1974–1994* (Lewisburg and London, 1997), 31–74.
45 This view has been advanced by a variety of critics but Ellen Sapega's formulation of it is especially direct. See 'No longer alone and proud: notes on the rediscovery of the nation in contemporary Portuguese fiction', in Kaufman and Klobucka (eds) *After the Revolution*, 168–86.

9 Longing for war

Nostalgia and Australian returned soldiers after the First World War

Stephen Garton

In 1895 Oliver Wendell Holmes, arguably America's foremost jurist, addressed the Memorial Day Graduating Class of Harvard University. His was not the usual bland entreaty to be worthy, wise and committed citizens, but an impassioned reflection on war, death and honour. One can imagine the slight discomfort of these young graduates, eager to embrace the world, as they listened to words about terror, the splat of bullets, feet slipping on dead bodies, oblivion and the ultimate silence, from the thrice wounded veteran of the American Civil War. But Holmes's reflections on war were far from melancholy. For him war may have been 'horrible and dull' but it also contained a 'divine message': soldiers 'have shared the incommunicable experience of war; we have felt, we still feel, the passion of life to its top'. Indeed it was the soldier's duty to advance against all odds, and in that moment they triumphed even in death. It was heroism that was the defining feature of the 'soldier's faith', and in this revelation Holmes sought to ground a system of values for a new generation to embrace.[1]

Historians have long noted the impact of Holmes's war experience on his life and even his later judgements as a member of the American Supreme Court.[2] War was the upholstery stud of his moral world, something that gave shape to his sense of 'being'. But there are complex undertones to Holmes's graduation address; not just triumphalism, the embrace of a warrior tradition he saw stretching back to Beowulf, moralism, a sense of spiritual transcendence in war or even the melancholy of remembrance and the acknowledgement of the grim finitude of existence. There was also a longing, a nostalgia for times of abandonment, the élan of a selfless, mad, heroic rush into oblivion, and the noble impetuosity of youth in contrast to the decline of old age. Giving rhetorical shape to these ideas and feelings was a pertinent trope – the truth of war in contrast to the mundane and suffocating world of civil life (particularly the Jeffersonian vision of an ideal society). For Holmes, 'who of us could endure a world, although cut up into five acre lots and having no man upon it who was not well fed and housed, without the divine folly of honor?'[3] War meant everything to Holmes.

Holmes was not alone in these feelings. There is abundant evidence for what George Mosse has aptly called the 'myth of war experience' – from combatants on all sides, in many modern wars, in many places.[4] Returned soldiers were active promoters of the 'divine' and 'transcendent' significance of war, an experience which set them apart from ordinary citizens. They established organizations not only to fight for entitlements and adequate pensions but also to fan the flames of war remembrance – the Order of the Cincinnati, the Grand Army of the Republic, the British, American and Canadian Legions, the Returned Services League in Australia, more recently the Vietnam Veterans' Association and various Gulf War associations. Even noted critics of the folly of war, such as Siegfried Sassoon, could look back, in time, to war as a defining moment, a pinnacle of experience, something that only those who fought could share.

The emotional tenor of Holmes's speech, indeed this larger emotional structure embodied in the 'myth of war experience', is echoed in the autobiographies, private writings and memories of ordinary soldiers. Here we can glimpse a deep current of longing and nostalgia for war among those who survived. This is evident in the writings and actions of returned Australian soldiers. In 1932 Arthur Graham Butler, a medical officer at Gallipoli and in France, witness to the physical agonies of war and later official historian of the Australian Medical Services, wrote to a colleague: 'I should love to go to the front line again and stay there'.[5] Three years later Raymond Lording (writing under the pseudonym A. Tiveychoc) wrote of his feelings on returning home almost twenty years previously: 'he felt not the glamour of heroics but rather the pathos of eventide that signified the passing of so much – the venture, the great experience had ended – the sun had set'.[6] In a more pathetic vein E.J.H. Joseph felt many years after the event that with war he 'had completed his life's work', all was decline thereafter.[7] More pathetic still was Oswald Bates who had returned from the trenches in France. In 1928, weary of life, he swallowed poison, turned to a friend asking him not to get help but light him a cigarette, stand him up against the fence, and let him fall on his face 'like the good lads on the other side'.[8]

These memories evoke a powerful and striking nostalgia for war, a phenomenon which this chapter seeks to explore. Our task is a difficult one. There are very few such accounts, although I would suggest that their rarity should not lead us to dismiss them as necessarily unrepresentative, singular or exceptional. In some senses the selection of memories is always arbitrary – a particular country, a particular war, not to mention the vagaries of how such testimony is gathered into an archive. The question of its representativeness can never be answered satisfactorily. Moreover there are powerful factors militating against the creation of a 'satisfactory' archive. Some of the familiar sources for such an investigation are more than usually partial in this context. This in itself is illuminating. Australian soldiers, for instance, were avid diary-keepers and many hundreds of these diaries exist in libraries and archives. The vast majority, however, cease with the return home. This might be due to a belief that libraries and archives were only interested in collecting war memorabilia but so many of them finish mid-volume it is just as likely that soldiers actually believed that their war experience was the one significant thing in their life worth recording. Many, it would seem, began to keep a diary as they enlisted for war and stopped, never to write one again, on demobilization.

Equally, pension case files contain many letters from returned men proclaiming the failure of governments to acknowledge their sacrifice. But this is a very specific context where complaint was a structured and highly ritualized mode of interaction. Thus such sources are not particularly useful for exploring the more general structures of memory of concern here. Far more illuminating are the oral histories of Australian returned soldiers, although most focus on the memories of men nearing the end of their lives, when they were anxious to tell a story.[9] Here I want to take up the problem posed by men like Butler, Lording and Bates. How was it possible to long for war in the first two decades after the cessation of hostilities? This requires a close reading of some very partial evidence, and in this instance I will concentrate, in particular, on Lording's autobiography. It also requires an effort to put this partial evidence into a wider context and thus infer a broader *mentalité* among Australian returned soldiers. In doing so I hope to raise some questions about the more general problem of nostalgia.

But this task is even more difficult because of the intricate relationship between these individual memories and the national narrative that has evolved in Australia since 1915. Australia was a nation which used the experiences of soldiers during the First World War to articulate crucial elements of national culture. In numerous reports, celebrations, marches, memorials and stories, the exploits of the Anzacs (Australian and New Zealand Army Corps) were proclaimed as the moment of realization of the potential of the former British colony of Australia to become a nation. Anzacs 'proved' that a vigorous and brave people inhabited the land, that in shedding blood for the nation the final right to claim the land for colonists had been won (thus symbolically dispossessing indigenous peoples) and they now had earned the right to stand on the world stage. The anniversary of the landing at Gallipoli became the major day of national celebration (and remains so today). And the extent to which Australian returned soldiers enrolled in veterans' organizations to perpetuate the memory of the war (a third of all returnees – a higher rate than any other country) points to the crucial investment of many veterans in this myth. In contrast to other allied nations, notably Britain, France and Canada, where memorialization of war was a solemn process of remembrance of the war dead, for Australians it was not only a day of remembrance but also a proud moment of national foundation. Many Australian war memorials recorded the names not just of those who died but all those who fought, to acknowledge the role of all soldiers in 'founding' the nation.[10]

The current public debates (and conflicts) in Australia over the 'nation' – encompassing such questions as whether Australia should become a republic (breaking its final constitutional ties to Britain), recognition of prior Aboriginal occupation and sovereignty (through such issues as native title, formal apologies for past practices of stealing indigenous children and other practices aimed at dispossession), and the emergence of active efforts to suppress such reform efforts (notably right-wing, white settler society proponents such as One Nation), all themes explored in greater depth in Ann Curthoys's contribution to this volume – highlight in new ways the contested nature of national culture. This chapter works within this context. In one sense it seeks to highlight the 'universal' nature of the war experience that has come to mark the specificities of Australian nationhood. This is an essential part of the critical work necessary to render the presumed self-evidence of nation and national identity problematical, particularly as questions of globalization and post-colonialism emerge as ones of vital significance in contemporary political culture. (It has the added virtue, in the context of this international and comparative volume, of making the Australian case an example of something of more general relevance for readers.) In other ways, however, it asserts the locality of its investigations by arguing that the Australian case is different in one important respect, namely that a 'universal' experience of soldiers in war was turned, uniquely, into a national narrative, something of considerable significance for those soldiers who returned and faced the difficult task of 'remembering' war.

* * *

Memory and nostalgia are now familiar problems within history. Much of this scholarship, most notably Raphael Samuel's idea of 'memory work' or Michael Frisch's idea of memory as a verb, analyses memory as an active and dynamic process, one where selection, both remembering and forgetting, make and re-make a narrative of the past.[11] In this memory work historians have argued that nostalgia is a 'romanticized' version of the past. Not all memory work leads to nostalgia but it is a type of remembering that can thrive in particular circumstances and conditions. Studies of nostalgia have defined it as the construction of an 'imagined' *Gemeinschaft* arising from 'present discontents', a context of dissatisfaction which shapes the moment of recollection or a process that works with 'distance' from the events remembered.[12] The function of distance has been explored in relation to war by Al Thomson, when war stories had hardened into well-structured but diverse performances such as amusing anecdote, the battler or the radical awakening. Here nostalgia wove its way into and out of other types of narrative.[13]

In contrast, my interest is in whether 'present discontents', in the sense of the post-war factors which shaped memory, is a sufficient concept to explain nostalgia in Australian returned soldiers in the two decades after the First World War. Here I want to suggest that war experience itself played a larger part than post-war 'discontents' in the generation of particular tropes of nostalgia. This is put forward to supplement rather than replace existing approaches. Thomson, more than most, is aware that the complex relationship between memory, discontents and public narrative is something not fixed, but a process. His interest in how a public memory helps to structure, 'from above', individual accounts is a vital insight. But in the case of Australian national narratives it is equally important to realize that the national narrative was, in part, a compilation of individual memories working their way up 'from below' to become something that was embraced by a national collectivity. The qualification 'in part' is crucial here. Elements of the fully developed Anzac legend drew on a longer, well established iconography, stretching back to the late nineteenth century when the laconic, radical nationalist bush worker came to embody a national type. But the 'bush legend', despite its continuing currency (although now a rather kitsch one), lacked, I would argue, a vital emotional content. It was too specific, too situated in a particular social class, to work effectively as a 'national' symbol, despite the efforts of important journals such as *The Bulletin* to promote the bush legend. Moreover it was always in competition with other images ('little boy from Manly', 'Minerva' and so on).[14]

The Great War, and more specifically Gallipoli, offered an opportunity to turn the bush worker into a truly national figure (because it was an experience that cut across most classes) and, even more importantly, it gave this figure a deeper emotional significance. The staggering loss of life and extent of casualties in the Australian armed forces (at nearly 64 per cent of the forces either killed or wounded, higher than any other allied force) affected most families in Australia in a very direct way.[15] It could incorporate a national experience, and, through the widespread sense of grief, loss and the need to memorialize, it gave a depth and resonance lacking in previous incarnations of the national type. Returned soldier organizations played a crucial role in cementing the Anzac legend as a national narrative, and their investment and the values they promoted had as much to do with their individual experiences as with national imperatives. Thus the national narrative which developed in Australia was a complex process of imperatives 'from above' finding accommodation with those 'from below' (accommodations continually remade since then). It is here that I would question some of the theoretical work on nostalgia, discontents and 'public memory'. All of these things are significant but it is also important to investigate how the shape of public memory takes on particular contours from the moment of its making, that discontents are 'present' things but that in memory they have to work with material, structures and stories that are 'old' things. And here the function of memory and nostalgia amongst Australian veterans offers an insight into some specific problems. The incorporation of the experience of the Great War (with its significant loss of life) into a national narrative was not an easy process. It required the 'repression' (conscious and unconscious) of much of that experience, and yet some of the emotional, fraternal and political structures that sustained soldiers at the front returned in subtle ways to mark out a particular type of national narrative, a narrative which in turn structured the individual remembering of war. Here I hope to explore some of these processes.

In this regard Australian returned soldier narratives stand in marked contrast to the more familiar response of English war poets, painters and writers such as Owen, Nash, Sassoon and Graves. Many historians have explored the diverse English and European cultural responses to war, although there are some marked differences of emphasis and interpretation. Where Fussell and Eksteins argue that the experience of the trenches was a potent impetus to a cynical and detached modernist aesthetic, Hynes and Winter have stressed the conservative and classical orientation of post-war culture.[16] But whether modernist or classical, war poets in particular romanticized a rural Arcadian past, a time forever lost through the horrors of war. What is remarkable about the narratives of Australian veterans is the extent to which the horror of war is elided. Men such as Bates, Joseph and Butler had experienced the worst of battle but there are very few references to the terror of fighting or the grisly horror daily experienced in the trenches. Most memoirs or accounts of war experience either ignored the battles entirely or brushed off the actual fighting with phrases such as 'brave engagement', 'sterling efforts' or 'noble sacrifice'. Lording is a notable exception. His is one of the few accounts that tries to describe an actual battle scene in grim detail. This makes his case a particularly interesting one. But even his efforts to incorporate a sense of horror, as we shall see, are marked by evasions and distancing techniques that serve to undermine any real sense of the realities of warfare. There is very little in the corpus of Australian depictions of war by soldiers to match the grim evocations of Sassoon, Nash or Owen. Even the diaries of front line Australian soldiers describe battle in a curiously distanced language, one marked by understatement or cliché.[17]

There were obvious psychic investments in the glory of fighting for one's nation that constrained rejection of war. Important studies, such as Klaus Theweleit's exploration of the characteristics of 'soldier males' and Graham Dawson's analysis of 'soldier heroes', position war as central to the formation of a glorified and heroic martial masculinity; one defined against domestic femininity and promoting, in Dawson's words, a 'projective disavowal of other aspects of the [war] experience'. It was a vulnerable and anxious masculinity but one that promoted stories of heroic soldier exploits rather than the 'horror of battle'.[18] But why some Australian soldiers found emotional succour in war itself rather than some Arcadian past, or more particularly how that idealized past became war itself, requires closer analysis. More importantly, our focus is on the forms of expression for nostalgia, and how those forms arose out of a particular front line culture. Thus, while I would not discount 'present discontents' in fostering nostalgia, my aim here is to analyse how nostalgia, and its particular modes, had deeper roots in war experience itself.

Part of the answer lies in the power of public memory. Australian soldiers, unlike veterans from other combatant nations in the Great War, faced a double repression. Not only did they have to divest themselves of the characteristics of 'soldier males' on returning to civilian life, they had to adjust to a culture which was making their exploits the foundation of national myth. This peculiar context made many Australians particularly antipathetic to efforts to question the 'glories' of war. There were extraordinary efforts by soldier groups and their supporters to crush modernism in inter-war Australia.[19] Australia was a culture particularly antagonistic to any aesthetic which queried the virtue and nobility of war. Thomson has explored the effects of this public context on the private memories of returned men. Over time their memory of war merged with the national narrative – horror was repressed or submerged, and glory, bravery, larrikinism and mateship became the dominant tropes of recollection. More tellingly still, actual memories were constructed out of the official histories and reports of Anzac exploits (returned men reported incidents as their own when in fact they had read about them not participated in them). Thomson has given us a very sophisticated insight into the diverse and complex ways in which Anzac memories were worked through in a context of a national narrative of heroic sacrifice and glory.[20]

Here, as suggested, I am seeking to open up another aspect of this memory work. It is one which played a part in the formation of returned soldier narratives. A common feeling amongst returned soldiers from most modern wars, was that in war they had reached the peak of their existence, had experienced life at its greatest intensity, and in doing so had fully realized themselves. We can glimpse some part of this feeling in Oliver Wendell Holmes's statement on having 'lived life to its top' or Lording's reflections on 'eventide' and the end of the 'great adventure'. Even Joseph's idea of a 'life's work completed' is part of the same ensemble of ideas and emotions. But what is curious about the Australian rememberings is that these men returned to a culture that proclaimed their exploits. Unlike veterans from other forces who had to struggle with cultures manifestly ambivalent about their exploits or largely uninterested, Australian soldiers found themselves at the centre of national iconography. And yet they still found themselves longing for war in later years. Was this due to post-war discontents? Undoubtedly this was the case, and in many cases there were real instances of material and personal hardship in the inter-war years which fostered these feelings, but this sense of discontent was also, in part, due to a view of the world made during the war. Most of these soldiers had returned home as young men (even if many believed that they had grown to maturity and lived a full life already), destined to face a number of years as ordinary civilians. For those few who put their memories into written form in the inter-war years, war was the crowning moment of life and everything afterwards decline – mundane, routine, comfortable, solid, respectable but somehow lacking. Thereafter life was a pale shadow of war, an afterglow – or, worse, carried an intense sense of loss of a 'true' self that had only been found in combat. So what was it that allowed the horrors of war to be turned into an enriching, comforting, poignant memory, the climax of the narrative and even, in some cases, an aching nostalgic, longing for return to the front?

* * *

Raymond Lording's autobiography *There and Back: The Story of an Australian Soldier*, published in 1935, was one of the most widely read accounts of war and return in the late 1930s. It is a complex and at times contradictory text and one that needs to be approached with some caution. Lording himself had a brief but traumatic battle experience. He enlisted as a young man, serving in both Egypt and France. At the Battle of Fromelles, Lording was wounded by machine-gun fire while charging a German trench and evacuated to a clearing station behind the lines. He hovered between life and death for a number of months. He survived but his wounds were so severe that over the next twenty years he underwent more than fifty major operations and consequently developed a serious addiction to morphine. He eventually overcame his addiction, married, became a father and suburban accountant, and wrote his story. It is a familiar and conventional narrative, part Bildungsroman and part tale of ultimate triumph over adversity. But the book also has an interesting publication history. Lording was closely associated with a number of important 'boosters' of the Anzac legend, including A.W. Bazley, assistant to the official war historian C.E.W. Bean; Frank Dunne, journalist with *Smith's Weekly*, the major returned soldier newspaper of the inter-war years; Major Ken Prior, Managing Director of the pro-Anzac and popular conservative journal *The Bulletin*; and prominent members of returned soldier organizations. *There and Back* was actually published by the Returned Sailors' and Soldiers' Imperial League of Australia. The timing is also interesting. In the early 1930s returned soldier groups were engaged in a struggle for a new type of pension, for those veterans who were chronically unemployed and struggling to meet the demands of life even though they had no discernible pensionable disability. Returned servicemen's organizations, in their campaign for the service pension, were anxious to reaffirm the bravery and sacrifice of soldiers and, more importantly, their plight since returning home. Thus Lording's autobiography was seen and promoted as an 'exemplary' tale suitable for a general audience. (The campaign for the service pension was successful soon after the publication of Lording's book, and, while there is probably no direct connection, this possibly explains why Lording's was one of the few books published by returned soldier groups during this period.)[21] This is an obvious constraint on the book as an 'impartial' or purely 'personal' reminiscence. But it does raise intriguing questions about what was it that made this account 'exemplary' and in what ways it sought to remember war.

There are a number of literary peculiarities in Lording's autobiography. First, it is written in the third person, through a persona Ted (shortened from Edward). This is Lording's middle name, the name of his father and the name of his son. It is certainly a distancing technique, one which objectifies the personal narrative, in an attempt to give it greater authority. But it is more than this. In an extraordinarily self-conscious fashion Lording states that by choosing this persona he hopes to link the generations in a seamless web, to blur the personal specificities of his tale in order to render it timeless and universal. His central character becomes an Australian 'everyman'. Thus its function as an exemplary tale arises from its specific mode of address. Second, the book is divided into three distinct parts, differentiated by genre. The first part is childhood memoir, where innocent Ted is enthused by the idea of war and seeks, although he is under age, to enlist. A series of influential adults (doctors, recruiting officers, a colonel, his batman, eventually his father) guide and facilitate his quest to flout the age regulations. Other men guide him through the pitfalls of training and camp life, ensuring that he becomes a part of the company. In the second part, there is a sharp shift in genre and register. Lording's war experiences are told in the diary format, although again it is in the third person and focused on the persona of Ted, suggesting that it was in part a 'fictional' diary, not merely a reprinting of an actual one, or more properly a memoir reconstructed in diary form. The final part returns to a more usual memoir form, beginning with the actual battle in which Ted is wounded and removed to hospital. The remainder of the account is an 'addict's tale' – of serious injury, the demon of addiction and the overcoming of that demon through perseverance and mateship.

Through these three set pieces war is remembered in different ways. It is adventure, laughs, mateship and a means of becoming a man. This is certainly the tone of the first part. Much of this continues in the more sparing episodic structure of the second part. But added now is the 'strangeness' of war. Here there is the anxious waiting, the curious flights of the mind as it contemplated what was to come, the thunderous noise, the near scrapes with death, the lucky escapes and then the final rush into battle – fire, sharp explosions, flying dirt, shrieks, the 'zipzipzip – zip machine gun bullets' and then 'another twisted heap of khaki hits the ground'. There is a certain exhilaration in the account of battle which is brought to a shocking and abrupt halt with 'It is Ted. He does not move.'[22] Working against this horror, however, is the frequent deployment of the disruptive trope of the laconic, cool under fire, joking soldier. Here are the devil-may-care attitudes, the amusing moments and interludes, all of which conjured up the conventional image of the Australian 'digger' central to the national legend – a great fighter, rugged individual, larrikin and fearless defender of a 'mate'. This is a very familiar characterization. But cutting against this representation are more unsettling moments: the exploding shells, the eerie sense of impending doom, the stark images of men blown to pieces, the death of good friends and the general effort to conjure a 'hell'. This last word is Lording's and while his literary skill is not sufficient to make this effort at a contrapuntal narrative of 'mateship' and 'hell' work successfully, the effort is noteworthy.

What seem to be operating here are two very different representations of war. A very conventional one of the nobility of soldiers and their transcendent character and spirit, and a more subversive one of the meaninglessness and horror of war. How does this autobiography grapple with these divergent strands? In one sense the horror appears to triumph because of the third part of the book on the aftermath of war. Lording's account of his injuries, the taxing recovery, his addiction, his struggle for survival and his frequent collapse, works to confirm a very moving, graphic and personal depiction of the tragedy of war. But Lording works very hard to resist this imperative, attempting to wrest a very different memory of war from his experiences. He constructs his war as a moment of great sublimity. Perhaps this is the perspective of someone whose life has been cut in half – a youthful, healthy, manly half, and a shadowy, painful, invalid half. War may have been the cause of the division but his war experience is decidedly, until the fateful day, in the first half. Thus war can be remembered as the crowning moment of his 'true', uninjured, youthful self. But this is too simple. Time and again Lording reasserts the 'hell' and 'futility' of war before his wounding. But after his evacuation, as the story threatens to lapse into a genuine 'hell', he punctuates his narrative with fond recollections and strikingly jocular memories.

The key theme here is mateship. The source of Lording's life force seems to flow from the community of soldiers and former soldiers. In hospital it is the 'yarn with a cobber [friend], joke with a wardsman [all ex-soldiers]' which meant more to him than anything else.[23] He thus eventually finds his way out of the wilderness of illness, thereafter devoting himself to the returned soldier cause – obviously something that makes his case a useful example for others. But the underlying tone of emotional solace is crucial here. At remembrance ceremonies 'they were all there – fallen comrades, veterans, youths. It was one great reunion – a reunion of souls.'[24] And at the end there is the ringing hymn of 'We will remember them'. This book is a massive effort to chart this remembrance, and the very difficulty of the task of rescuing such a message through the life of Lording is an obvious impetus to its publication. In the end it is the memory of a transcendent comradeship that is rescued and made the key feature of Lording's story. Here, in the final pages, the horror of Lording's experience serves not to undercut but to reinforce the narrative of Anzac triumph over adversity.

But there are more curious facets to this autobiography that work against the manifest message of the text. The focus on mateship works through an almost complete disavowal of family. Nowhere is Lording's wife mentioned in his account, although we know from the introduction that he is married and has three children. In this effort of remembrance the private world is entirely erased, as if it played no part in his recovery. More than this there is a dismissal of civilian intrusion into the sacred communion of returned men. Repatriation efforts are worthy but it is the returned soldiers who smooth over civilian ignorance and the well-meaning ineptitude of bureaucrats and doctors. What marks this text is a radical disjuncture between Anzac and civilian. This works largely through a persistent forgetting of civilian assistance, most notably that of family. There seem to be two worlds here that are almost unbridgeable and, although Lording is forced to return to civilian life, this process is relegated to a few perfunctory sentences. Thus the powerful memory of war is strengthened by an equally powerful disavowal of home. This memory and forgetting parallels a strong message of many returned soldier organizations. Lording's focus on mateship and disavowal of home is a key theme of much returned soldier discourse and it is this, as much as his stress on the good work of returned soldier organizations, which makes his autobiography an exemplary one.

* * *

The circumstances of the writing and publication of Lording's autobiography also suggest problems with theories of nostalgia as a product of 'present discontents'. This book was written at the moment of Lording's triumph over his addiction. He has come through an ordeal and forged a meaningful life in the face of the severest adversity. Far from a romanticized recollection born of bitter resentment and current unhappiness, *There and Back* constructs a memory of war as transcendent and sublime at a time of great personal triumph. There are two great moments in this life – war and survival of addiction. But there is an interesting imbalance in their presentation. Lording's experience of trying to enlist, actual enlistment, training and then war constitutes two-thirds of the book. His evacuation and early hospital experiences take up much of the rest of the book. The triumph over addiction and the forging of a new life are a mere thirty pages, just over a tenth of the length of the book (although in reality it constituted almost half his life to that point). In the scheme of life it is war that bulks largest. And this is hardly explained by the concept of 'present discontents'.

There is a useful contrast here with one of the other returned men who put their thoughts and recollections into writing between the wars. A.G. Butler, the official historian of Australia's wartime Medical Services, remained a passionate advocate of the Anzac legend all his life. To this end he gave up a potentially lucrative medical practice to publish his three-volume history, a labour of love that took him nearly twenty-five years. Butler's circumstances offer some evidence for 'present discontents'. His salary as official historian was small and the growing frustration of others, notably Bean, at his slowness was the source of great emotional burden. By the late 1930s he was being hammered from all sides – subjected to repeated demands to write and struggling financially. There may have been an element of 'present discontent' in his love of the Anzac legend. But these discontents were of his own making. He could have returned to medical practice at any time. Even after his official task was completed in 1943 he continued to work, publishing a popular hymn of praise to the Anzacs, *The Digger: A Study in Democracy* (1945), comparing the Anzacs to Winstanley and the Levellers of seventeenth-century England. There is something much more than dissatisfaction and discontent propelling the romanticized and nostalgic reminiscences of men like Butler and Lording. Both exhibit a fundamental refusal to give up their war experience. Overriding everything else in their life – family, work, church, friends – was war, the glory of war and the community of comrades to whom they felt forever joined.

It is this blurring of past and present, the erasure of a sense of time, that kept the Great War alive in the minds and memories of these men. What is striking here is the ways in which this personal memory work paralleled and echoed the public pronouncements and aims of prominent returned soldier groups. The key here was the poignant entreaty 'Lest We Forget', and returned soldier organizations dedicated themselves to keeping the flame of remembrance alive. This was central to the emerging national and public narrative of the Anzac legend in Australia. Of course this was hardly unique to Australia. Nor was the mateship that was so highly prized in this legend. As historians of other combatant forces have made clear, comradeship was a key and sustaining feature of war experience for all combatant forces.[25] But, as many historians have suggested, most notably Ken Inglis, Australians took this further than any other nation in weaving this into the central thread of the national story.[26] Moreover, as suggested, Australian returned soldiers flocked to those organizations which perpetuated such ideas more than any other veteran force.

So in public organizations and the published and private memories of

Australian returned soldiers there was a marked refusal to let go of war experience, an essential continuity, one that asserted the spiritual brotherhood of the front line and worked through a disavowal of civilian life and family. Here we can see that such 'memories' were fostered by public forms of commemoration. Men consciously chose to align themselves with a public 'comforting' tradition. But the deep emotional resonance, the passion of the commitment of many to the perpetuation of such memories points to deeper psychic roots for the 'legend'. Mourning for lost friends, desire for a lost youth, guilt at survival and perhaps even deeper masculine investments in the murderous orgy of war may have worked in complex ways to embed profound attachments to the war commemoration.[27] The ways war experience found deep emotional roots, both publicly and privately, is beyond the scope of this chapter. My point here is that such structures of feeling found certain 'types of expression'. And while present discontents may have sharpened nostalgic memories, the forms of that nostalgia, its modes of expression and the tropes for its recollection, were found in the culture of front line experience.

There is abundant evidence to suggest that the ensemble of ideas and emotions underpinning later nostalgia was present from a very early stage. It did not evolve slowly with the passing years or gain lustre and force with distance and discontents. Rather, the essential tropes of front line and home front were present in returned soldier literature even before the war finished. The sense of identity forged on the front line actually found its force and shape in relation to those at home. Australian soldiers in particular, thousands of miles from home, developed a complex demonology of home front betrayers, fanned by newspapers, letters from home, stories in journals and magazines and the reports of new troops, recently arrived from Australia. The focus of their enmity were all those seen to be failing in their duty to support the soldiers at the front. They saw home through the lens of the front, with its peculiar perspectives. And at that distance particular objects of hatred included those politicians who failed to deliver the promised manpower and resources; profiteers, who withheld crucial supplies to drive up prices and make excessive profits at the expense of soldiers and their families; strikers and members of trade unions who stopped or slowed down industries at home for their own personal gain, regardless of how necessary those industries were for the war effort; and shirkers or stay-at-homes, men who refused to enlist, thus undermining morale and reducing manpower at the front.[28]

One of the most corrosive anxieties afflicting soldiers on the front line was their sense that women at home were not fully committed to their cause. As a number of historians have noted, the division between front line and home front is, in part, a gendered one; a context for the representation of gender relations and a site for anxiety and the disruption of these relations.[29] This is evident in popular journals and magazines of the war and post-war years. Battle in many ways was represented as an assertion of manhood, and the symbolic prize of that assertion was women. Soldiers were seen as having a rightful expectation of faithfulness from their own women and, equally, a reasonable expectation that they would be desired by all women. This dichotomous representation of woman as both faithful wife/mother and desiring lover/whore drew on a long heritage of masculine images of women, but war brought them into sharp focus. Both images pervade the pages of Australian journals read by soldiers. In the official journal of the Australian Returned Soldiers' Association, for instance, the motto of the token women's column was 'A mother's heart is always with her children, a wife is the key to the house'.[30] But in front line journals women were the objects of a male sexual gaze, ever available, willing and compliant – ' " Diggus Feminus", easily domesticated … lives chiefly on chocolates, cakes and ice cream … very fond of theatres, movies, moonlight excursions and picnics' – and variously known as 'The Tabbie', 'The Bint', 'The Clinger' and 'The Skirt'.[31]

While such images and stories spoke of male desire unshackled in war, they also undermined male confidence in the faithfulness of Australian women. The fickle and selfish woman was a recurring figure in Australian popular culture in the war and immediate post-war years. Such a figure was of course as much an entreaty to women to act honourably as it was a direct articulation of soldiers' fears, but the two streams produced a bitter and scarifying portrait of women on the home front. In one cartoon in *The Bulletin*, one of Australia's most popular journals, two military wives decked out in all their finery, discuss the effect of war:

Mrs Lieutenant Smith – You were always complaining before the war about Dick's late hours, but now he's away you seem happy as a cricket.

Mrs Captain Dick Brown – Ah, yes. It's such a comfort knowing he was safe at the front.

In a similar vein, a cartoon pictured a young flapper talking to an old man:

Old Thing – It is gratifying to find you young ladies still persisting with a cheerful heart in your efforts to drive the war on …

Young Thing – Oh, I just love my bit! I've seven fiancés in Palestine and twelve in France, and if the war hadn't fizzled out till Easter I'd have made two dozen.

In these and many similar depictions women joined the pantheon of forces who, for their own selfish ends, were eager for men to continue fighting and dying.[32]

The object of greatest soldier contempt was the woman who consorted with shirkers and profiteers. War, instead of rewarding 'real' men with the love and affection of women, allowed the shirker to exploit their absence. In 1919 *The Soldier*, published a particularly savage drawing: 'The Girl Who Didn't Wait'.[33] Here a distraught soldier is restrained and comforted by a mate as he sees his former sweetheart on the arm of a rich, older man (thus linking many pervasive soldier antagonisms – the shirker, the profiteer, the older generation and the faithless woman). Such depictions preyed upon pervasive cultural anxieties and, for some, became reality. We can glimpse the complex intertwining of life and representation in the frequent news reports of outraged returned soldiers abusing civilians. In 1919 Eric Bannister, a returned soldier, was charged with assault after he saw John Price kiss a woman in George Street. He broke up the kiss shouting 'What right have you to kiss a girl when you haven't been to war?' and punched Price in the face. Other newspaper reports harped on the numerous cases of soldiers who returned to find that they had been 'cuckolded' by shirkers and profiteers.[34]

At the front line soldiers had come to expect betrayal as much as glory. And the experiences of many returned men suggests that they found both. Certainly glory was very much in evidence. Returning soldiers were greeted by tumultuous and joyous crowds, feted by dignitaries, embraced by wives, lovers, friends and family, and proclaimed by newspapers and politicians alike as the men of 'whose deeds of valour the whole world had sung paeans of praise'. They had made 'the immortal name of Anzac' and later they came to be seen as the exemplars, even the founders, of a distinctive national ethos.[35] But it wasn't long after the shouts of the crowds had died down that many returned men came to view the sentiments of the banners, speeches and the crowds as rather hollow. Returned soldier representatives capitalized on and fanned this discontent in their arguments that the men were not receiving what they had been promised. Politicians had legislated to provide a series of rights and entitlements for returning soldiers – pensions for the disabled, free hospital and medical care, training for future careers, housing assistance, employment preference and a land settlement scheme. There were inevitable delays in implementing these policies, and these angered men anxious to get their just reward. The Federal Minister overseeing most of these assistance schemes, Edward Millen, earned the nick name 'Millen-nium' (because it was clear that his promises would not be filled until then) by exasperated returned soldier groups. Men at the front had come to despise politicians for placing them in the mire of France. And many found their low opinions confirmed by bureaucratic delay and political compromise. In numerous returned soldier journals politicians are depicted as corrupt, thieving and conniving traitors. After the war there were calls for a 'soldier parliament' as returned soldier groups despaired of the 'do-nothing' attitudes of traditional party politics. They claimed that only the men who had fought had proved themselves worthy of full citizenship, and with their reputation for action, were the only ones capable of alleviating the mess of post-war society.[36] Such attitudes fuelled the frequent returned soldier riots after the war and efforts to impose 'loyalty' on those deemed 'disloyal', as well as fostering the growth of returned soldier secret armies in the inter-war years, faithfully imagined in D.H. Lawrence's *Kangaroo* (1923).[37]

What is striking here is the extent to which the memories and reminiscences of individual returned men, such as Lording and Butler, were structured by tropes that were an integral part of returned soldier literature from the war years. Central to the formation of the returned soldier identity was an assertion of the importance of the spiritual brotherhood of Anzacs. This was sustained by a prevalent returned soldier antagonism to those at home or an equally poignant erasure of family from recollections of their lives. The bitter divisions of the front line and the home front, which plagued soldiers in war, also sustained many of them after the war. There were complex tensions, then, in the emergent returned soldier identity in inter-war Australia. On the one hand a national narrative proclaimed their significance to the nation. On the other hand persistent antagonisms of war helped sustain this identity in opposition to civilians. Thus disavowal of home worked in intimate concert with a refusal to let go of war experience. These emotional, ideological and, in the hands of more formal returned soldier organizations, political, frameworks helped shape the experience of a number of returned soldiers and even their actual modes of remembrance. Thus the individual and specific reminiscences of men like Lording and Butler were part of a broader stream of memory, not born out of distance or discontent, but present in the basic structure of a returned soldier identity.

* * *

The memories of Australian veterans of the First World War only rarely touched on the brutal realities of combat. Such events were a disturbing memory and many returned soldiers sought to forget them entirely. Some, quite a few in fact, refused to participate in post-war celebrations or returned service organizations, preferring to expunge all memory of war. But a substantial number did seek to maintain their front line friendships in such organizations and many more participated in the annual memorial day for Anzac exploits. There is a powerful cultural imperative to memorialize bravery, sacrifice, courage, fortitude and mateship rather than the underside of war experience – death, crippling injury, bitterness, dependency and discontent. And while solemn remembrance is now more to the fore, the celebration of a national ideal remains. It is not surprising that individual memory and national narrative have come together as one in such public celebrations of the Anzac ethos. Here a powerful nostalgia pervades the accounts of war experience.

What is more surprising is how embedded this nostalgia is, even in the private memories of ordinary veterans. Away from the crowds and national narratives they have come to live the Anzac ethos, although, as Thomson has shown, it is sometimes a fragile and difficult accommodation. The national narrative is a powerful agent. But here I have tried to suggest that the process of return played an integral part in the formation of a certain type of memory work in relation to war. This nostalgia, which relied so much on forgetting, was rooted in discontents. Discontent was the suppressed supplement that allowed, indeed shaped, a narrative of wartime glory. But these discontents were not just those of later time and circumstance. They were in fact inextricably intertwined in the experience of war itself. The front line was a charged emotional environment of deep masculine bonds, bonds of such intensity that it is not surprising that men hungered for them after the war. But these intensities were also fuelled by anger, resentment and frustration, much of it directed at the home front. Many of these men, while delighted to be home, could not shake off the culture of the front. They expected to find unsympathetic civilians at home and, like many expectations, lived experience worked to confirm them. Many returned men found ample reason to be dissatisfied (and many of these dissatisfactions were justified) and out of this arose a hunger for due acknowledgement, evident in the desire to maintain the Anzac flame on the one hand, and, on the other, to become dependent on a society they thought owed them. Discontent, then, was forged on the front line and found its confirmation on the home front. Indeed, for a few, it was home that was constructed as 'hell'. In such a troubled emotional context, men forgot the horror of war and remembered the mateship, the camaraderie, the excitement, the thrill, the fun and the laughter of an innocent time when they lived life to the full.

Notes

1 'The Soldier's Faith' in Mark De Wolfe Howe (ed.) *The Occasional Speeches of Justice Oliver Wendell Holmes* (Cambridge, MA, 1962), 73–83. I am indebted to Michael Grossberg, Indiana University, for this reference to the work of Holmes.

2 See for example Sheldon M. Novick, *Honorable Justice: The Life of Oliver Wendell Holmes* (Boston, 1989), 29–89 and Richard Maxwell Brown, *No Duty to Retreat: Violence and Values in American History and Society* (New York, 1991), 30–7.

3 'The Soldier's Faith', 76.

4 George Mosse, *Fallen Soldiers: Reshaping the Memory of Modern Wars* (New York, 1990). Mosse, however, makes it clear that this 'myth' develops with the emergence of volunteer citizen armies from the late eighteenth century.

5 A.G. Butler to Dr H. Pern, 24 September 1932, A.G. Butler Papers, Australian War Memorial (AWM), Series 41, box 68.

6 A. Tiveychoc, *There and Back: The Story of an Australian Soldier* (Sydney, 1935), 224. The pseudonym is both an affirmation and an ironic commentary. Lording had been assigned to a battalion under the charge of General Tivey which had not seen action at Gallipoli, instead being stationed in Egypt. Australian soldiers generally referred to those who had not seen the heat of battle as 'chocolate soldiers', hence the pseudonym refers to being one of General Tivey's chocolate soldiers. The irony is that the battalion saw a great deal of action in France, with devastating consequences, as we shall see, for Lording.

7 E.J.H. Joseph, *The Long Road Back* (Hobart, 1985), 14.

8 *Sydney Morning Herald* 4 April 1928.

9 Alistair Thomson, *Anzac Memories: Living with the Legend* (Melbourne, 1994). Other useful works which use oral history sources, usually alongside more conventional sources, include Marilyn Lake, *The Limits of Hope: Soldier Settlement in Victoria 1915–38* (Melbourne, 1987), Patsy Adam-Smith, *The Anzacs* (Melbourne, 1991), and A. Jacomos and D. Fowell, *Forgotten Heroes: Aborigines at War from the Somme to Vietnam* (Melbourne, 1993).

10 The literature on the Anzac myth and its importance for national identity, in contrast to other combatant countries, is voluminous. See for example K.S. Inglis, 'The Anzac tradition', *Meanjin*, 24.1 (1965), 25–44; Geoffrey Serle, 'The digger tradition and Australian nationalism', *Meanjin*, 24.2 (1965) 149–58; Richard White, *Inventing Australia: Images of Australia 1688–1980* (Sydney, 1981) 125–39; and Stephen Garton, *The Cost of War: Australians Return* (Melbourne, 1996), 66–9.

11 See Raphael Samuel, *Theatres of Memory* (London, 1994) and Michael Frisch, *A Shared Authority: Essays on the Craft and Meaning of Oral and Public History* (Albany, 1990). The

literature on memory and history is now voluminous. For some other useful recent contributions see Elizabeth Tonkin, *Narrating our Pasts: The Social Construction of Oral History* (Cambridge, 1992); Raphael Samuel and Paul Thompson, *The Myths We Live By* (London, 1990); David Thelen, 'Memory and American history', *Journal of American History*, 75.4 (1989), 1117–29; and Paula Hamilton and Kate Darian-Smith (eds) *Memory and History in Twentieth-Century Australia* (Melbourne, 1994).

12 A very useful collection of essays on nostalgia which argues forcefully for the origin of nostalgia in 'present discontents' is Christopher Shaw and Malcolm Chase (eds) *The Imagined Past: History and Nostalgia* (Manchester, 1989). Michael Frisch has also discussed the way 'romanticized' versions of the past develop with distance from the events, in Frisch, *A Shared Authority*, 15–28. For other relevant studies see Stephanie Coontz, *The Way We Never Were: American Families and the Nostalgia Trap* (New York, 1992) and Patrick Wright, *On Living in an Old Country: The National Past in Contemporary Britain* (London, 1985).

13 Thomson, *Anzac Memories*, 1–22.

14 For studies of this older legend see Russel Ward, *The Australian Legend* (Melbourne, 1958) and more recently White, *Inventing Australia*, 85–124. White also explores how the bush legend became the Anzac legend, something that had been analysed in an early article by Serle, 'The digger tradition'.

15 The comparable figures for the armed forces for the United Kingdom were 47.1 per cent, Canada 49.7 per cent and New Zealand 58.6 per cent. See A.G. Butler, *Official History of the Australian Army Medical Services in the 1914–18 War*, vol. 3 (Canberra, 1943), 880.

16 See Paul Fussell, *The Great War and Modern Memory* (Oxford, 1975), Samuel Hynes, *A War Imagined: The First World War and English Culture* (New York, 1990), Modris Eksteins, *Rites of Spring: The Great War and the Birth of the Modern Age* (New York, 1990), and Jay Winter, *Sites of Memory, Sites of Mourning: The Great War in European Cultural History* (Cambridge, 1995).

17 For a very important analysis of these front line diaries see Bill Gammage, *The Broken Years: Australian Soldiers in the Great War* (Melbourne, 1975).

18 See Klaus Theweleit, *Male Fantasies*, vol. 1 (Minneapolis, 1987) and Graham Dawson, *Soldier Heroes: British Adventure, Empire and the Imagining of Masculinities* (London, 1994).

19 See for example Robin Gerster, *Big-Noting: The Heroic Theme in Australian War Writing* (Melbourne, 1992) and John Williams, *The Quarantined Culture: Australian Reactions to Modernism 1913–39* (Melbourne, 1995).

20 Thomson, *Anzac Memories.*

21 See Tiveychoc, *There and Back.*

22 Ibid., 172.

23 Ibid., 276.

24 Ibid., 281.

25 See Eric J. Leed, *No Man's Land: Combat and Identity in World War I* (Cambridge, 1979); Eksteins, *Rites of Spring*, 95–135; John Ellis, *The Sharp End of War: The Fighting Man in World War Two* (London, 1980); C.E.W. Bean, *The Story of Anzac: The Official History of Australia in the War of 1914–18*, vol. I (St Lucia, 1988, first published 1921), 605–7.

26 See Inglis, 'The Anzac tradition', 25–44. See also John Laffin, *Digger: The Legend of the Australian Soldier* (Melbourne, 1959) and Adam-Smith, *The Anzacs.*

27 Some recent provocative accounts of war and mourning suggest dimensions of these investments. See Joanna Bourke, *An Intimate History of Killing* (London, 1999) and Joy Damousi, *The Labour and Loss: Mourning, Memory and Wartime Bereavement in Australia* (Melbourne, 1999). Although the exploration of the psychology of such public and private memories is beyond the scope of this chapter Graham Dawson's study of *Soldier Heroes* stands out as a pioneering effort to develop an historical understanding of such processes.

28 For an excellent analysis of the Australian soldier mentality and their views on life at

the front and home see Gammage, *The Broken Years*. See also Garton, *The Cost of War*, 13, 189. For accounts of similar frameworks amongst other soldiers see Eksteins, *Rites of Spring*, 209–38.

29 See for example M. Higonnet, J. Jenson, S. Michel and M. Weitz (eds) *Behind the Lines: Gender and Two World Wars* (New Haven, CT, 1987); H.M. Cooper, A.A. Munich and S.M. Squier (eds) *Arms and the Man: War, Gender and Literary Representation* (Chapel Hill, NC, 1989); and M. Cooke and A. Woollacott (eds) *Gendering War Talk* (Princeton, NJ, 1993). For an analysis of the gendered character of the Anzac ethos see Marilyn Lake, 'Mission impossible: how men gave birth to the Australian nation – nationalism, gender and other seminal acts', *Gender and History*, 4.3 (1992), 305–22. A very interesting analysis of the gendered character of the Anzac legend and the ways in which the 'feminine/domestic' was devalued by front line soldiers is Annabel Cooper, 'Textual territories: gendered cultural politics and Australian representations of the war of 1914–1918', *Australian Historical Studies*, 25.100 (1993), 403–21.

30 *The Soldier*, 2 November 1917.

31 'The Tabbie', *Aussie*, 10 (January 1919).

32 See 'Relieved from anxiety', *The Bulletin*, 21 November 1918 and 'Her heart was in her work', *The Bulletin*, 23 January 1919.

33 'The Girl Who Didn't Wait', *The Soldier*, 27 June 1919.

34 'Peace and war', *Truth*, 17 November 1918. For a typical example of the 'cuckold' story see 'Currans cuckolded', *Truth*, 15 December 1918.

35 For a characteristic newspaper account of return see *The Age*, 25 November 1918.

36 See for example *The Soldier*, 10 January 1919. For the call for a soldier's parliament see *The Soldier*, 15 November 1918.

37 For post-war riots see Raymond Evans, *The Red Flag Riots: A Study of Intolerance* (St Lucia, 1988) and Fiona Skyring, 'Driving the mongrels out: riots against disloyalty in Australia 1918–21', unpublished PhD thesis, University of Sydney, 1997. It has been argued that Lawrence modelled some of his central characters on actual people. See Robert Darroch, *D.H. Lawrence in Australia* (Sydney, 1981). For broader discussions about secret armies see Andrew Moore, *The Secret Army and the Premier: Conservative Paramilitary Organizations in New South Wales 1930–32* (Kensington, 1989) and Michael Cathcart, *Defending the National Tuckshop: Australia's Secret Army Intrigue of 1931* (Melbourne, 1988).

10 Involuntary commemorations

Post-traumatic stress disorder and its relationship to war commemoration

Jo Stanley

This chapter focuses on combat-related post-traumatic stress disorder (PTSD) and argues that it is an involuntary commemoration of war. Because of the memories they bring home, those who go through wars can, and often do, become involuntary walking memorials to that experience. The tragedy lies in the involuntary nature of their remembering. They do not choose to be walking war memorials but are helpless victims of their memories. Rather than mentally browsing old war zones by choice, as if on an armchair commemorative tour recalling fallen comrades, they submit to memory's visits. They are chronically possessed by certain overwhelming memories so strong that the war is re-experienced with all its nightmarish force. Their remembrance is neither chosen nor gracious, but imposed upon them by the particular ways their brain processes personal memories of collective combat.

War's legacy is, however, not only indelibly inscribed in their lives, but in the lives of the people around them. PTSD sufferers' carers also are forced by memory – another's – to experience the effects of combat. Perhaps with even less volition than the sufferer, partners, family and professional mental-health staff vicariously 'remember' disabling aspects of war. Indeed some carers become bearers of the other's problematic memories, suffering secondary traumatization (STSD).

PTSD, this involuntary private commemoration, has a complex and often contradictory relation to public memory and commemoration. Ironically the clinical illness may well be triggered by public commemorations such as Remembrance Sunday. In the very havens where they go to get clinical support in burying – or at least managing – the troubling memories, veterans may seek out popular culture's versions of wars such as SAS novels. Ex-combatants' personal needs to keep faith with dead comrades may keep them suffering in a way that fallen friends would never have wished. The nominal public respect for veterans on anniversaries is undermined by the actual daily lack of a supportive infrastructure such as pensions and appropriate health care for those privately bearing the disabling legacies of their service. Above all, carers such as wives and nurses bear the private burden of a public event, but are not themselves commemorated.

A personal endurance rather than a public commemoration, PTSD is not a glorying, orderly and social remembrance event with bands, tank parades, bunting, well-marshalled columns and choreographed formal movements on fixed dates. Instead this illness that was once called shell shock or battle fatigue, is often experienced chaotically at home in the depths of the night. Clinical experience reveals that PTSD is a total bodily, intellectual and emotional experience. And this privatized and non-chosen remembrance of combat can have more to do with deeply private feelings of shame, guilt and a personal sense of defeat than with pride – however glorious the battle was, or is said to have been, in public histories and popular memory.

This chapter analyses the tensions between the privatized experience of PTSD and public remembrance of war. First, it summarizes the nature and treatment of PTSD memories. Second, it discusses the responses of PTSD sufferers to the representation of war in film and official commemorations, and their efforts to fashion alternative forms of public remembrance. Third, it explores the situation of the carers who privately handle the involuntary commemorations of public wars in which they were not directly involved.

Primarily I use reflective oral testimony as my source. The information was gathered in autumn 1999 during an investigatory stay at Audley Court, a Shropshire respite-care home for PTSD sufferers. It is run by the principal charity in the field: the Ex-Services Mental Welfare Society, now more snappily called Combat Stress. Audley Court, with its neat red brick and proud hanging-baskets, is a now-unrecognizable old workhouse in a Salopian village's suburban avenue. The newest Combat Stress home, which has thirty-one beds, was set up in 1996 and functions as a safe holding-space for people who cannot speak easily of their memories and the resulting symptoms in the outside world. This particular Combat Stress home is seen by some sufferers as a loved haven.

My principal informants were three professional PTSD carers and three PTSD sufferers.[1] Several of them subsequently spoke to me at length by telephone and in person, commented on drafts of this chapter, and sent me documents and videos to enhance my understanding. This chapter is thus co-authored, to the extent that was possible given the time available and the professional constraints on what carers felt able to say publicly. The sufferers' and carers' insights are backed by primary and secondary reading of PTSD literature, especially from the Imperial War Museum's sound and documentary archives; secondary reading on life-story composition; telephone interviews with leading PTSD doctors; and repeated interviews with four wives of sufferers. The value of repeated face-to-face interchange was that I could tap into the privatized experiences of the sufferers and their carers, and gain information that has not previously been made public, particularly not in oral-history contexts.

That reciprocity was assisted by my being the ex-wife of a PTSD sufferer and having had advanced training in counselling. I approached the subject as a writer and oral historian, but primarily as an involved life-story worker who shared some background with my informants. My energy came partly from a need retrospectively to make sense of my own private suffering during that 1970s marriage. Combat-related PTSD was not then a phenomenon recognized by lay people. Eighteen years ago I travelled three times a week from my own analysis at the Tavistock Clinic to the British Library to read First World War shell shock expert, William Rivers, trying to make connections with my situation as, by then, an ex-carer and a person wrecked by over-doing her duty. PTSD's emergence into public knowledge later in the 1980s, and the wider acknowledgement of wives' problems, brought an end to my lonely ignorance. For the past six months I have been writing fiction based on the wives' stories. That personal motivation connects with a sense of political responsibility. As a feminist and socialist, who wears both white poppies and red each November, I am both intellectually and emotionally interested in investigating how this crucial legacy of war affects the psyche and subsequent lives both of combatants and of the women who pay the private price.

PTSD and memory

PTSD is now recognized, not without controversy, as a real illness. However there is an uneven awareness of it in society at large. This causes problems to sufferers and carers. The American Psychiatric Association's definition of PTSD – as a development of characteristic symptoms following a psychologically traumatic happening outside the range of normal human experience – is now widely accepted.[2] Fundamentally the nature of PTSD symptoms indicates a disturbance to memory – and raises questions about how people handle the effects of involuntary memories of an event that is not assimilable.

This disturbance to memory and its related effects now occupy significant time in popular and in academic thinking. In popular culture, the number of websites on PTSD reaches into the hundreds and refers to civilian as well as military memories, especially of sexual abuse and events such as mass tragedies in football stadiums. The websites indicate the extent to which 'traumatic memory' is now a familiar, even domestic, term. People know about it in relation to incest as much as, if not more than, shelling on the Somme. The illness can be seen as resulting from wrongs done to people who may or may not be 'innocent victims' but certainly did not have the power to stop the traumatizing event from happening. It is not in any way a voluntary commemoration of that event.

Combat-related PTSD is just one sub-category. Those few members of the public who think about it probably understand it to be a by-product of war but also of life. Most public knowledge is limited, vague and received indirectly, for example through television documentaries or novels. Combat-related PTSD is still not easily assimilated in public understanding in the UK. Lack of realistic and direct knowledge of its realities brings loneliness, misunderstanding and stigma to returned combatants with PTSD. Sufferers and their partners report that their neighbours consider them 'loony'. Their families shun them. Symptomatic behaviour like habitually searching for explosives beneath the family car is ridiculed. Such ignorance arises because in Britain wars are no longer events that directly affect the whole population, however much selected news footage and popularized aspects of war glimpsed on television may mean that, more than ever before, we imagine we know what war is and does to people.

The treatment of PTSD as an illness is not a late twentieth-century development but dates back to the First and Second World Wars, although differently named: as, for example, 'war neurasthenia' and 'shell shock.'[3] PTSD has changed in status from being pooh-poohed as scrimshanking or dismissed as simple funk in 1915 to being recognized as an extremely complex illness. This development was largely bankrolled by Uncle Sam's military, especially after the Vietnam War with its very visible and militant psychic casualties. Such socio-medical concern is less about understanding memory and more to do with securing public popularity for the military, and avoiding the waste of expensively trained fighting units: people. In the UK, by contrast, a less vociferous population of veterans together with a reduced fighting role for the armed services has meant that Britain has had less of a social problem to deal with; and our National Health Service (NHS) has anyway subsumed the medical problem under general health care, to some extent. Today, by cruel irony, economic priorities may mean that PTSD veterans of the First and Second World Wars are not offered recognition in the form of health care. Many area health services in the UK operate a cut-off age of between 65 and 70 years, meaning that community mental-health teams cannot access help for ex-combatants. Lack of training may mean that PTSD is not detected; instead it can be misread as simply a symptom of ageing. Thus the problems of handling disturbances of memory are compounded by practical difficulties.

Monetary support to address ex-combatants' sufferings is one way that they can feel their contribution is properly recognized and that they are not 'forgotten warriors.'[4] The state finances some PTSD care, which implies a degree of acceptance of this responsibility for sufferers. Yet divisions exist between the central bureaucracy and those suffering personally from their memories of situations that took place allegedly in the public interest. While the state has a moral duty to support financially the legacy of previous governmental decisions, it finds the charge on the public purse is great and thus acts with the zealous selectivity common to most payers of benefits. The War Pensions Agency awards different levels of pensions for PTSD sufferers depending on the severity of their case.[5] Cases are reviewed every few years. Combatants going up a few percentage points after periodic reviews can feel some limited, if long-delayed, compensation for the memories that cannot be exorcized. But most feel bitter about this relatively small recognition of what the war has done to them. The acrimony is all the more sour because the state position is seen as reflecting the changed symbolic meanings of war, not least a decrease in patriotism and respect for veterans.

Generally the clinical attention given to combat-related PTSD is motivated by pragmatic concern to handle the symptoms and feelings, in some cases so that sufferers are well enough to go back into combat or at least manage their own daily living needs. What clinical research does show, however, is that PTSD arises from people's experience of a moment that is too upsetting to be coded and stored normally in the memory. Memories of the trauma-generative moment won't lie down. To summarize Hunt and Robbins, oral historians of Second World War and Korean War PTSD sufferers: severe shocks can mean that people's capacity to synthesize sensory impressions into a coherent and conscious whole is partly destroyed.[6] Freud believed that memories of such traumatizing events exist outside consciousness and only emerge later into consciousness, through therapy or when defence mechanisms somehow allow them to do so.[7] PTSD takes over the here-and-now life that the person is trying to live. Symptoms include stammering and inability in speaking, nightmares, temporary blindness, weeping fits, tremors, starting at sudden noise and anxiety about failing in duty by not showing courage. Sufferers at home turn to alcohol and substance abuse, which leads to an increase in flashbacks. Personality changes mean marriages fail and many PTSD sufferers have trouble holding down jobs. Degrees of severity vary.

Many ex-combatants suffering less severe PTSD, says Audley Court PTSD nurse, Claire Evans, use a metaphor to describe the memories of combat. In therapy sessions they say, ' "It's like there's a video playing and I can't find the pause button … " You hope to give them the faith that they will find it.' However, this videotape analogy only refers to hearing and vision. In fact, she notes, some veterans experience the memories through their whole bodies, causing uncontrollable shaking and lack of mobility; every sense is affected. The Director of Clinical Services at Combat Stress, Linda Winn, points out that the more severely traumatized do not see the memories as a show but actually re-experience the moments of trauma. 'They are actually *there*', as if the sufferers had an irremovable virtual-reality machine attached to their heads. This agonized way of remembering war is termed 'abreaction'.[8]

One of the key new neurobiological ideas is that memories of the trauma-causing event are laid down and encoded differently to other memories. In studying a range of PTSDs, not just combat-related versions, Siegel found 'emotional states bias the interpretation of stimuli, encoding and retrieval of memories in a process known as state-dependent learning'.[9] In the body's high-adrenalin preparations to deal with a crisis there is a narrowing of the visual field, so that only vital information is taken in. Claire Evans sees her ex-combatant patients as having experienced it in the following way:

> You may be involved in a situation or series of events which demand that all your resources – physical, psychological and spiritual – are directed towards one end, often survival … It feels [to me] as though these memories are experienced on a more 'gut' level than in a processed, intellectual way.

Longer-established explanations of trauma link its effects to meaning-making and narrative. Nineteenth-century trauma psychologist, Pierre Janet, argued that memory was essentially 'the action of telling a story'.[10] By recounting a memory people seek a way to place it in the right position in their recountable life history and their idea of themselves. Allan Young, a social anthropologist studying

PTSD today, argues that PTSD sufferers are affected by cognitive dissonance; that is, the gap between their preferred version of their life and self, and what they have actually experienced.[11] For example, combatants who believe themselves to be moral or brave people might have acted in a way they think is immoral or cowardly. Young points to three main ways in which sufferers try to handle the dissonance. The first method is to reframe the memories in order to make the content fit pre-existing cognitive schema. So, for example, the story might be that: 'I've always been a kind person who believes in justice; he was a terrorist full of hate. I had to stop him before he hurt people.' Second, they might empty the memories of their power to hurt by assimilating them or denying them; for example, by telling how: 'We all have to do terrible things in war. I was no different from anyone else'; or: 'He was probably already dead before I stuck the bayonet in.' Third, they might revise their cognitive schema; for example, by remembering that: 'I used to think it was wrong to kill, but I was ignorant then. The war taught me that you have to sometimes.'

Tensions often exist between the damaging nature of personal traumatic memories and ex-combatants' desires to hold on to such memories, even when traumatic. To retell a traumatic memory can be a way, however agonized, to keep faith with dead comrades. Most PTSD sufferers feel guilt about their comrades' deaths and their own survival. To lose or dispossess themselves of the traumatic memory can feel like deserting someone with whom they have an interwoven identity, whom they have been trained to protect like a family member, and who has already suffered enough loss and insult. Additionally, says Claire Evans:

> It is sad when the traumatic memories eclipse the rest of the veteran's service experiences, as for many veterans this was a rewarding time also, [giving them] a realization of potential, a sense of comradeship unknown in other spheres. A consequence of trauma 'eclipsing' other experiences may mean that clients have further difficulties 'letting go' or 'moving on' as [this] may mean they will then leave or lose the good as well.

For many, war was a high point in their life when they travelled, had a little job security and a common purpose, and excitingly discovered their strengths – just as do civilians on Outward Bound training courses today. Then, war may have felt worthwhile, perhaps an opportunity for working-class pride in contributing to the nation's greatness. For some it was a time of exhilarating challenge when combatants felt more alive than ever before or since, in the face of multiple deaths and carnage. Remembering the good aspects of the war can give meaning to its other aspects. To no longer have the memory would be to make the past dead and to face maybe a still lonelier void.

PTSD treatment in the UK varies according to site practices, individual sufferers' needs, and whether or not they are still serving personnel. Provision is by the military, the state (through a range of NHS in-patient and out-patient care), and ex-services charities such as Combat Stress and the British Legion.

PTSD sufferers still serving in the armed forces have access to a world-wide network of community nurses working in cooperation with doctors and community psychiatrists, or go to the tri-services hospital at Catterick in Yorkshire. This deals with 100 outpatients and has six in-patient beds for people living far away.

Those who were in the services from the First World War, the Second World War, Korea, Northern Ireland, the Falklands, the Gulf and Bosnia, but are no longer serving personnel, are usually treated by home support plus short hospital or nursing-home stays of two to six weeks per year. This is delivered by charities such as Combat Stress, which has been helping such sufferers since 1916.[12] It organizes home support and repeated fortnightly sojourns in its four homes: Audley Court, one in Ayr and two in Surrey. Ex-services personnel usually self-refer to these havens, say twice a year. The procedure is that they decide they feel ready for a stay and contact the Combat Stress Regional Welfare Officers who are already handling their cases. These officers, often ex-services officers who have a similar role to social workers, visit the sufferers and decide if they may benefit from attending. If so, and if the clients, their doctors and their psychiatrists agree, then the sufferers apply to the War Pensions Agency to see if it will fund treatment and travel. A substantial part of the cost of the homes is thus borne by the state through the War Pensions Agency. This is a recognition that is deeply appreciated.

Homes such as Audley Court function for sufferers as an interface between the public and the private. They are a safe haven, a longed-for opportunity to speak out freely in a particular kind of public context, both in groups and with individuals. The private difficulty and agony, previously confined to solitary bedrooms or dealt with by resort to the booze cupboard at home, can be moved into a sanctuary where every single person knows – directly or indirectly – about war and about PTSD. They are refuges where staff can handle the extreme behaviour, tackle it, and even help the sufferers themselves to do the same. Further support comes from other sufferers who can offer empathy and tips on how to cope. The sanctuary may be perhaps all the more treasured because it offers constructive, non-judgemental attention both from other sufferers who know about war (maybe even *your* war) and health professionals. Here there are friendly young women, on duty throughout the night, who bring wisdom gained from accompanying so many other veterans through their rememberings of war.

Treatment at Audley Court focuses especially on memory, symbolization and narrativization. Linda Winn reports that:

> Nine Registered Mental Nurses work with ex-combatants from 19 to 96. Nursing is one-to-one, and group work. [It] comprises dramatherapy including collage and spectrograms – because metaphor and distanciation can help expression – anxiety management, PTSD education,[13] reflexology, and sometimes Tai Chi as well as basic counselling and physical care.

Winn is an internationally recognized expert on dramatherapy and PTSD but by no means all clients are thought able to benefit from dramatherapy, which is only one of the treatments. Audley Court's techniques involve learning the self-management skills that many civilians suffering from stress can use too. They also work cognitively, specifically with the *content* of memories and finding the right storage places for each of them. Nurse Claire Evans uses the metaphor of putting an untidy garden shed in order. In one-to-one sessions she asks clients, 'Do you want to get everything out, sort it, then put it back? Or will you just tidy round while you are in there?' Those who choose the first option are seen as being potentially able to benefit from deeper therapeutic treatment. Memories are explored at a level and in ways that enable sufferers to be able to put them at rest, at least temporarily. Other memories may be symbolically buried or even burned. A key part of the recovery process is integration and the creation of coherence.

But putting a memory to rest in a wrong way may mean that sufferers then stop commemorating important aspects of the war. Therefore part of the recovery process involves them in consciously handling the memories differently, to commemorate their lives, their dead comrades and their own contributions in a less self-damaging manner. Claire Evans encourages them to gain perspective, recognize and acknowledge what has been of value, and create a healthy and satisfactory 'storage system'. This enables them to feel more empowered and in control. Because so many of the clients are fans of information technology, Audley Court staff use a computer metaphor: 'The bad memory is like a file that you can save. You don't have to keep it on screen all the time. But it will be safely stored, and still there if you want to let it out.' Candles are important. Many ex-combatants use a line from the Chris de Burgh song, 'Carry her like a flame in your heart', as inspiration for this process of selective retention of memories that can commemorate the war without damaging the rememberer.

PTSD expert, Daniel Siegel, defines abreaction as:

> a term referring to the reinstatement of a memory with its concomitant emotional responses from the time of the event … a therapeutic abreaction emphasizes both emotional and cognitive processing of disassociated elements of a previously inaccessible or only partly accessible memory. What is different in then recycling from the actual experiencing is that the patient can now reflectively process the experience and co-construct a narrative with the therapist.[14]

At Audley Court recovering PTSD sufferers, in groups and one-to-one sessions with therapists, recount their version of their experience. According to Claire Evans, speaking about it and finding their commonality after lonely experiences of suffering (as well as sharing information about what helps them handle their symptoms) leads to many saying: 'It's the first time I've been able to talk about it.'

Group discussions by PTSD sufferers enable combat memories to be dealt with in two ways that are significant to oral historians. First, hearing many others' stories enables people new to speaking about their memories to compose their narratives from a range of possible narrative examples. It helps them own what may have been previously inadmissible truths, such as a veteran's shame at not being able to save a Chinese woman from being raped by Japanese guards in front of his hut. This owning may act as commemoration for the victims, a turning of the involuntary memory into a voluntary one. Second, sufferers may feel rage and rivalry at others' stories. Rage occurs if they cannot articulate their own memories as coherently and without breaking down. The rivalry is often intergenerational, and particularly common between the Second World War and Northern Ireland veterans; it is expressed in the form of an accusation such as: 'I had a worse (or a more dramatic) war than you. You had it cushy', and a riposte such as: 'You weren't even bloody there to see. How do you know? It's all different now.' Sometimes staff members have to separate physically battling ex-combatants. Other PTSD sufferers argue over 'what really happened'. Such competition is familiar in oral historiography dealing with contesting memories, especially family stories, in which each family member assumes that their own memory is the True One. However, the ex-combatants need to construct a particular version of their war memories – both to commemorate colleagues and victims, and maybe as an act of respect towards an earlier self. Such commemorations also include the pilgrimages and memorials detailed later in this chapter.

As Siegel points out, neurobiological studies of development of memory show, amongst other things, that 'the manner of *post-event dialogue* regarding an experience can influence the manner and probability of recall of aspects of an experience'.[15] As oral historian, Alistair Thomson, discovered in interviews with First World War veterans in Australia, those living in towns and with radical companions remembered and discussed the war more critically than those living in rural, conservative areas.[16] One effect of this for PTSD sufferers could be that someone remembering, say, how he failed to save his mate, Jim, might experience less unrealistic expectations of his own performance and less isolated, guilty remembrances in a socially critical context, such as an urban Labour Party ex-servicemen's gathering, than if he were to be going over the event with colleagues in a conservative British Legion club. The veteran in the former case may also gain more awareness of the total traumatic situation, for example, that there were actually many others who could equally have been expected to save Jim, or that Jim was not savable. In this context the remembrance may become a more realistic and controlled one, less guilt-driven and less involuntary. After therapy, some sufferers move from private, involuntary agony to chosen public statements about war, as did Audley Court client, Frank Capacity, whose story is outlined below.

Public commemoration/private memory

Complex and paradoxical interrelations exist between, on the one hand, personal memories of trauma, and on the other hand, the range of public memories and commemorations of war. Even in safe havens like Audley Court, veterans are not hermetically sealed off from such public commemorations and their impact. Each public commemoration impacts on survivors in different ways. One of the most surprising aspects of the complex relationship is that veterans are not straightforwardly pleased by a commemoration such as Remembrance Sunday. Some value these occasions, some are traumatized by them, others have both responses. A further tension is that, while non-combatants' notions of heroism can make a veteran feel unrecognized as a person who has been through trauma, to be unpopular because of public attitudes to that particular war may also cause or exacerbate trauma. Additional tensions exist between self-organized commemoration by PTSD sufferers and public commemoration, and in the range of veterans' reactions to secondary representations such as war movies.

There are three main types of commemoration in which these complexities and tensions are manifested. First, there are public commemorations, whether local or national, which are officially organized and include representatives of the state as well as ex-combatants. Second, there are secondary commemorations such as films and books made about a war. And, third, unofficial, collective commemorations are set up in which personal commemoration by veterans plays a key part.

In the first case, official commemorations, veterans may appreciate the ceremony as an honouring of dead comrades. They may also want their own contribution recognized as meaningful. Many veterans, including those in the Falklands War, gripe about not being welcomed home by official ceremonies. 'We came in through the back door', they complain to Linda Winn. Medical research at the Veterans Administration Medical System's National Centre for PTSD demonstrates that people who come back from a 'popular war' – and therefore may have some degree of heroic status – cope better than those who do not. Evidence from Vietnam veterans shows that ex-combatants who suffer most after a war are those who sense that their contribution was unwelcome or futile.[17] This is one of the factors that makes it feel most difficult to overcome depression. When Claire Evans began her work at Audley Court, her expectations of the experiences of ex-services personnel were based on what is demonstrated by public commemorations, that is pride, comradeship, patriotism and a sense of common purpose. She soon became aware that this was not necessarily reality. 'One client in particular said that he felt no pride in what he had been part of. He uses phrases to the effect that it was "a dirty little war – it was meaningless".' She believes that 'when it feels as though there is no value or meaning to the action, it can also feel that there is no value (or meaning) to their (wounded) condition now … But a greater number [of veterans] feel a sense of pride than do not.'

If official commemoration can provide an important form of recognition, at the same time many PTSD sufferers know that they feel at their worst on such occasions. Rather than being a safe memorial, the event can become the flash which ignites the still-active gunpowder, that has been so carefully packed away for safety it may have been 'forgotten'. After the fiftieth anniversary of D-Day, over 1,000 ex-services personnel contacted Combat Stress, because they were experiencing PTSD-style symptoms.[18] Staff stand by for a high response after any such event. Some veterans make prior bookings for Combat Stress home places during the Remembrance Day period or the anniversary of their particular trauma-generating conflict, says Linda Winn. Why? 'Because they feel safe here.'

Safety means feeling safe, above all, from their own memories and feelings, thinks Audley Court nurse, Judith Jones. It also means safety from the company of people who do not understand, such as wives and civilians. For some veterans, the existence of the nurse-station at the end of their corridor means that the enemy cannot get them. It is a helpful barrier, albeit a limited and temporary one. It is not just official ceremonies that may prove difficult, however. Linda Winn reports that traumatized Falklands combatants often felt alienated from the homecoming street parties organized in their honour. These veterans did not want to join in the loud festivities outside, with the party poppers that reminded them of gunfire. Rather, they wanted only to sit quietly in their bedrooms re-adjusting to their families and remembering their dead comrades who were not being feted. Sometimes these ex-combatants felt that the person who was being welcomed home no longer existed. Instead it was an untrue version of themselves as a hero, or a pre-war version of themselves as a family member. Such interactions were profound moments of private suffering in the midst of public festivity by those ignorant of modern warfare. One PTSD nurse sees the welcome-home parties she hears about as 'an affront to the dignity of all the people concerned – and completely without understanding of the death and destruction caused by war.'

Handling public commemoration may be especially difficult when the rememberers' memories constitute a more complex or contradictory version of what happened than that offered in the commemorative event itself, but they have no satisfactory narrative with which to voice this difference to themselves or others. Veterans can feel distressed and alienated because the public knowledge of an event is often very different to what they know really happened: it is usually less spectacular and more regrettable. The popular notions of heroic sacrifice freely made, flattering though they may be, can contradict the subjective experience of coercion, collective powerlessness at wrong-headed top-brass decisions, crises of masculinity, and all the lonely lacunae between pomp and fear – between the ceremonial city of today and the bloody battlefield of yesterday. PTSD nurse, Judith Jones, points out: 'There is no place in official remembrance for looking at the justification for wars. Some clients have been searching for some justification for certain actions for fifty or more years.' The very character of the discourse surrounding official commemorative events – reverent, tear-jerking and, at worst, pompous, uncritical of war and focusing on heroism not fear – may trigger stressful memories. Veterans therefore are divided as to whether they want public commemorations. Such markings can belatedly give *gravitas* and dignity to a situation, even though veterans suspect the motivation is a concern for national and armed-forces' public relations, and that reverence for the dead varies in sincerity and wisdom.

A second type of public commemoration comes in the secondary form of popular culture's products: films, plays, novels and television programmes about war, usually created by those who have never fought in one. The library at Audley Court contains hundreds of war paperbacks donated by clients, even though initial policy was to eschew such books. SAS novels far outweigh the Sidney Sheldons and *Reader's Digest* compendiums. A popular day trip for clients is to the RAF aerospace museum at Cosford. And they watch war programmes on television at home. Created for popular consumption rather than as a salute to the fallen, the meanings produced by such cultural artefacts are usually qualitatively different from veterans' own subjective experience of war. Yet such products appear to offer them some comfort in that at least their war is spoken of, however erroneously. The conflict in which they fought is not forgotten. Indeed it is available for 'remembering' at the touch of a button. Carers' explanation for the apparent contradiction here is that the service world is so live and relevant to the ex-combatants that material about it makes them feel at home, even though it is associated with anguish.

A further benefit of media representations such as these is that they perpetuate war in public memory, and may educate the public about the private price of war. In Britain wars are no longer events that directly affect the whole population. In the First and Second World Wars so much of the country's population was mobilized that anyone in any corner shop could talk about battlefields such as Ypres, Gallipoli and Dunkirk, as well as the effects of war; for example Dad's amputated leg, or a brother's bouts of the shakes. Physical and mental war wounds were a familiar public reminder of what half the men in the street could not help but remember. People knew about war at first hand, or through stories vividly told to them by people they knew well, as well as through the mediation of populist representations such as Pathé Gazette newsreels or *Picture Post* photographs.

However, one of the additional complexities of the relationship between private memory and public commemoration is that such entertainment, like official commemorative events, can also trigger PTSD. *Saving Private Ryan* (1998), the film acclaimed for offering viewers the opportunity to feel as if they were actually in the Second World War, brought about the onset of PTSD for some Second World War combatants. One UK veteran went to the film to show himself that he was all right, found that he was not, and had to contact Combat Stress. Similarly in the USA in July 1998 the US Department of Veteran Affairs set up a website offering support to those suffering the movie's impact. Over 15,714 people visited the site in the first thirteen months.

A third type of public commemoration is self-organized and largely unofficial. It is created by veterans as a direct and personal, but public, way to recognize what happened. Veterans may create counters to the popularized versions of war they see on the silver screen. For example, Audley Court client, Frank Capacity, a retired publican and an ex-prisoner of war who survived three-and-a-half years on the Burma–Thailand 'Railway of Death', created a home video. In sharp contrast to the well-known heroic story of *The Bridge on the River Kwai* (1957), his video is a compilation of recorded television programmes about present-day confrontations with Japanese soldiers, mixed with his own family's pilgrimage to the Far East. Shots of war cemeteries are inter-cut with swimming pool scenes; 'Railway of Death' news cuttings lie next to tourist shots of Bangkok. In such ways PTSD sufferers insert agency into the marking of a past situation over which they did not have enough control.

Such initiatives reflect a wider shift in veterans' and civilians' relations to war remembrance. In the 1950s the aim was to forget war. Veterans of earlier wars report that, even if they met in the British Legion or Ex-Servicemen's Clubs, they might eschew war talk. It was easier and more manly to 'forget'. Indeed, crucially, this kind of bluff may have meant they could get a job. Such avoidance tactics were officially encouraged, reports Frank Capacity. However the post-1970s popularity of researching one's family roots, plus the availability of cheaper travel, has contributed to enabling battlefield tourism by veterans and their families. Today the UK's veterans are more like the American Vietnam veterans who are actively involved in memorial creation, the production of books and other activities in their own war commemoration groups. Many Far East prisoners of war (FEPOWs) now use the world-wide web and become engrossed in electronic contacts with other veterans, which seems to have a beneficial effect rather than dragging them into distress. The private experience and memories are helped by the collective action and sharing of information – especially for educative purposes.

Frank Capacity's situation illustrates some of these changes. From 1946 Frank spent several decades avoiding thinking about the war and his time on the Kwai. Then several years ago his PTSD became so severe that he now frequently dreams that the Japanese have come after him, and 'hears' them shouting as they come up the stairs at home. But to his surprise, for the past four years he has found himself happily involved in setting up Kwai Railway Memorial (Three Pagodas Group), by subscription. This meant 'I was living with it [the war] but not *re-living* it.'[19] The initiative arose from Frank's 1995 pilgrimage to Singapore and Thailand, to commemorate the fiftieth anniversary of the ending of the war there. With his daughter, granddaughter, and son-in-law, he visited cemeteries including the 7,000 graves at Kanchanaburi War Cemetery; his old POW camp at Changi; and the opening of a new memorial at Hellfire Pass, Konyu.[20]

That memorial was built and paid for by the Australian government to honour the Australian veterans and war dead. Seeing that there was no involvement from the British government in the commemoration of war in the Far East, Frank and other members of the pilgrimage formed themselves into a fund-raising charity, which he now chairs.[21] He was upset when both the then-Minister of Defence, Earl Howe, and subsequently the Prime Minister, Tony Blair, refused to fund such memorials. 'They brushed it under the carpet. They don't want to know.'[22] It was especially upsetting as he remembers Blair, when in opposition, promising that there would be a Labour government spokesperson or Department for Veteran Affairs, just as the Australians have a Ministry. His feelings worsened when he heard that £5,000 had been given for a Liverpool memorial to the victims of the Irish Potato Famine. 'I asked Tony Blair: if they could make an exception for that, why couldn't they make an exception for us too?' It was fruitless.[23]

Despite such disheartening confrontations, the attempt to organize a memorial has given Frank private pleasure. And it is a socially responsible pleasure, too. 'It's not for the past but it's for the future and for me now, for their memory. It's for understanding. This must never happen again, so that everyone has got to know.'[24] Frank now also has a new, much more personal memorial project. He has collected over fifty photographs of his comrades' graves in Burma, which he will give to those of their relatives he can find. 'That, to me, is the real commemoration.'[25] His project will restore the dead to a place – on domestic mantelpieces or in family albums – from which they have been symbolically missing. It may even facilitate narrative closure for the families, enabling memories of the dead to be more properly located within private psyches.

Unusually for PTSD sufferers, Frank moves not just from personal memory to public commemoration but also from subjective memory to public education. He does so as a means of winning political change, to prevent others ever having traumatic combat memories. The memory is transformed into a pedagogical tool with peace as the goal. But it still besets him at night. His activism and a sense of the politics of commemoration does not resolve or end his PTSD. The two coexist, illustrating that the involuntary commemoration can continue despite the existence of adequate memorials. Public commemorations, however collectively created, appear to be in no way a final cure although they can assist the better re-arranging of personal memories.

Sufferers from these traumatic memories require another sort of public commemoration too: material support. A number are campaigning, personally and collectively, to get fuller state recognition of, and compensation for, their illness, in the form of adequate pensions. Such an initiative demonstrates that commemoration must take many forms, including material compensation with all its complex implications of value and approval. As this chapter left my computer, the British Legion were raising in Parliament[26] the question of the British government compensating FEPOWs for what the Japanese did to them. After the war Frank Capacity was paid just £78 from sequestered Japanese assets, plus a total of £178 soldier's wages, for being a prisoner of war for three-and-a-half years. Now he supports the British Legion's move to redress a problem caused by the post-war UK government's waiving of compensation claims in return for trade agreements with the Japanese. A sense of injustice keeps the memories alive. For a government to put its money where its mouth is, is qualitatively different from honouring the dead with a paper poppy at the Cenotaph. Material compensation may not be able to stop the veterans' memories, guilt and grief. However it can create the key feeling of being respected for effort and suffering.

Witnesses/commemorators/carers

Those who care for PTSD sufferers are often the involuntary 'witnesses' of traumatic memories, and, as such, suffer doubly from war's side-effects. Such carers are positioned in between the sufferer and a society that may well not understand PTSD or may want to deny it. Both private and public carers are walking second-hand war memory banks – banks from which no-one wants to make a withdrawal because these memories are not desirable commodities. A carer's role is a complex, undervalued, and under-supported one. While ex-combatants are to some degree commemorated, their carers are usually barely known about, let alone memorialized. The structural relations of carers – to war, to its commemoration, to the health system and to the sufferers – illustrate the ways the *care* of combat-related PTSD exists even further back in the hinterland of public knowledge than does PTSD itself. However, the relationship of carers to this public knowledge depends largely on whether they provide support in a private, domestic context, or whether they are paid professionals. Carers of both sorts can suffer from secondary or vicarious traumatization (STSD).[27] The effects may be worse for private carers, however, who have neither the support systems and the formal training in distancing, nor the opportunities to debrief and the time off that professional carers can enjoy.

Private carers, usually wives or partners, are overwhelmingly female, unpaid and usually unable to do paid work outside the home because of the demands of their caring duties. One full-time PTSD worker, Lieutenant-Colonel John Booth, Combat Stress's Welfare Officer for Worcester, writes of visiting the home of 78-year-old Cyril, a Burma Star medal holder: 'His wife tells me he goes back to Imphal nearly every night. She has to apologize to the neighbours some mornings after he has been "howling like an animal on the floor of his room." '[28] Most private carers report that they nurse their husband or partner; act as gaolers to keep him from the bottle; escort him in public situations where he might crack up; 'translate' his behaviour for the public, such as neighbours; act as his spokesperson and advocate at the doctor's surgery and War Pensions Agency reviews; function as his secretary, and as his public-relations officer to those alienated by his extreme behaviour. They do almost everything for him. And when he is healthy and able to do things for himself, they stand by for the next bout. It may be that the more carers are trusted by the sufferer, the greater their burden. By contrast, those sufferers who keep their symptoms to themselves – to protect themselves and their partners – create a state of psychic divorce that has different strains for their partner. Many marriages sunder. Over 40 per cent of Vietnam veterans have divorced at least once.[29]

No vicarious glory accrues to the private carers of 'heroes'. These women, often isolated, unrecognized war 'casualties', do not march in the parades of ex-services personnel. They may have repeatedly endured harrowing battle scenes by proxy, often in sleepless suburban bedrooms after public parades. The public commemoration only brings more private labour for them. It may indeed bring public shame if he cracks up at the Cenotaph and they have to escort him home on buses where his state is visible and not understood by other passengers. Among the many tensions in carers' positions is that, despite their secondary suffering, a number of wives remain patriotic and pro-war, even though their damaged husbands may have become anti-war, and vice-versa.

Private carers often have no one to whom they can narrate their stories as partners in involuntary remembering. Indeed, private carers may have only a very incoherent knowledge of his war. Their information may come mainly from television versions of war, coupled with any fragments the sufferer may or may not let out, including involuntarily during his nightmares. This was my own experience. I bought books on *his* wars to fill in all the gaps between the nightmare fragments. But so different were the public and private versions that almost nothing added up.

Writers on traumatic remembering, Shoshana Felman and Dori Laub, ask 'What is the relation between speech and survival?'[30] We could add, what is the relation between being actually involved in an event and involved in it by proxy as carer? To whom can she speak in order to survive? How does she negotiate between her reactions to the war stories given by the media and the ones she is given – or protected from – by her partner? Because PTSD carers have not yet been studied,[31] and because their lives are often so cloistered, the answers can only be speculated upon. Many do not speak to others about their burden, not even to doctors. There is not even a specific support group for them within the organizations where one might be expected: MIND (the UK's National Association for Mental Health), the Carers' National Association and Combat Stress. Professional PTSD carers are probably the main experts on PTSD sufferers' private carers. For example, an Audley Court nurse might telephone a wife to tell her how her partner's treatment is going. In doing so the nurse discovers the wife's feelings and can answer requests for information. Similarly, a partner may reveal her situation in passing as she comes to pick up the veteran after a stay at a haven.

Carers are not primarily interested in the historical event that caused the PTSD in susceptible subjects, the nature of memory or the historiographical questions that concern us here. Their preoccupation is with the personal well-being of the product of that history – and their own survival as human beings and carers. Professionals may, in Felman and Laub's terms, 'bond with the narrator in a common struggle to release the testimony which, in spite of inhibitions on both sides, will allow the telling of the trauma to proceed and to reach its testimonial resolution.'[32] But the wives I know report more pragmatic goals: to just end the rememberer's pain somehow; to snatch a decent night's peace; or to evade the tedium and horror of an endlessly repeated story. Their experience raises the question of how do such 'witnesses' debrief from all those different levels? The answer is that we do not yet know.

Professional carers, by contrast, at least belong to a public sphere, where the illness is recognized and understood. They have access to books, training courses, supervisors and international resources. Because they care for not one, but a stream of hundreds of clients per year in public and private havens, their experience of PTSD is intense in a different and far more social way. More than any other group of people, professional PTSD carers know about war's personal after-effects and the private costs of the involuntary remembering. Like private carers, professional carers do not yet appear to have been studied except in relation to STSD. Many professional carers can be reticent informants to outsiders such as myself. Their caution is motivated by social concerns: those of mental-health workers concerned to behave ethically in relation to their clients, but also those of people intimately bound up with the armed forces and their security preoccupations. Services personnel sign the Official Secrets Act, and therefore PTSD sufferers may well be breaking it by talking of their experiences. Combat Stress staff do not sign the Act. In care situations for serving personnel, such as Catterick hospital, nurses are themselves in the forces whereas Audley Court tends to have civilian workers. But certainly most professional carers feel personally connected to sufferers.

Referring to the psychic costs to professional carers, Audley Court nurse, Judith Jones, feels that they deal with it 'sometimes with difficulty and pain'. She is helped by both official and private support: having clinical supervision at work and an outside spiritual director/friend. Linda Winn at Audley Court sometimes encourages staff to recover after therapy sessions by burning a candle as a tribute to the dead they have heard about, or bathing away the painful stories that may otherwise obsess them. Others have developed their own strategies in the face of a general, modern, Western lack of funerary rituals. Female nursing culture may offer some support. The memories of Astrid Ortega, an American nurse in the Vietnam War, are recorded in Shirley Dick's anthology of interviews with post-Vietnam PTSD sufferers. When Ortega went home to New Jersey she experienced the public hostility against the war, but her PTSD personally benefited from a feminine culture of sharing feelings.

> I didn't experience the worst kind of trauma because I had someone to verbalize with who had been in Vietnam also. That is the thing about nurses. They have nurse friends that they could tell how they felt and talk to, but the men didn't have that.[33]

Implicitly she is suggesting that the masculine habits of not unburdening feelings may worsen trauma.

Even so, the emotions aroused by witnessing may become overwhelming. Nurse Cecelia Christie many times witnessed psychiatric abreaction treatment under pentathol in the Second World War, and later wrote a thesis on those most likely to be affected by it. She told her Imperial War Museum interviewer that even in 1993, and as a skilled counsellor: 'I can't see a soldier or hear a military band without being choked ... and I wasn't in the front line. I myself am [now?] an almost uncontrollable emotional.'[34] Some carers respond by simply getting on with the job until things get too unbearable. Nurse Judith Jones tellingly believes: '[I have] no strong feelings (most of the time).'

Professional carers have an additional, unrecognized task on top of performing and surviving their care work: that is, selecting the ways in which they represent war to themselves. The sheer volume of sufferers witnessed – at least 500 per year, maybe 10,000 in a career – means that professional carers develop personal opinions about war, as well as wider professional expertise, to a greater extent than a private carer dealing with only one sufferer. As the repositories, like museums, of these involuntary commemorations, how do carers themselves commemorate war? For example, what do they metaphorically put in the main showcases? How do they phrase the introductory panel? Do they metaphorically picket the very museum's existence, becoming anti-war as some of their clients have done? The answer is, it varies. Not all professional carers share the hegemonic notion that war is a necessary evil, especially if they themselves are not ex-services personnel. Some may have class-related sympathy for the rankers' view of themselves as ill-understood by the civilians whose dirty work they have done. Some may make political connections with war and become increasingly anti-war because of what they repeatedly hear. Judith Jones can't help but see war as 'a breakdown in the process of what it means to be civilized'. When sufferers represent war to her and her colleagues, 'I think we hear the very personal aspects that wouldn't be admitted to other vets, and which are painful to admit to themselves.'

* * *

This chapter has described the ways PTSD sufferers and their carers, often themselves secondarily traumatized, handle memories of war. Sufferers and carers alike are living, private and involuntary commemorators of old wars that still have their effect in ex-combatants' minds and bodies. The ways that carers and sufferers handle the memories offers us insights into an invisibilized history.

Professional carers notice memory-related aspects of PTSD that would benefit from public research; for example, the way memories of trauma are screened, encoded and then articulated. PTSD nurse, Claire Evans, finds that sometimes veterans speak of a memory with such profundity that they feel no need to repeat this. Other sufferers do need to repeat their stories, given in the same form, in a manner that is distanced from emotion or profundity. She remembers the qualitatively different feel to some veterans' memories that, in contrast to such 'war stories', may be only mentioned once:

> For example, one POW made his experiences of daily life [in captivity] suddenly very personal and clear [to me] by describing how he thought of his family sitting down to dinner on a Sunday. This is not the kind of memory recounted in history books and it feels different to his descriptions of the conditions in the camp, the illness, suffering, and fear. Those memories are shared – his thoughts of his family were his alone.

Laub and Felman ask what can traumatic testimony teach us and how? We learn from the remembering processes of combat-related PTSD that the crucial embodied experience of PTSD involves a complex history – of sufferers and carers – that goes far beyond what is yet recorded. Such a history can help commemorate the literally and metaphorically fallen in multi-dimensional and respectful ways.

Notes

1 I thank the Director of Clinical Services at Combat Stress, Linda Winn, and the Audley Court nurses, Claire Evans and Judith Jones, for their exceptional courtesy, lively intellectual interaction and hospitality during my visit in August 1999. All non-footnoted quotations are from our discussions during this visit and subsequent checkings of their remarks by telephone and letter. I value their knowledge and sensitive expertise. Their views and mine are not necessarily the same and they are not responsible for my errors. I thank, too, the war veterans and PTSD sufferers, John and Dave, and especially Frank Capacity for his promptness and generosity; and Margaret, Katherine, Kathy and Ruth, the wives and carers of PTSD sufferers. Some names are changed at participants' request.

2 American Psychiatric Association, *Diagnostic and Statistical Manual of Mental Disorders* (Washington, DC, 1987: 3rd edn). PTSD experts, both clinicians and those concerned with more social aspects have formed no consensus about the condition. For example, US anthropologist, Allan Young, has caused strong reactions among those thinking more philosophically by arguing that PTSD is an illusion 'constituted through a researcher's techno-phenomenon and styles of scientific reasoning', in *The Harmony of Illusions: Inventing Post-Traumatic Stress Disorder* (New Jersey, 1995), 10. In the clinical world Stephen O'Brien recently caused controversy by arguing that PTSD can occur even when there is no traumatic *incident.* See S. O'Brien, *Traumatic Events and Mental Health* (Cambridge, 1998).

3 For an accessible basic summary see A. Babbington, *Shellshock: A History of the Changing Attitudes to War Neurosis* (London, 1997).

4 The term comes from J.P. Wilson, *Forgotten Warriors* (Washington, 1979).

5 For example, someone with 50 per cent disability currently (Nov. 1999) gets £60 per week tax free, in addition to regular benefits such as old-age pension.

6 N. Hunt and I. Robbins, 'Telling stories of the war: ageing veterans coping with their memories through narrative', *Oral History*, 26.2 (1998), 57–64.

7 S. Freud, 'Introduction to psychoanalysis and the war neuroses' (1919), in J. Strachey (trans. and ed.) *The Complete Psychological Works of Sigmund Freud*, vol. 17: *1917–19* (London, 1955), 209–10.

8 D.J. Siegel, 'Memory and trauma', in D. Black, M. Newman, J. Harris-Hendricks and G. Mezey (eds) *Psychological Trauma, A Developmental Approach* (London, 1997), 50.

9 Ibid., 45.

10 P. Janet, *Psychological Healing* (1925), cited in Young, *Harmony of Illusions*, 35.

11 Young, *Harmony of Illusions*, 8.

12 The Combat Stress archives are in the British Library.

13 PTSD education offers information in very accessible, everyday language but the very term PTSD is psychologized. Nikolas Rose argues that 'when our culture provides us with life narratives couched in psychological terms, our lives really do become psychological in their form'. See N. Rose, 'Assembling the modern self', in R. Porter (ed.) *Rewriting the Self: Histories from the Renaissance to the Present* (London, 1997), 237. But it may be that framing war trauma in clinical terms makes it easier to legitimate and less humiliating to express.

14 Siegel, 'Memory and trauma', 50.

15 Ibid., 45.

16 A. Thomson, *Anzac Memories: Living with the Legend* (Oxford, 1995), 118–28.

17 See A. Fontana and R.A. Rosenheck, 'Post-traumatic stress disorder among Vietnam theater veterans: a causal model of etiology in a community sample', *Journal of Nervous and Mental Disease*, 182.12 (Dec. 1994), 677–84.

18 Telephone interview with Combat Stress, Aug. 1998.

19 Interview with Frank Capacity (a pseudonym), Audley Court, 17 Aug. 1999. He gave further background information in a number of telephone interviews, Sept.–Nov. 1999.

20 For a summary of this event see T. Wright, 'Return to the Railway of Death', *The Age* (Melbourne), 25 April 1998.
21 The Kwai Railway Memorial (Three Pagodas Group) can be contacted via 3 Gulpher Road, Felixstowe, IP11 9DP.
22 Telephone interview with Frank Capacity, 10 Oct. 1999.
23 Ibid.
24 Interview with Frank Capacity, 17 Aug. 1999.
25 Ibid.
26 On 20 Oct. 1999.
27 See C.R Figley (ed.) *Compassion Fatigue: Coping with Secondary Stress Disorder in Those Who Treat the Traumatized* (New York, 1996).
28 J. Booth, 'Life on the streets: a day in the life of a welfare officer', *Combat Stress News* (Spring 1999), 7.
29 S. Dicks (ed.) *From Vietnam to Hell: Interviews with Victims of PTSD* (Jefferson, NC and London, 1991), 11.
30 S. Felman and D. Laub, *Testimony: Crises of Witnessing in Literature, Psychoanalysis, and History* (London, 1992), xiii.
31 There are two excellent popular works from the USA: A. Matsakis, *Vietnam Wives* (Lutherville, MD, 1996) and C. McCarty-Gould, *Life with the Combat Veteran* (Commack, NY, 1998).
32 Ibid., xvii.
33 Dicks, *From Vietnam to Hell*, 107.
34 Cecelia Helen Christie, Imperial War Museum Sound Archive interview, #13121, recorded 1993.

Part III

Debates and reviews

11 War commemoration in Western Europe

Changing meanings, divisive loyalties, unheard voices

T.G. Ashplant

Martin Evans and Kenneth Lunn (eds) *War and Memory in the Twentieth Century* (Oxford: Berg, 1997)

Sarah Farmer, *Martyred Village: Commemorating the 1944 Massacre at Oradour-sur-Glane* (Berkeley: University of California Press, 1999)

Adrian Gregory, *The Silence of Memory: Armistice Day 1919–1946* (Oxford: Berg, 1994)

Alex King, *Memorials of the Great War in Britain: The Symbolism and Politics of Remembrance* (Oxford: Berg, 1998)

Alessandro Portelli, *The Battle of Valle Giulia: Oral History and the Art of Dialogue* (Madison: University of Wisconsin Press, 1997)

One of the central tasks of the nation-state in war commemoration is to maintain or secure the unity of the 'imagined (national) community', and its associated narratives and rituals, in the face of sometimes acute social divisions. In twentieth-century Western Europe, this has often proved a difficult task, given shifts in national borders, class conflicts within societies, ideological allegiances cutting across nationality, occupation by foreign powers and separation from colonized territories. The recent publication of significant studies dealing with the politics of war memory and commemoration in Britain (Gregory, King), France (Farmer) and Italy (Portelli), together with contributions to the synoptic collection edited by Evans and Lunn, can, when brought together, reveal much about these tensions and the means by which they can be explored.

What emerges strikingly from the studies by Adrian Gregory and Alex King is how unproblematical for the state was First World War commemoration in England, Scotland and Wales.[1] Britain had emerged victorious from the war, with its territory and prestige undamaged, and endured relatively limited internal conflicts over the next twenty years. These circumstances are apparent in the forms which war commemoration took between 1919 and 1939. Gregory shows how such remembrance had to address two constituencies in particular: the bereaved, who came to be figured as predominantly female and civilian; and the veterans. The dominant focus of the most solemn public commemoration soon centred on the bereaved and their (supposed) concerns. Armistice Day with its public ceremonial and Two-Minute Silence, which usually fell on a weekday, drew a predominantly female attendance; the Tomb of the Unknown Warrior offered a place for grieving for all those wives and mothers whose dead lay in unmarked graves. Remembrance Sunday, by contrast, a less central event, became the occasion for veterans. The balance was restored by Poppy Day, in which the whole population could express its gratitude for the sacrifices of the veterans in the most practical way, by raising funds for those most in need. In this way, the needs of both major constituencies were recognized.[2]

This picture is reinforced by King's account of the building of local war memorials (symbolic or utilitarian). The dominant theme is continuity. To establish the memorials, the existing mechanisms of the local state, in conjunction with relevant agencies in civil society, proved entirely adequate. Key features of the forms and symbolic language of war commemoration which had been tentatively initiated by the military in the later nineteenth century, and developed by various localities to honour their own volunteers after the Boer War, were deployed, though on a vastly greater scale and modified where necessary to recognize the unprecedented nature of the First World War.[3]

Gregory's account of key national ceremonies shows how the dominant public meaning attributed to such commemoration, while remaining broadly consensual, subtly changed over those two decades. Its core proved remarkably constant: the war was figured as a redemptive sacrifice, not a militaristic triumph. Nevertheless, an important shift occurred. In the immediate post-war years, this sacrifice was seen to have been secured through the victory of British arms; while the defeated foe was specifically Prussian militarism. From the mid-1920s, this interpretation began to weaken as circumstances changed. Post-war reconstruction had not built a land fit for heroes, domestic social and economic tensions were increasing, and revelations about pre-war diplomacy had challenged the view that Germany alone was to blame. In the new understanding which began to crystallize, war commemoration now marked a sacrificial victory over militarism *tout court*; the onus was now as much on Britain as any other nation to realize the motto 'never again'.[4]

Within this broad interpretative framework, there was room for vigorous discussion about specific aspects of commemoration, turning in part on whose needs were to be given priority. Gregory examines two major debates. In 1925 there were complaints that the charity concert on the evening of Armistice Day showed a lack of respect for those mourning. Both veterans and bereaved were divided over the question, and the two leading popular daily newspapers took divergent lines, one emphasizing a moral conservatism which stressed domestic and religious values, the other a libertarian conservatism which would permit people to commemorate or celebrate as they saw fit. On this occasion, the proponents of a more sober celebration won (thereby strengthening the association of the day with the grief of the bereaved rather than the relief of the survivors).[5] Throughout the inter-war years, there were also occasional demands to reduce or end the military presence in the Armistice Day ceremonies. As Gregory shows, both sides argued within the shared assumption that what was being marked was a victory over militarism. Debate was over whether the continuing presence of soldiers contradicted that message or not.[6] Similar debates over the proper forms of commemoration, though generally fairly muted, are apparent in the localities studied by King (the most violent examples of disagreement he cites arose from religious or petty political differences, having little to do with the meanings of war or loss).[7] King suggests that the resolution of such disputes as did occur over the form and symbolism of memorialization required, not the imposition or acceptance of a dominant meaning, but simply a continuing practice of reverence by all sides for the sacred space and time of memorial sites and their commemorative rituals. These remained a shared locus within which meaning could continue to be contested.[8]

These overviews of the meanings associated with British memorial practice after 1918 can be supplemented by studies in Martin Evans and Kenneth Lunn's volume which examine in greater depth the experience of more specific social groups. Barry Doyle offers a case study of responses to the death of a single officer, Harry Jewson, among the tightly knit Liberal nonconformist bourgeoisie of the city of Norwich from whom he came. At various occasions of commemoration from 1917 to 1923, his contemporaries and seniors sought to reaffirm the values of their community. Such reaffirmation might be backward-looking, seeking to vindicate the breach with tradition which nonconformist support for the war in 1914 had represented; here, fortuitously, Jewson's death confronting the Turks before Jerusalem (rather than German Protestants in Europe) eased the task. Or it might be present-oriented, stressing the continuing need for the cross-class cooperation represented by his pre-war voluntary efforts to alleviate poverty and help youth. A double process of conservation of values was at work here. Jewson's death was to be vindicated as the outcome of the high values which had motivated his whole life; while the continuing relevance of those values had been underlined by the courageous manner of his death.[9]

What this suggests is a powerful but narrow consensus in the British case: that the dead had not perished in vain, and that the war had a meaning – as the war to end wars. This set firm limits to the contest over meaning. As Catherine Moriarty comments, on the range of adjectives which preceded the word 'memory' in lapidary inscriptions:

> Memory was always described in positive terms: the loss of a loved one was to be viewed as a glorious and meaningful sacrifice rather than as a painful, pointless loss. Thus we never read of 'regretful memory', 'angry memory' or 'broken-hearted memory'.[10]

Those who stepped outside that consensus, such as women Cooperators selling White (anti-war) Poppies, were liable on occasion to threats and casual violence.[11] Gregory shows how it was the growing threat of a new war in the later 1930s which put this dominant message under strain. At first, the reaction was to reaffirm more strongly the need to defeat militarism, to ensure that the sacrifice had indeed been redemptive; this reaffirmation worked alongside the newly emergent pacifist movement. But once war broke out again in 1939, this meaning had been irretrievably broken; the Great War had not, after all, been the war to end wars. The celebration of Armistice Day was immediately suspended; when it returned after 1945, it had been assimilated into Remembrance Sunday, and its meaning thereby altered.[12]

In other, more traumatized or divided national contexts, commemoration was considerably more complex and conflicted. Even within the British Isles, the bitter independence struggle leading to the creation of the Irish Free State in 1922 occluded the memory of those Irishmen who had volunteered to fight for Britain in 1914. Many of those Irish veterans whom Jane Leonard (and others) interviewed had been ambivalent, retaining loyalties to both Britain and Ireland; they were, though, especially angered by the use of demobilized British troops (the 'Black and Tans' and the Auxiliaries) to fight the Irish Republican Army during the independence struggle.[13] Some even supported the IRA; but despite this, all were to a degree 'suspect' in the new state in those years because of their war service. A high proportion emigrated, to Britain or further afield. Nevertheless, though there was considerable official hostility to any First World War commemoration between the wars, British Legion clubs for ex-servicemen operated freely and poppies were sold. With the re-eruption of armed conflict in Northern Ireland from 1969, a new hostility to war commemoration emerged and was sustained for two decades; many Legion clubs closed down. But the shifting political context of the late 1980s and 1990s brought a reversal. The developing peace process in the North with its accompanying ceasefire, together with questioning of the monolithic nationalist narrative of the Republic's history, opened up a space in which for the first time the Irish state could give official recognition to those Irishmen who died in the First World War.

France, like Britain, emerged victorious from the First World War, and was able to reinforce its national-collective memory with a picture of having successfully defended the Fatherland via republican values. However, even here there were fissures within the body of the nation. Alsace-Lorraine had been annexed to a victorious Germany after the Franco-Prussian War of 1870–71. William Kidd's study of commemoration in Lorraine notes that, although the territory was recovered in 1918, the French state faced the problem that more of its men had fought for the German rather than for the French army.[14] Hence many standard elements of French war memorial language were inappropriate, and alternative symbols had to be found. This was made easier by the fact that the separation of French Church and state in 1905 had not extended to Lorraine; hence specifically Christian imagery, which could not have been used in the rest of France, was available for this purpose.

The nature and ideological complexities of the Second World War posed far more profound difficulties in the way of any composure of national memory for most Western European countries.[15] All the major continental combatant nations underwent foreign (usually German) occupation; collaborationist regimes staffed by sections of the existing political elite came to power; some subordinated nationalities were offered the opportunity to form an independent nation-state; the international ideologies of communism (and to a much lesser extent fascism) evoked adherence and loyalty across the boundaries of existing (or aspiring) nation-states. Hence at the end of the war, there could often be no simple commemoration and (re-)legitimation of the nation. Instead, nations constructed post-war collective memories via complex processes of negotiation and repression.[16] Furthermore, in such a situation it is not only the nation-state as a whole which must maintain legitimacy; so too must the major competing political forces within it. Each of these will seek to inflect the memory (and hence the identity) of the 'nation at war' in such a way as to secure its own place as a legitimate element within the nation-state. To identify oneself with the nation remained, even for oppositional groups, the most effective mode of legitimation.

In France (as in Italy) after the Second World War a dominant narrative was constructed which celebrated the Resistance as representing the true France; while positioning Vichy as an alien regime, a foreign imposition. This allowed two of the major power blocs within the nation – the Gaullist (Free French) right and the Communist left – to identify themselves with the French regime which emerged from the war. To do this, Vichy – which involved leading political personalities and drew on widespread ideological themes of the pre-1939 French political world – had to be designated as fundamentally foreign.

Sarah Farmer's detailed study of the official designation and preservation of Oradour-sur-Glane as a 'martyred village' throws light on both the extent of, and the limits to, the silences and exclusions employed to fit the events of June 1944 into the developing French official memory of the war.[17] Oradour, in the Limousin, came to be singled out as symbolic in part due to the extent and character of the events themselves. On 10 June, Waffen-SS soldiers killed 642 civilians and torched the village. Women and children had been herded into the church which was then set on fire; those who tried to escape were shot down.[18] Yet, Farmer notes: 'The events of the massacre had to be removed from their historical context and dramatized, visually and in narrative, to be rendered suitable for telling the archetypical story of innocence and victimization.'[19] The German forces had perpetrated several massacres of civilians during the occupation. These had become more frequent during 1944, in response to the increasing military activities of the *maquis*.[20] It was in this context, of a policy of reprisals for French resistance, that the initial accounts of Oradour in 1944–45 situated it.[21] Quite soon, however, the story of Oradour began to be removed from this frame in response to what Farmer terms 'a persistent tension ... between historical facts, context, and the commemorative message'.[22]

The purpose was to play down the divisiveness of the activities of the Resistance, through both their left-wing politics and the risks to which they exposed local civilians. Such decontextualization was made easier because the Oradour massacre, unlike some of the others, had no precise and obvious Resistance action as its trigger. As it came to be commemorated, the village blurred the distinction between resisters and bystanders: all were at risk. It represented France not as a nation of resisters, but as 'a nation of victims, martyred regardless of political choice or wartime activity'.[23] This reframing suited the winding down of the process of *l'épuration* (the post-war purge) which occurred in 1947, amid a growing desire to put aside the divisive legacy of Vichy. As well as uniting the French themselves, as common victims of a cruel enemy, Oradour could also offer a message directed to the allies: that France had been the 'hapless victim of Nazi barbarism'. To see France as martyr, or victim of overmastering force, avoided confronting aspects of its wartime experience which might be humiliating or dishonourable.[24]

However, while the forms taken by commemoration at the site were successful in repressing the internal conflicts of wartime France, the legal process was eventually to reveal them only too starkly. Among the troops involved in the massacre, and still alive and available to be brought to trial in 1953, were fourteen Frenchmen, all Alsatians.[25] Following the German re-annexation of Alsace-Lorraine in 1940, the men of the provinces were made liable to conscription from 1942. Some Frenchmen (termed the '*malgré-nous*') had been compelled to fight in the German army (some after having fought in the French army in 1939–40); while others had fought in the Resistance, and others still had been civilian victims of reprisals.[26] The trial was held in 1953 at the end of a long national debate about the appropriate punishment for collaboration which had resulted in amnesties and shortening of sentences.[27] More immediately, the presence of the Alsatian conscripts among the accused led to vehement demands from their native region that its specific wartime situation be recognized. Fear that this rich province, only recently reintegrated into the motherland, might be alienated, swayed the French parliament into granting an amnesty immediately after sentence. The inhabitants of the Limousin felt themselves to have been betrayed in turn, a poor, rural, leftist region with less political leverage sacrificed to placate Alsace.[28] In reaction to this perceived betrayal, the survivors of Oradour themselves turned away from the state; they returned the medals awarded to the village and the survivors' association, banned government officials from the annual commemorative service and refused to transfer the ashes of the victims into the government-sponsored crypt.[29]

In Italy too, the Resistance, which followed the collapse of Mussolini's fascist regime in 1943 and the German occupation, became a post-war source of legitimation, the root of Italian republican democracy. It allowed the nation to see itself, in Alessandro Portelli's words, as the subject and not the object of its own liberation. In particular, the communist left's role in the Resistance vindicated its national identity (though never sufficiently to permit it a place in government). The advent of the Cold War tended to freeze official memory, discouraging attempts to modify it which might destabilize contemporary political alignments. Nevertheless, the official narrative of the Italian Resistance was never uncontested; and challenges from the right have grown more powerful since 1989.[30] Where Farmer focuses primarily on the place of Oradour in the creation of official French war memory, Portelli is particularly concerned with these later re-negotiations of Italian war memory. His essays comprise an intervention from the left against what he sees as a too easy capitulation to efforts to delegitimize the Resistance. Portelli endorses the argument of Claudio Pavone that the Resistance in Italy has to be seen as comprising not one singular conflict, but three overlapping wars: a patriotic war against the Germans, a class war between powerful and powerless, and a civil war between left and right.[31] He approaches war memory through the close analysis and comparison of discourses (the memoirs and oral testimonies of partisans and survivors), examined in relation to the arenas in which they have been articulated.

He explores the complexities of legitimation through the example of the battle of Poggio Bustone (March 1944), where a partisan battalion killed a group of Italian fascists. Such violence, played down by the left after 1945 to reinforce its democratic identity, was later to be used by the right to redefine the Resistance fighters, not as Italian patriots but as murderous communists. Through an examination of contradictions between partisan accounts of the battle (whether the fascists were killed indoors or outside in the square, whether they had surrendered first), and mythic elements within them (what he terms the primal scene of the Western, in which the bad guys draw their guns first but are nevertheless killed), Portelli uncovers various ways in which these oral narratives and songs reassert the legitimacy of the Resistance. In its early months, when the partisans controlled certain territory, executions of fascists were carried out publicly, in a manner precisely intended to stress the legitimacy of the struggle. Later, as the Resistance was forced back underground, killings of fascists had to be carried out covertly. Such actions, Portelli argues, are retrospectively placed within a narrative frame in which the Resistance as a whole is figured as an act of self-defence. The contradictions refer not so much to what happened as to what was *felt*: we *had* to defend ourselves, they got their just deserts; but also – we *wanted* to kill them. These stories, told by active Resistance partisans, encode recognitions suppressed within the sanitized left version of official war memory: that the conflicts were multiple; and that violence was necessary, perhaps even of revolutionary value, but that it makes men beasts inside.[32]

Portelli also offers a detailed analysis of one specific counter-memory, that of the massacre of 115 civilians by German troops at Civitella in June 1944, ostensibly in retaliation for the killing of three German soldiers by partisans a few days earlier.[33] While the dead were, after 1945, celebrated as martyrs of the Resistance, the Civitella survivors blamed, not the Germans who carried out the massacre, but the partisans whose earlier killings had provoked it. Portelli's essay is a reflection on a conference held on the fiftieth anniversary, entitled 'In Memory: For a European Memory of Nazi Crimes after the End of the Cold War'. His response to the powerfully expressed testimony of survivors is to insist that their shared/common memory, put forward in opposition to official memory, is itself no pure record of the original events, but one with its own evidential, narrative and value structures.[34] He explores the ways in which the values embedded in this deeply felt Civitella memory are rooted in the town's pre-existing class and ideological structure (with its marked segregation between town dwellers, and the rural peasants and workers from whom most partisans were drawn), and its dominant wartime stance within what he terms the 'grey zone' of non-involvement between victim and killer. This memory of the survivors has also been modified in the fifty years since the massacre, with immediate post-war anger at the German perpetrators giving way to powerful hostility against the partisans. In particular, a key moment seems to have been the fortieth anniversary, when, for the first time, survivors (who had hitherto simply boycotted the civil celebration and attended a religious ceremony) mounted their own public counter-commemoration, attended by national figures from the political right. New elements were now incorporated into their account, which had previously been absent or discounted. These reinforced the themes of political quietism and Christian virtue which the townspeople hold to characterize their community. Sectional memory was becoming oppositional, addressing itself to a national, rather than a local community, arena.[35] It gained greater recognition in the following decade – a process culminating in the 1994 conference – as Italian politics shifted with the demise of international communism, enabling the neo-fascists to reposition themselves as democrats while weakening the legitimacy of the Resistance.

Oral history methodology allows Portelli a means to explore the conflicts which run through individual and shared/common memories, even if they have been allowed no expression within the public narratives of sectional or oppositional memory. It is therefore worth juxtaposing his approach with a valuable comment made by Gregory in his 'Introduction'. He points out that he had 'paid a great deal of attention to how the ritual was explained by contemporaries and which tropes they used', and that since 'the discourse surrounding the ritual was so rich and so diverse in its political implications, I was content to accept that all of the memory of war would be shaped and reflected by the languages I have presented'. (King's study similarly rests on extensive analysis of public statements about the purposes of memorials.) However, Gregory continues, Geoffrey Moorhouse's study of First World War memory in one town, Bury, 'suggests that beneath all the varied public discourses there remained a "hidden transcript" about the war, a memory preserved at familial and personal level which was at odds with the rhetoric about the war.' Hence, he suggests, the need for local studies to uncover this memory, even if 'the problems of uncovering it are methodologically profound'.[36] The work of Portelli on Italian war memory, like that of Alistair Thomson on Australia, suggests some of the ways in which such hidden transcripts might be uncovered.[37] Paying attention to the distinctions between different arenas, agencies and discourses of articulation of war memory and commemoration can only increase the value of the rich body of work now being produced in this area, by making more explicit what sorts of memory are being explored in each case.

Notes

1 A. Gregory, *The Silence of Memory: Armistice Day 1919–1946* (Oxford: Berg, 1994); A. King, *Memorials of the Great War in Britain: The Symbolism and Politics of Remembrance* (Oxford: Berg, 1998).

2 Gregory, *The Silence of Memory* (Oxford: Berg), Chs 1, 3. D.W. Lloyd, *Battlefield Tourism: Pilgrimage and Commemoration of the Great War in Britain, Australia and Canada, 1919–1939* (Oxford, 1998), 49–83, shows how the events surrounding the inauguration of the permanent Cenotaph and the grave of the Unknown Warrior, from July 1919 to November 1920, were crucial in first establishing an official war narrative which

centrally addressed the sacrifices and losses of these two groups. See further Ashplant, Dawson and Roper (Chapter 1, in this volume, section 2.1).

3 King, *Memorials of the Great War*, 20–44, 184–7.

4 Gregory, *The Silence of Memory*, 34–8, 84–6, 118–42, 149–72.

5 Ibid., 61–84.

6 Ibid., 118–31, 150–2, 159–72.

7 King, *Memorials of the Great War*, 86–99.

8 Ibid., 210–12, 228–39, 246–9. Catherine Moriarty's account of the erection of local war memorials tends rather to emphasize the representation of a dominant official meaning which was rarely challenged: 'Private grief and public remembrance: British First World War memorials', in M. Evans and K. Lunn (eds) *War and Memory in the Twentieth Century* (Oxford: Berg, 1997), 125–42.

9 B.M. Doyle, 'Religion, politics and remembrance: A Free Church community and its Great War dead', in Evans and Lunn (eds) *War and Memory in the Twentieth Century*, 223–38. These commemorative efforts did not, however, prevent the fracturing, in Norwich as elsewhere, of the religio-political bloc of pre-war Liberalism. Indeed, this had been partly foreshadowed by Jewson's own political stance before 1914 – something not dwelt on in the tributes paid to him.

10 Moriarty, 'Private grief and public remembrance', 137. Lloyd, *Battlefield Tourism*, 90–1, similarly suggests that the Cenotaph 'could refer to the dead without invoking images of the Allied victory' and thus that 'it appealed to people who did not accept the traditions which that victory affirmed'. Hence it was possible to leave political wreaths there, as did a march of the unemployed in 1921; however some wreaths were censored by the police, such as that bearing the slogan 'To the dead victims of Capitalism from the living victims of Capitalism'.

11 Gregory, *The Silence of Memory*, 152–9, 163–4; King, *Memorials of the Great War*, 234–6; Lloyd, *Battlefield Tourism*, 70.

12 Gregory, *The Silence of Memory*, 118–33, 149–78, 191, 215–22.

13 J. Leonard, 'Facing "the finger of scorn": veterans' memories of Ireland after the Great War', in Evans and Lunn (eds) *War and Memory in the Twentieth Century*, 59–72.

14 W. Kidd, 'Memory, Memorials and Commemoration of War in Lorraine, 1908–1988', in Evans and Lunn (eds) *War and Memory in the Twentieth Century*, 143–59.

15 For Britain, undefeated and unoccupied, the war proved much easier to render within a united narrative; so much so, that it was to become probably the central moment in the national imaginary. Even so, Peter Sjølyst-Jackson (in Chapter 6 above) shows how popular war films in 1950s Britain negotiated political, class and gendered divisions so as to sustain a dominant picture of a nation united against the enemy. Lucy Noakes, 'Making histories: experiencing the Blitz in London's museums in the 1990s', in Evans and Lunn (eds) *War and Memory in the Twentieth Century*, 89–104, shows how, even so, the Imperial War Museum's 1990s presentation of the London Blitz was sanitized, omitting all reference to particular disasters and keeping painful memories low-key and marginal. The emphasis was on how 'We was all in it together' (101), rather than on the inequalities very apparent at the time.

16 For Finland's and Norway's handling of these problems, see Raivo and Sjølyst-Jackson (Chapters 5 and 6 above).

17 S. Farmer, *Martyred Village: Commemorating the 1944 Massacre at Oradour-sur-Glane* (Berkeley: University of California Press, 1999).

18 Ibid., 20–8.

19 Ibid., 55.

20 Ibid., 37–40, 44–54.

21 Ibid., 37–41, 55–6.

22 Ibid., 57.

23 Ibid., 49–50, 56–8, quoted at 58.

24 Ibid., 59–65, 78–88, quoted at 87.

25 Ibid., 136. There had also been seven refugees from Alsace and forty-four from Lorraine among the dead: 223 n.22, 252 n.46.

26 Ibid., 139–40. One of the fourteen put on trial was a volunteer, the remainder all *malgré-nous*. Kidd, 'Memory, memorials and commemoration of war in Lorraine', 152–5, explores the problem of memorialization in the recovered provinces after 1945, even more complex than it had been after 1918. In some cases, new symbols were developed, which sought to express these complexities of a fractured history while refusing any easy transcendence. 'Never again' became a powerful theme.

27 Farmer, *Martyred Village*, 136–9.

28 Ibid., 140–8, 152, 158–68.

29 Ibid., 167–70, 178–9. They also persisted in displaying at the site the names of the deputies who had voted for the amnesty, to the latter's continuing discomfort: 179–81.

30 A. Portelli, *The Battle of Valle Giulia: Oral History and the Art of Dialogue* (Madison: University of Wisconsin Press, 1997), 139, 142, 149. Chs 8–10 of this book deal with Italy during the Second World War.

31 Portelli, *The Battle of Valle Giulia*, 127–8. A. Prost, 'The Algerian War in French collective memory', in J. Winter and E. Sivan (eds) *War and Remembrance in the Twentieth Century* (Cambridge: Cambridge University Press, 1999), 174, argues that *l'épuration* 'made it impossible to face the fact that this war had been a civil war in some respects'. He cites the example of a young French leftist politician murdered by the French collaborationist police in 1944; his tombstone reads 'killed by the enemies of France', the school named after him bears the legend 'victim of Nazi barbarity'.

32 A. Portelli, 'The Battle of Poggio Bustone: violence, memory and imagination in the partisan war', in his *The Battle of Valle Giulia*, 126–39.

33 A. Portelli, 'The massacre at Civitella Val di Chiana (Tuscany, 29 June, 1944): myth and politics, mourning and common sense', in his *The Battle of Valle Giulia*, 140–60.

34 For discussion of the specific role and responsibilities of historians in this debate, see Ashplant, Dawson and Roper (Chapter 1, in this volume, section 3.5).

35 Portelli himself, in 'The Massacre at Civitella Val di Chiana', 156–7, talks of there being 'two official memories'.

36 Gregory, *The Silence of Memory*, 6, referring to G. Moorhouse, *Hell's Foundations: A Town, its Myths and Gallipoli* (1992; London: Sphere, 1993).

37 A. Thomson, *Anzac Memories: Living with the Legend* (Melbourne: Oxford University Press, 1994).

Index

The following index headings are used to group concepts, or clusters of related entries: battles and military operations; films; novels; television series; veterans; veterans' organizations; war commemoration (commemorative events; memorials; organizations and publications); war memory; wars.